Highway Capacity Analysis

HCM & HCS

William M. Sampson

cognella® | ACADEMIC PUBLISHING

Bassim Hamadeh, CEO and Publisher
Jennifer McCarthy, Acquisitions Editor
Gem Rabanera, Project Editor
Christian Berk, Associate Production Editor
Jess Estrella, Senior Graphic Designer
Brian Fahey, Licensing Coordinator
Don Kesner, Interior Designer
Natalie Piccotti, Senior Marketing Manager
Kassie Graves, Director of Acquisitions and Sales
Jamie Giganti, Senior Managing Editor

Printed in the United States of America.

ISBN: 978-1-5165-2525-6 (pbk) / 978-1-5165-2526-3 (br)

cognella® | ACADEMIC PUBLISHING

Highway Capacity Analysis

The preliminary edition of this text was published the week between my wife's birthday and our forty-second anniversary—what wonderful reminders of her inspired wisdom and our daughters' unconditional support. Without my family, I could not possibly be living the happy and fulfilled life I have been so blessed to enjoy—I owe everything to them.

ACKNOWLEDGMENTS

The author would like to thank the Transportation Research Board (TRB) and its Highway Capacity and Quality of Service (HCQS) Committee for producing the Highway Capacity Manual (HCM) and the National Academies Press for allowing the use of selected exhibits and equations; the University of Florida for its support in this work and for permission to use references and images from the Highway Capacity Software (HCS); and Samantha Taningco for her analytical expertise and comprehensive editing effort on all example problems.

CONTENTS

Introduction

Highway Capacity Analysis is performed following the methodologies in the Highway Capacity Manual (HCM), typically applied using the Highway Capacity Software (HCS). The HCM 6th Edition (HCM6) was published and implemented in HCS Release 7 (*HCS7*) at the end of 2016, and this text covers these latest versions.

These procedures are divided into two categories: Uninterrupted Flow (Freeways and Highways) and Interrupted Flow (Intersections and Streets). This text is organized to cover Uninterrupted Flow first in Part I, including Basic Freeway Segments, Freeway Weaving Segments, Freeway Merge and Diverge Segments, Freeway Facilities, Multilane Highways, and Two-Lane Highways. Interrupted Flow follows in Part II to include Stop Control, Roundabouts, Signalized Intersections, Urban Streets, Interchanges, and Alternative Intersections. Multimodal Analysis and Travel Time Reliability are included for the applicable procedures. Part III extends the procedural analysis into broader applications, including Signal Timing Optimization and Traffic Impact Studies.

Each procedure is defined in detail, including step-by-step through applicable equations, tables, and exhibits to essentially be able to perform theses analyses manually with examples following each method. Since applications use software in practice, the coding and results of each example are illustrated using *HCS7*.

The goal of this text is to guide the use of the HCM and application in the HCS in a practical, succinct, and logical manner to facilitate a complete understanding of the procedures and proficiency in applying them to any analysis scenario.

Highway Capacity Analysis permeates many disciplines within traffic engineering and transportation planning for modeling freeways, highways, streets, and intersections. These research-based and peer-reviewed procedures defined in the Highway Capacity Manual and implemented in the Highway Capacity Software have served to define level of service for over thirty years. These methods are used in short- and long-range planning; highway and intersection design; signal design and timing; and traffic impact analysis by cities, counties, states, and consulting firms nationally and internationally. Understanding highway capacity analysis and its applications translates to one of the most marketable skills in our profession.

HCM

The HCM was first published in 1950 with a second edition in 1965, but the third edition in 1985 brought these concepts and procedures into the traffic engineering and transportation planning practice. The third edition was developed with underlying research and panel peer review before facing the scrutiny of the Highway Capacity and Quality of Service (HCQS) committee that oversees the HCM for the Transportation Research Board (TRB). Beyond making these procedures extremely defendable with this level of review, software was developed to automate the analysis on personal computers that were relatively new on the scene in 1985. This combination of confidence and efficiency quickly made these methods a standard in several areas of the transportation profession, including operations, design, planning, signal timing, and traffic impact studies.

Since 1985, there have been two major updates in 1994 and 1997, the three new editions in 2000, 2010, and 2016 (HCM6). Each update has served to add to the accuracy and breadth of the methodologies, expanding to oversaturated conditions on both freeways and streets; into multimodal applications for pedestrians, bicycles, and transit; and most recently to alternative intersections and travel time reliability.

HCS7

The HCS was first introduced in 1986 as an application developed at Polytechnic University on an Apple IIe, then converted to DOS before releasing it to the public through the Mc*Trans* Center at the University of Florida in 1987. From that time, Mc*Trans* has distributed, supported, and improved the HCS through seven releases: *HCS2* in 1995; *HCS3* in 1998; *HCS2000* in 2000; *HCS+* in 2005; *HCS2010* in 2011; and *HCS7* in 2016. Each release has implemented the procedures in the corresponding HCM and improved the interface and functionality to match the DOS and Windows standards at the time.

The underlying architecture of the HCS has gone from Quick Basic to Visual Basic, through C++ and onto C# Object Oriented in the current *HCS7*. Every minor change to the HCM as approved by the HCQS committee in most of its two meetings annually has been implemented in the HCS and made available to users as downloadable updates to maintain consistency with the HCM.

Alternative Tools

There are situations that go beyond the ability of the HCM procedures to analyze. In those cases, the process for using alternative tools (like microsimulation) are defined to ensure the results are HCM-compliant. This is done by prescribing values for underlying parameters in these tools to be consistent with the HCM methods—otherwise the density and delay results would not be appropriate for defining level of service, since those scales assume the HCM-produced service measures. The process for analyzing many situations for all methodological chapters in the HCM is defined.

As an example, the left-turn queue that is longer than the left-turn storage lane "spills over" into the adjacent through lane, affecting the capacity of that through movement, sometimes quite significantly. The HCM procedure predicts the length of that queue and compares it to the coded storage length to compute the queue storage ratio, which exceeds one for the spillover condition. However, the effects on the adjacent through lane are not modeled—meaning results are incorrect for those movements and that approach using the HCM procedures. Modeling this in a microsimulation tool is recommended, and *HCS7* automates that conversion.

Capacity

Capacity is defined as the (sustainable) hourly rate during a particular (normally 15-minute) period and is used in every procedure. The basis for capacity is almost always headway, the average time between vehicles (front bumper to front bumper) passing a given point. The shorter the headway, the higher the capacity; two seconds between vehicles would be a capacity of 1800 vehicles per hour with 3600 seconds in an hour. So, many of the procedures focus significantly on what would affect a driver's headway.

Think about your own driving:
Approaching a signalized intersection in the rightmost lane, you feel the friction of parked cars on the street next to you. You are concerned over a door opening or a car pulling out, so you compensate with a slightly increased headway. If the normal 2.0-second approach headway increases by only two tenths of a second to 2.2 seconds, the capacity would decrease by 10%.

The HCM consistently uses the peak fifteen-minute period for computing service measures to determine level of service. Hourly volumes are converted fifteen-minute flow rates using the peak hour factor (PHF), expressed as an hourly rate. The PHF is computed by dividing the hourly volume by four times the peak fifteen-minute volume. This ratio is used by dividing the hourly volume by the PHF to generate the peak fifteen-minute flow rate, essentially the same as multiplying the peak fifteen-minute volume by four.

Time Period	Volume (veh)	Rate of Flow (veh)
5:00-5:15	1,000	4,000
5:15-5:30	1,200	4,800
5:30-5:45	1,100	4,400
5:45-6:00	1,000	4,000
5:00-6:00	4,300	

$$PHF = V/(4 \times V_{15})$$

$$v = V/PHF = 4300/0.90 = 4800 = 4 \times 1200$$

$$PHF = 4300/(4 \times 1200) = 4300/4800 = 0.90$$

Level of Service

Level of Service (LOS) is quantifying as a service measure to indicate the quality of the trip over a roadway or through an intersection and is defined for every procedure. Quality is the relative comfort and convenience provided to the user—essentially quantifying something that is qualitative.

Density (in passenger cars per mile per lane) is the service measure for most uninterrupted flow. On the surface you might think this would be speed for a freeway or highway, but density is more indicative of the driver's level of comfort and convenience.

Think about your own driving:
At midnight on a freeway, your speed is likely to be about 75 mi/h on a freeway and you are not affected by other vehicles going slower in front or pushing you from behind because the traffic is so light. At noon on a freeway, your speed is still likely to be about 75 mi/h, but you can't change lanes whenever you want and other vehicles are affecting you. While your speed is the same, you are much less comfortable and afforded much less convenience at noon than at midnight, so it's the density of vehicles that affects your level of service in these two scenarios, not speed.

Delay (in average seconds per vehicle) is the service measure for most interrupted flow. How long you must wait at an intersection (signalized, stop-controlled, or roundabout) is normally what affects your comfort and convenience navigating the intersection.

System Element	HCM Chapter	Service Measure(s) by Mode: Automobile	Pedestrian	Bicycle	Transit	Systems Analysis Measure
Freeway facility	10	Density	--	--	--	Speed
Basic freeway segment	12	Density	--	--	--	Speed
Multilane highway	12	Density	--	LOS score[a]	--	Speed
Freeway weaving segment	13	Density	--	--	--	Speed
Freeway merge and diverge segments	14	Density	--	--	--	Speed
Two-lane highway	15	Percent time-spent-following, speed	--	LOS score[a]	--	Speed
Urban street facility	16	Speed	LOS score[a]	LOS score[a]	LOS score[a]	Speed
Urban street segment	18	Speed	LOS score[a]	LOS score[a]	LOS score[a]	Speed
Signalized intersection	19	Delay	LOS score[a]	LOS score[a]	--	Delay
Two-way stop	20	Delay	Delay	--	--	Delay
All-way stop	21	Delay	--	--	--	Delay
Roundabout	22	Delay	--	--	--	Delay
Ramp terminal, alternative intersection	23	Experienced travel time	--	--	--	Travel time
Off-street pedestrian–bicycle facility	24	--	Space, events[b]	LOS score[a]	--	Speed

Scales define level of service by matching a quantity of each service measure to a corresponding quality as a letter grade from A (best) to F (worst). The scales vary by control type with signalized intersections on an eighty-second scale and unsignalized intersections on a 50-second scale. This is because of your required driving tasks (work) and your relative level of concern (worry).

Think about your own driving: At a signal with the light red and ten cars back from the stop bar, you have no driving tasks and you are confident you will get through in a relatively short time when the light turns green. At a stop-controlled intersection as the first car, you must watch every conflicting vehicle to make a decision every few seconds, and it doesn't take long to feel like you are "never going to get to go." At a signal, you are not "working" or "worried," but at a stop sign, you are both "working" and "worried." This comparison is the reason the scales differ with stop-control level of service values deteriorating more quickly than signals.

PART I Uninterrupted Flow

This part deals with procedures for analyzing uninterrupted flow segments and facilities with each HCM chapter and HCS module described in detail. Each methodology is defined step-by-step, culminating in exercises to work problems by hand to demonstrate a full understanding and code them in software to learn how it will be done in practice.

Chapters and modules in uninterrupted flow analysis include: Basic Freeway Segments; Freeway Weaving Segments; Merge and Diverge Segments; Freeway Facilities; and Multilane Highways and Two-Lane Highways.

Basic Freeway Segments

This procedure provides an understanding of converting traffic volumes to flow rates (including use of the peak-hour factor and the new passenger-car equivalent values) and of adjusting free-flow speed (toward computing the average travel speed from the speed-flow curves and equations, converging on the calculation of density for the determination of level of service (LOS). New methods for defining density and LOS for managed lanes are also defined.

Freeway Weaving Segments

This procedure explains the lane-changing rate process that is needed to compute the weaving intensity factor for determining weaving speed. The two required checks of capacity (based on demand and density for the segment and for the ramps) and maximum weaving length (if exceeded dictates a basic segment analysis) are defined. Combining weaving and non-weaving speeds, the overall speed is used with flow rate to compute density and find LOS. New methods for quantifying the cross-weave to manage lane reduction in general-purpose lane capacity are also defined.

Freeway Merge and Diverge Segments

This procedure focuses on the flow rate in the two lanes nearest the on- or off-ramp toward computing the density for determining LOS. Adjacent ramps can influence these values and ramp capacities must also be checked. Adjustments are made for left-side ramps and to model these junctions on ten-lane freeways. New equations for computing density beyond the influence area to include all freeway lanes are also defined.

Freeway Facilities

This procedure models the combination of segment analyses across an entire freeway facility to generate a composite density for a facility-wide LOS. New procedures for modeling the effects of work zones on capacity and speed are defined. This method is the only way to model oversaturated conditions across multiple segments over multiple time periods to capture the interactive effects of both time and space. The interaction of bottleneck segments upstream (generating queues) and downstream (protecting from flow) are modeled to generate results beyond density to include queuing and delay. Procedures are defined for modeling travel time reliability with demand, weather, incidents, work zones, and special events varying for a distribution of results for hundreds of scenarios to produce travel time reliability indices at three different percentiles.

Multilane Highways

This procedure is very similar to the Basic Freeway Segments method as now being combined in the same HCM chapter. Differences in undivided option, total lateral clearance definitions, access point (travel direction only) adjustments, base free-flow speed values, free-flow speed ranges, and consistent break points are all discussed. Procedures are defined for determining bicycle level of service.

Two-Lane Highways

This procedure includes methods for three highway classes with two service measures as average travel speed and percent time spent following with different scales for each highway class. Grade and percent no-passing zones are variables in computing the service measures, which are further modified for the presence of a passing lane defined by the lengths of four analysis regions, including truncated segments. Procedures are defined for determining bicycle level of service.

CHAPTER 1 Freeways

Freeways are divided into segments for analyzing one direction at a time, consisting of Basic Freeway Segments, Freeway Weaving Segments, and Freeway Merge and Diverge Segments. A Basic Freeway Segment is a portion of the freeway outside of any ramp influence areas. A Freeway Weaving Segment consists of either one with one-sided (on-ramp followed by an off-ramp with a continuous auxiliary lane between) weaving or two-sided (a left-side on-ramp followed by a right-side off-ramp or a right-side on-ramp followed by a left-side off-ramp) weaving. A Freeway Merge or Diverge Segment models the on-ramp (merge) or off-ramp (diverge) interaction with the adjacent two freeway lanes. This chapter deals with the analysis of individual freeway segments, while Chapter Two defines combining multiple segments into a Freeway Facility to be able to model multiple time periods for analyzing oversaturated conditions as well as travel time reliability (TTR). Since the segment procedures do not model oversaturated conditions, any segment analysis resulting in a failing level of service must be modeled within a facility analysis.

Basic Freeway Segments

The analysis of a Basic Freeway Segment defined in HCM Chapter 12 involves estimating the free-flow speed and the flow rate for computing the travel speed, which is used to calculate density for determining level of service. This is likely the simplest of all procedures and can be done manually quite easily.

Required Data

To compute the free-flow speed, the geometric data required include number of lanes, lane width, right-side lateral clearance, and total ramp density. If the free-flow speed is measured in the field as the average speed in low-volume conditions, it is assumed to be equal to the adjusted free-flow speed (since these geometric attributes would have influenced the field-measured average).

To compute the flow rate, the demand data required include the prevailing volume (with driver familiarity information), peak hour factor, truck percentage, and the terrain type or specific grade (for which truck data must include the proportions of tractor trailers and single-unit trucks).

Free-Flow Speed

The process for computing adjusted free-flow speed (*FFS*) begins with a base free-flow speed (*BFFS*) that defaults to 75.4 mi/h. This value can be changed with justification like field data to defend using a locally derived speed. Ramp density is computed by counting the ramps within a 6-mile distance, 3 miles upstream to 3 miles downstream of the midpoint of the analysis segment. Adjustments are made if the average lane width is below 12 feet, the right-side lateral clearance is less than 6 feet, and/or the total ramp density (*TRD*) is greater than zero ramps per mile, according HCM Equation 12-2 and HCM Exhibits 12-20 and 12-21.

These adjustments reflect how drivers react to geometric restrictions with slower speeds. Field measured free-flow speed is recommended if there are geometric situations that are not being modeled—most specifically if left-side lateral clearance (like a guardrail within 6 feet of

Figure 1.1 *HCM Equation 12-2*

$$FFS = BFFS - f_{LW} - f_{RLC} - 3.22 \times TRD^{0.84}$$

f_{LW} = adjustment for lane width, from Exhibit 12-20 (mi/h),

f_{RLC} = adjustment for right-side lateral clearance, from Exhibit 12-21 (mi/h), and

TRD = total ramp density (ramps/mi).

Figure 1.2 *HCM Exhibit 12-20*

Average Lane Width (ft)	Reduction in FFS, f_{LW} (mi/h)
≥12	0.0
≥11–12	1.9
≥10–11	6.6

Figure 1.3 *HCM Exhibit 12-21*

Right-Side Lateral Clearance (ft)	Lanes in One Direction			
	2	3	4	≥5
≥6	0.0	0.0	0.0	0.0
5	0.6	0.4	0.2	0.1
4	1.2	0.8	0.4	0.2
3	1.8	1.2	0.6	0.3
2	2.4	1.6	0.8	0.4
1	3.0	2.0	1.0	0.5
0	3.6	2.4	1.2	0.6

the travel lane) potentially influences free-flow speed.

Think about your own driving:

Your speed will be higher on a freeway with a 60-foot grass median to your left, a 30-foot clear zone on your right, on 12-foot lanes and level terrain than it is with guardrails within a few feet on both sides, and on narrow lanes with curves and hills.

Free-flow speed can be adjusted using the Speed Adjustment Factor (*SAF*) for the effects of weather and/or driver familiarity, following HCM Equation 12-5 with HCM Exhibits 11-21 and/or 26-9.

Figure 1.4 *HCM Equation 12-5*

$$FFS_{adj} = FFS \times SAF$$

Think about your own driving:

Have you ever noticed someone driving more slowly for no apparent reason? Then, you realize you are in Orlando and everyone is looking at their phones trying to find Mickey Mouse. While not too common, since we typically are modeling rush hours with very familiar commuters, driver unfamiliarity can affect both speed and capacity.

Flow Rate

Flow rate (v_p) is estimated by modifying the hourly volume in vehicles per hour to adjust for the presence of heavy vehicles and to convert the value to a peak fifteen-minute rate using the peak hour factor, following HCM Equation 12-9. The heavy vehicle adjustment factor (f_{HV}) is determined by the passenger-car

Figure 1.5 *HCM Exhibit 11-21*

		Speed Adjustment Factors				
Weather Type	**Weather Event Definition**	**55 mi/h**	**60 mi/h**	**65 mi/h**	**70 mi/h**	**75 mi/h**
Medium rain	> 0.10 – 0.25 in/h	0.96	0.95	0.94	0.93	0.93
Heavy rain	> 0.25 in/h	0.94	0.93	0.93	0.92	0.91
Light snow	>0.00 – 0.05 in/h	0.94	0.92	0.89	0.87	0.84
Light–medium snow	>0.05 – 0.10 in/h	0.92	0.90	0.88	0.86	0.83
Medium–heavy snow	>0.10 – 0.50 in/h	0.90	0.88	0.86	0.84	0.82
Heavy snow	>0.50 in/h	0.88	0.86	0.85	0.83	0.81
Severe cold	<–4°F	0.95	0.95	0.94	0.93	0.92
Low visibility	0.50 – 0.99 mi	0.96	0.95	0.94	0.94	0.93
Very low visibility	0.25 – 0.49 mi	0.95	0.94	0.93	0.92	0.91
Minimal visibility	< 0.25 mi	0.95	0.94	0.93	0.92	0.91
Non–severe weather	All conditions not listed above	1.00	1.00	1.00	1.00	1.00

Figure 1.6 *HCM Exhibit 26-9*

Level of Driver Familiarity	SAF_{pop}
All familiar drivers, regular commuters	1.000
Mostly familiar drivers	0.975
Balanced mix of familiar and unfamiliar drivers	0.950
Mostly unfamiliar drivers	0.913
All or overwhelmingly unfamiliar drivers	0.863

equivalents (*PCE*) as a function of terrain type or specific grade data in combination with the proportion of trucks. This proportion is defined as a split between tractor-trailers (*TT*) and single-unit trucks (*SUT*) that include buses and recreational vehicles. HCM Equation 12-10 is used with passenger-car equivalent values from HCM Exhibits 12-25, 12-26, 12-27, and 12-28 to generate the final flow rate value.

Interpolation and extrapolation are utilized for values not specifically listed in the provided tables. If the truck percentage is high along a long, steep upgrade, the mixed-flow model should be used in place of this equation and these tables. HCM Exhibit 26-4 illustrates the modified speed-flow curves that are required to model these situations.

Capacity

Capacity (c) is also defined as a function of free-flow speed in the equation from HCM Exhibit 12-6 between 2200 pc/h/ln at 55 mi/h and 2400 pc/h/ln at 75 mi/h.

Capacity can be adjusted using the Capacity Adjustment Factor (*CAF*) for the effects of weather, incidents, and/or driver familiarity, following HCM Equation 12-8 with HCM Exhibits 11-20, 11-23, and/or 26-9. For incidents, the adjustment is applied to the open lanes.

Figure 1.7 *HCM Equation 12-9*

$$v_p = \frac{V}{PHF \times N \times f_{HV}}$$

where

v_p = demand flow rate under equivalent base conditions (pc/h/ln),

V = demand volume under prevailing conditions (veh/h),

PHF = peak hour factor (decimal),

N = number of lanes in analysis direction (ln), and

f_{HV} = adjustment factor for presence of heavy vehicles (decimal).

Figure 1.8 *HCM Equation 12-10*

$$f_{HV} = \frac{1}{1 + P_T(E_T - 1)}$$

where

f_{HV} = heavy-vehicle adjustment factor (decimal),

P_T = proportion of SUT and TTs in traffic stream (decimal), and

E_T = passenger-car equivalent of one heavy vehicle in the traffic stream (PCEs).

Figure 1.9 *HCM Exhibit 12-25*

Passenger Car Equivalent	Terrain Type	
	Level	Rolling
E_T	2.0	3.0

Figure 1.10 *HCM Exhibit 12-26*

% Grade	Length (mi)	Percentage of Trucks and Buses (%) 2%	4%	5%	6%	8%	10%	15%	20%	>25%
−2	0.125	2.62	2.37	2.30	2.24	2.17	2.12	2.04	1.99	1.97
	0.375	2.62	2.37	2.30	2.24	2.17	2.12	2.04	1.99	1.97
	0.625	2.62	2.37	2.30	2.24	2.17	2.12	2.04	1.99	1.97
	0.875	2.62	2.37	2.30	2.24	2.17	2.12	2.04	1.99	1.97
	1.25	2.62	2.37	2.30	2.2	2.17	2.12	2.04	1.99	1.97
	1.5	2.62	2.37	2.30	2.24	2.17	2.12	2.04	1.99	1.97
0	0.125	2.62	2.37	2.30	2.24	2.17	2.12	2.04	1.99	1.97
	0.375	2.62	2.37	2.30	2.24	2.17	2.12	2.04	1.99	1.97
	0.625	2.62	2.37	2.30	2.24	2.17	2.12	2.04	1.99	1.97
	0.875	2.62	2.37	2.30	2.24	2.17	2.12	2.04	1.99	1.97
	1.25	2.62	2.37	2.30	2.24	2.17	2.12	2.04	1.99	1.97
	1.5	2.62	2.37	2.30	2.24	2.17	2.12	2.04	1.99	1.97
2	0.125	2.62	2.37	2.30	2.24	2.17	2.12	2.04	1.99	1.97
	0.375	3.76	2.96	2.78	2.65	2.48	2.38	2.22	2.14	2.09
	0.625	4.47	3.33	3.08	2.91	2.68	2.54	2.34	2.23	2.17
	0.875	4.80	3.50	3.22	3.03	2.77	2.61	2.39	2.28	2.21
	1.25	5.00	3.60	3.30	3.09	2.83	2.66	2.42	2.30	2.23
	1.5	5.04	3.62	3.32	3.11	2.84	2.67	2.43	2.31	2.23
2.5	0.125	2.62	2.37	2.30	2.24	2.17	2.12	2.04	1.99	1.97
	0.375	4.11	3.14	2.93	2.78	2.58	2.46	2.28	2.19	2.13
	0.625	5.04	3.62	3.32	3.11	2.84	2.67	2.43	2.31	2.23
	0.875	5.48	3.85	3.51	3.27	2.96	2.77	2.50	2.36	2.28
	1.25	5.73	3.98	3.61	3.36	3.03	2.83	2.54	2.40	2.31
	1.5	5.80	4.02	3.64	3.38	3.05	2.84	2.55	2.41	2.32
3.5	0.125	2.62	2.37	2.30	2.24	2.17	2.12	2.04	1.99	1.97
	0.375	4.88	3.54	3.25	3.05	2.80	2.63	2.41	2.29	2.22
	0.625	6.34	4.30	3.87	3.58	3.20	2.97	2.64	2.48	2.38
	0.875	7.03	4.66	4.16	3.83	3.39	3.12	2.76	2.57	2.46
	1.25	7.44	4.87	4.33	3.97	3.50	3.22	2.82	2.62	2.50
	1.5	7.53	4.92	4.38	4.01	3.53	3.24	2.84	2.63	2.51
4.5	0.125	2.62	2.37	2.30	2.24	2.17	2.12	2.04	1.99	1.97
	0.375	5.80	4.02	3.64	3.38	3.05	2.84	2.55	2.41	2.32
	0.625	7.90	5.11	4.53	4.14	3.63	3.32	2.90	2.68	2.55
	0.875	8.91	5.64	4.96	4.50	3.92	3.56	3.07	2.82	2.67
	1	9.19	5.78	5.08	4.60	3.99	3.62	3.11	2.85	2.70
5.5	0.125	2.62	2.37	2.30	2.24	2.17	2.12	2.04	1.99	1.97
	0.375	6.87	4.58	4.10	3.77	3.35	3.09	2.73	2.55	2.44
	0.625	9.78	6.09	5.33	4.82	4.16	3.76	3.21	2.93	2.77
	0.875	11.20	6.83	5.94	5.33	4.56	4.09	3.45	3.12	2.93
	1	11.60	7.04	6.11	5.47	4.67	4.18	3.51	3.17	2.97
6	0.125	2.62	2.37	2.30	2.24	2.17	2.12	2.04	1.99	1.97
	0.375	7.48	4.90	4.36	3.99	3.52	3.23	2.83	2.63	2.51
	0.625	10.87	6.66	5.79	5.21	4.46	4.01	3.39	3.08	2.89
	0.875	12.54	7.54	6.51	5.81	4.94	4.40	3.67	3.30	3.08
	1	13.02	7.78	6.71	5.99	5.07	4.51	3.75	3.37	3.14

Figure 1.11 *HCM Exhibit 12-27*

% Grade	Length (mi)	Percentage of Trucks and Buses (%)								
		2%	4%	5%	6%	8%	10%	15%	20%	>25%
−2	0.125	2.67	2.38	2.31	2.25	2.16	2.11	2.02	1.97	1.93
	0.375	2.67	2.38	2.31	2.25	2.16	2.11	2.02	1.97	1.93
	0.625	2.67	2.38	2.31	2.25	2.16	2.11	2.02	1.97	1.93
	0.875	2.67	2.38	2.31	2.25	2.16	2.11	2.02	1.97	1.93
	1.25	2.67	2.38	2.31	2.25	2.16	2.11	2.02	1.97	1.93
	1.5	2.67	2.38	2.31	2.25	2.16	2.11	2.02	1.97	1.93
0	0.125	2.67	2.38	2.31	2.25	2.16	2.11	2.02	1.97	1.93
	0.375	2.67	2.38	2.31	2.25	2.16	2.11	2.02	1.97	1.93
	0.625	2.67	2.38	2.31	2.25	2.16	2.11	2.02	1.97	1.93
	0.875	2.67	2.38	2.31	2.25	2.16	2.11	2.02	1.97	1.93
	1.25	2.67	2.38	2.31	2.25	2.16	2.11	2.02	1.97	1.93
	1.5	2.67	2.38	2.31	2.25	2.16	2.11	2.02	1.97	1.93
2	0.125	2.67	2.38	2.31	2.25	2.16	2.11	2.02	1.97	1.93
	0.375	3.76	2.95	2.77	2.64	2.47	2.36	2.20	2.11	2.06
	0.625	4.32	3.24	3.01	2.84	2.63	2.49	2.29	2.19	2.12
	0.875	4.57	3.37	3.11	2.93	2.70	2.55	2.33	2.22	2.15
	1.25	4.71	3.45	3.17	2.99	2.74	2.58	2.36	2.24	2.17
	1.5	4.74	3.47	3.19	3.00	2.75	2.59	2.36	2.24	2.17
2.5	0.125	2.67	2.38	2.31	2.25	2.16	2.11	2.02	1.97	1.93
	0.375	4.10	3.13	2.92	2.77	2.57	2.44	2.26	2.16	2.10
	0.625	4.84	3.52	3.23	3.03	2.77	2.61	2.38	2.26	2.18
	0.875	5.17	3.69	3.37	3.15	2.87	2.69	2.43	2.30	2.22
	1.25	5.36	3.79	3.45	3.22	2.92	2.73	2.47	2.33	2.24
	1.5	5.40	3.81	3.47	3.24	2.93	2.74	2.47	2.33	2.25
3.5	0.125	2.67	2.38	2.31	2.25	2.16	2.11	2.02	1.97	1.93
	0.375	4.89	3.54	3.25	3.05	2.79	2.62	2.39	2.26	2.19
	0.625	6.05	4.15	3.75	3.47	3.11	2.89	2.58	2.42	2.32
	0.875	6.58	4.43	3.97	3.66	3.26	3.01	2.67	2.49	2.39
	1.25	6.88	4.58	4.10	3.77	3.35	3.09	2.72	2.53	2.42
	1.5	6.95	4.62	4.13	3.80	3.37	3.10	2.73	2.54	2.43
4.5	0.125	2.67	2.38	2.31	2.25	2.16	2.11	2.02	1.97	1.93
	0.375	5.83	4.03	3.65	3.39	3.05	2.84	2.55	2.39	2.30
	0.625	7.53	4.92	4.38	4.01	3.53	3.24	2.83	2.62	2.50
	0.875	8.32	5.34	4.72	4.29	3.75	3.42	2.97	2.73	2.59
	1	8.53	5.45	4.81	4.37	3.81	3.47	3.00	2.76	2.62
5.5	0.125	2.67	2.38	2.31	2.25	2.16	2.11	2.02	1.97	1.93
	0.375	6.97	4.63	4.14	3.81	3.38	3.11	2.74	2.55	2.43
	0.625	9.37	5.89	5.16	4.68	4.05	3.67	3.14	2.88	2.72
	0.875	10.49	6.48	5.65	5.09	4.37	3.93	3.34	3.03	2.85
	1	10.80	6.64	5.78	5.20	4.46	4.01	3.39	3.08	2.89
6	0.125	2.67	2.38	2.31	2.25	2.16	2.11	2.02	1.97	1.93
	0.375	7.64	4.98	4.43	4.05	3.56	3.26	2.85	2.64	2.51
	0.625	10.45	6.45	5.63	5.07	4.36	3.92	3.33	3.03	2.85
	0.875	11.78	7.16	6.20	5.56	4.74	4.24	3.56	3.22	3.01
	1	12.15	7.35	6.36	5.69	4.85	4.33	3.62	3.27	3.05

Figure 1.12 *HCM Exhibit 12-28*

% Grade	Length (mi)	Percentage of Trucks and Buses (%)								
		2%	4%	5%	6%	8%	10%	15%	20%	>25%
-2	0.125	2.39	2.18	2.12	2.07	2.01	1.96	1.89	1.85	1.83
	0.375	2.39	2.18	2.12	2.07	2.01	1.96	1.89	1.85	1.83
	0.625	2.39	2.18	2.12	2.07	2.01	1.96	1.89	1.85	1.83
	0.875	2.39	2.18	2.12	2.07	2.01	1.96	1.89	1.85	1.83
	1.25	2.39	2.18	2.12	2.07	2.01	1.96	1.89	1.85	1.83
	1.5	2.39	2.18	2.12	2.07	2.01	1.96	1.89	1.85	1.83
0	0.125	2.39	2.18	2.12	2.07	2.01	1.96	1.89	1.85	1.83
	0.375	2.39	2.18	2.12	2.07	2.01	1.96	1.89	1.85	1.83
	0.625	2.39	2.18	2.12	2.07	2.01	1.96	1.89	1.85	1.83
	0.875	2.39	2.18	2.12	2.07	2.01	1.96	1.89	1.85	1.83
	1.25	2.39	2.18	2.12	2.07	2.01	1.96	1.89	1.85	1.83
	1.5	2.39	2.18	2.12	2.07	2.01	1.96	1.89	1.85	1.83
2	0.125	2.67	2.32	2.23	2.17	2.08	2.03	1.94	1.89	1.86
	0.375	3.63	2.82	2.64	2.52	2.35	2.25	2.10	2.02	1.97
	0.625	4.12	3.08	2.85	2.69	2.49	2.36	2.18	2.08	2.02
	0.875	4.37	3.21	2.96	2.78	2.56	2.42	2.22	2.11	2.05
	1.25	4.53	3.29	3.02	2.84	2.60	2.45	2.24	2.13	2.07
	1.5	4.58	3.31	3.04	2.86	2.61	2.46	2.25	2.14	2.07
2.5	0.125	2.75	2.36	2.27	2.20	2.11	2.04	1.95	1.90	1.87
	0.375	4.01	3.02	2.80	2.65	2.46	2.33	2.16	2.06	2.01
	0.625	4.66	3.35	3.08	2.88	2.64	2.48	2.26	2.15	2.08
	0.875	4.99	3.52	3.21	3.00	2.73	2.56	2.32	2.19	2.12
	1.25	5.20	3.64	3.30	3.08	2.79	2.60	2.35	2.22	2.14
	1.5	5.26	3.67	3.33	3.10	2.80	2.62	2.36	2.23	2.15
3.5	0.125	2.93	2.45	2.34	2.26	2.16	2.09	1.98	1.92	1.89
	0.375	4.86	3.46	3.16	2.96	2.69	2.53	2.30	2.18	2.10
	0.625	5.88	3.99	3.59	3.32	2.98	2.76	2.46	2.31	2.22
	0.875	6.40	4.26	3.81	3.51	3.12	2.88	2.55	2.38	2.28
	1.25	6.74	4.43	3.96	3.63	3.21	2.96	2.60	2.42	2.32
	1.5	6.83	4.48	3.99	3.66	3.24	2.98	2.62	2.44	2.33
4.5	0.125	3.13	2.56	2.43	2.34	2.21	2.13	2.01	1.95	1.91
	0.375	5.88	3.99	3.59	3.32	2.98	2.76	2.46	2.31	2.22
	0.625	7.35	4.75	4.22	3.85	3.39	3.10	2.71	2.51	2.39
	0.875	8.11	5.15	4.54	4.13	3.60	3.27	2.83	2.61	2.47
	1	8.33	5.27	4.63	4.21	3.66	3.33	2.87	2.64	2.50
5.5	0.125	3.37	2.69	2.53	2.42	2.28	2.19	2.05	1.98	1.94
	0.375	7.09	4.62	4.11	3.76	3.31	3.04	2.66	2.47	2.36
	0.625	9.13	5.68	4.97	4.49	3.88	3.51	3.00	2.74	2.59
	0.875	10.21	6.24	5.43	4.88	4.18	3.76	3.18	2.89	2.71
	1	10.52	6.41	5.57	5.00	4.27	3.83	3.24	2.93	2.75
6	0.125	3.51	2.76	2.59	2.47	2.32	2.22	2.08	2.00	1.95
	0.375	7.78	4.98	4.40	4.01	3.51	3.20	2.78	2.56	2.44
	0.625	10.17	6.23	5.42	4.87	4.17	3.75	3.18	2.88	2.71
	0.875	11.43	6.88	5.95	5.32	4.53	4.04	3.39	3.06	2.86
	1	11.81	7.08	6.11	5.46	4.64	4.13	3.45	3.11	2.90

Figure 1.13 *HCM Exhibit 26-4*

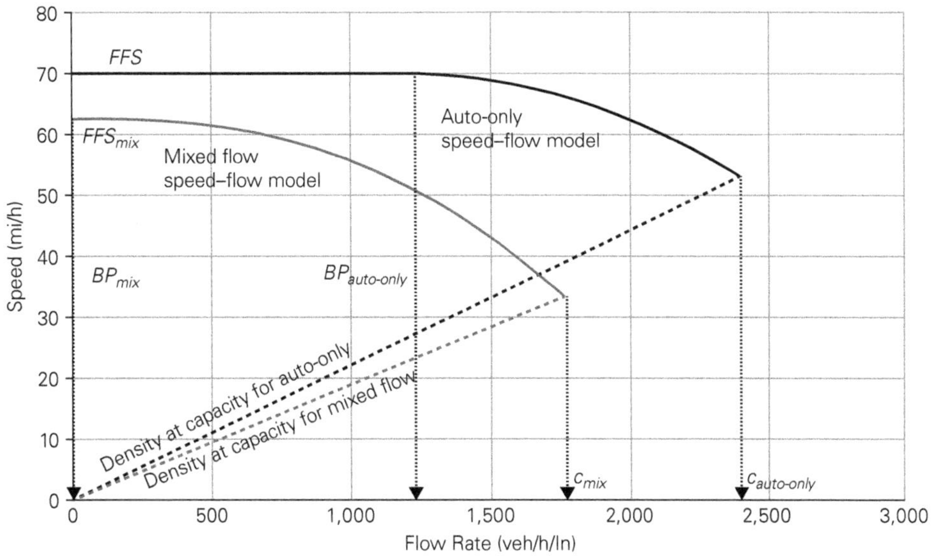

Figure 1.14 *HCM Exhibit 12-6*

$$c = 2,200 + 10(FFS - 50)$$
$$c \leq 2,400$$
$$55 \leq FFS \leq 75$$

Figure 1.15 *HCM Equation 12-8*

$$c_{adj} = c \times CAF$$

Figure 1.16 *HCM Exhibit 11-20*

		Capacity Adjustment Factors				
Weather Type	**Weather Event Definition**	**55 mi/h**	**60 mi/h**	**65 mi/h**	**70 mi/h**	**75 mi/h**
Medium rain	> 0.10–0.25 in/h	0.94	0.93	0.92	0.91	0.90
Heavy rain	> 0.25 in/h	0.89	0.88	0.86	0.84	0.82
Light snow	>0.00 – 0.05 in/h	0.97	0.96	0.96	0.95	0.95
Light-medium snow	>0.05 – 0.10 in/h	0.95	0.94	0.92	0.90	0.88
Medium-heavy snow	>0.10 – 0.50 in/h	0.93	0.91	0.90	0.88	0.87
Heavy snow	>0.50 in/h	0.80	0.78	0.76	0.74	0.72
Severe cold	<–4° F	0.93	0.92	0.92	0.91	0.90
Low visibility	0.50 – 0.99 mi	0.90	0.90	0.90	0.90	0.90
Very low visibility	0.25 – 0.49 mi	0.88	0.88	0.88	0.88	0.88
Minimal visibility	< 0.25 mi	0.90	0.90	0.90	0.90	0.90
Non-severe weather	All conditions not listed above	1.00	1.00	1.00	1.00	1.00

Figure 1.17 *HCM Exhibit 11-23*

Directional Lanes	No Incident	Shoulder Closed	1 Lane Closed	2 Lanes Closed	3 Lanes Closed	4 Lanes Closed
2	1.00	0.81	0.70	N/A	N/A	N/A
3	1.00	0.83	0.74	0.51	N/A	N/A
4	1.00	0.85	0.77	0.50	0.52	N/A
5	1.00	0.87	0.81	0.67	0.50	0.50
6	1.00	0.89	0.85	0.75	0.52	0.52
7	1.00	0.91	0.88	0.80	0.63	0.63
8	1.00	0.93	0.89	0.84	0.66	0.66

Figure 1.18 *HCM Exhibit 26-9*

Level of Driver Familiarity	CAF_{pop}
All familiar drivers, regular commuters	1.000
Mostly familiar drivers	0.968
Balanced mix of familiar and unfamiliar drivers	0.939
Mostly unfamiliar drivers	0.898
All or overwhelmingly unfamiliar drivers	0.852

Think about your own driving:
With weather, even a little rain generates a little more caution where you back off the vehicle in front slightly, increasing headway times and lowering overall speed and capacity. While this change is small for rain, almost all drivers do this and it takes just a little longer to get to work when it rains. For heavy rain or snow or reduced visibility, these effects can be quite significant.

Break Point

Speed is unaffected by flow rate until the flow rate exceeds the break point that is defined as a function of free-flow speed in the equation from HCM Exhibit 12-6. Generally, the break points are 1000, 1200, 1400, 1600, and 1800 (pc/h) for free-flow speeds of 75, 70, 65, 60, and 55 (mi/h), respectively. When the flow rate is lower than the break point, speed is equal to free-flow speed.

Figure 1.19 *HCM Exhibit 12-6*

$$BP_{adj} = [1{,}000 + 40 \times (75 - FFS_{adj})] \times CAF^2$$

Figure 1.20 *HCM Equation 12-1*

$$S = FFS_{adj} \qquad v_p \leq BP$$

$$S = FFS_{adj} - \frac{(FFS_{adj} - \frac{c_{adj}}{D_c})(v_p - BP)^a}{(c_{adj} - BP)^a} \qquad BP < v_p \leq c$$

Think about your own driving:
Speeds are generally unaffected by traffic until it reaches a certain level, where other drivers are in front slowing you down, or behind pushing you to speed up, and on either side preventing you from getting out of their way.

Speed

Speed (S) is computed using the adjusted flow rate and free-flow speed values following the speed-flow curves defined by HCM Equation 12-1 and illustrated in HCM Exhibit 12-7.

Density

Density (D) becomes simply adjusted flow rate divided by speed as shown in HCM Equation 12-11, following the speed-flow relationship illustrated by HCM Exhibit 12-16.

LOS

Level of Service (LOS) is defined by the thresholds shown in HCM Exhibit 12-15 with

Figure 1.21 *HCM Exhibit 12-7*

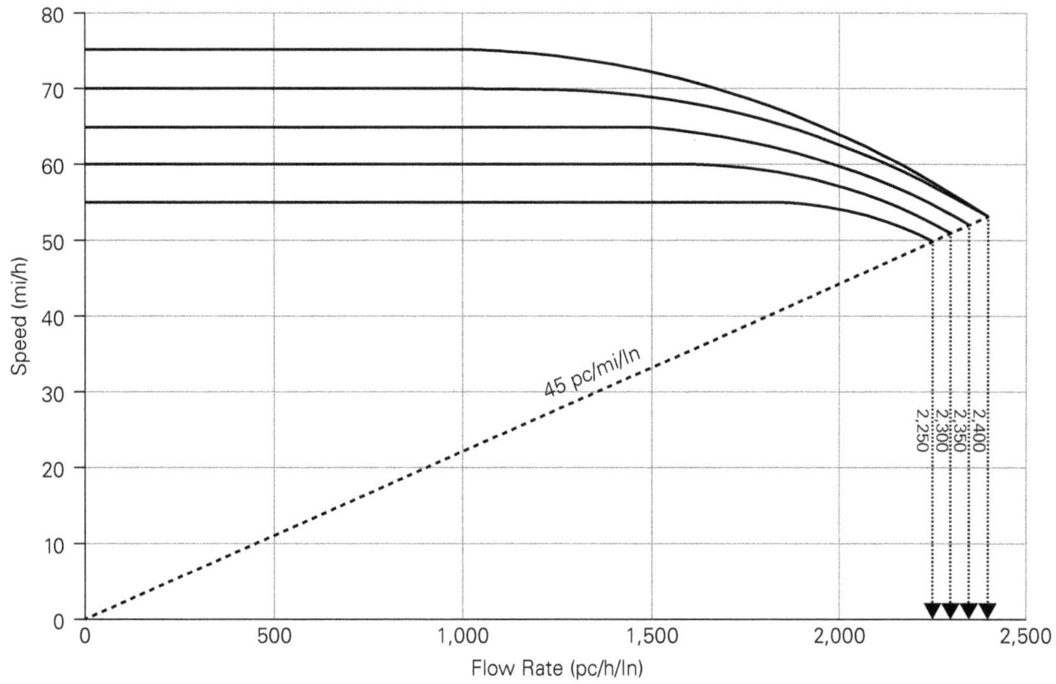

Figure 1.22 *HCM Equation 12-11*

$$D = \frac{v_p}{S}$$

where

D = density (pc/mi/ln),

v_p = demand flow rate (pc/h/ln), and

S = mean speed of traffic stream under base conditions (mi/h).

LOS F further dictated when demand exceeds capacity.

Think about your own driving:
At 2 a.m., you might be driving 75 mi/h essentially alone on the freeway, where you can change lanes at will with few other vehicles around. At 10 a.m., you might still be driving 75 mi/h, but now there are drivers to your left and right, in front and behind, restricting your freedom. This is why density—not speed—determines level of service as a measure of your comfort and convenience.

Managed Lanes

Managed lanes (ML) are modeled for five different operational designs: continuous access (single lane with a skip or solid stripe); buffer-separated (single lane); buffer-separated

Figure 1.23 *HCM Exhibit 12-16*

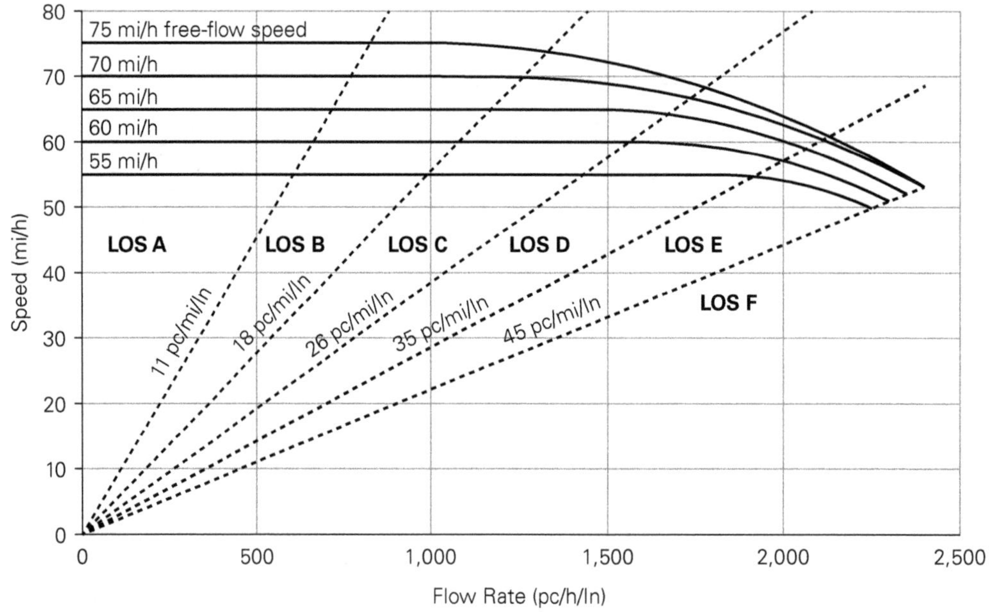

Figure 1.24 *HCM Exhibit 12-15*

LOS	Density (pc/mi/ln)
A	<11
B	>11–18
C	>18–26
D	>26–35
E	>35–45
F	Demand exceeds capacity OR density >45

(multiple lanes); barrier-separated (single lane); and barrier-separated (multiple lanes) as shown in HCM Exhibit 12-9.

The capacity values for managed lanes differ from those for general purpose lanes defined earlier as shown in HCM Exhibit 12-11. Even though the two-lane designs appear to be similar to two-lane freeway segments, the confinement felt by the driver tends to increase headways between vehicles and lower capacity values.

The speed-flow relationships for continuous access and buffer one designs can be altered by friction as illustrated for a continuous access design in HCM Exhibit 12-12. This friction is generated when the general purpose lanes reach a density of 35 pc/mi/ln, which is fairly crowded with capacity around 45 pc/mi/ln.

The procedure for computing density and level of service in managed lanes is similar to that defined earlier for general purpose lanes, but with adjustments to speed, break points, and capacity as appropriate for each managed-lane design after accounting for potential friction with general purpose lanes. HCM Equations 12-12 through 12-19 define the modifications with coefficients provided in HCM Exhibit 12-30.

Ultimately, a final density value is computed to determine level of service for the managed lane segment, separate from the results for the general purpose lanes. While the capacity and speed results will likely be worse than for the general purpose lanes, density and level of service could be better (and result in a better LOS) because of the lower flow rates.

Figure 1.25: *HCM Exhibit 12-9*

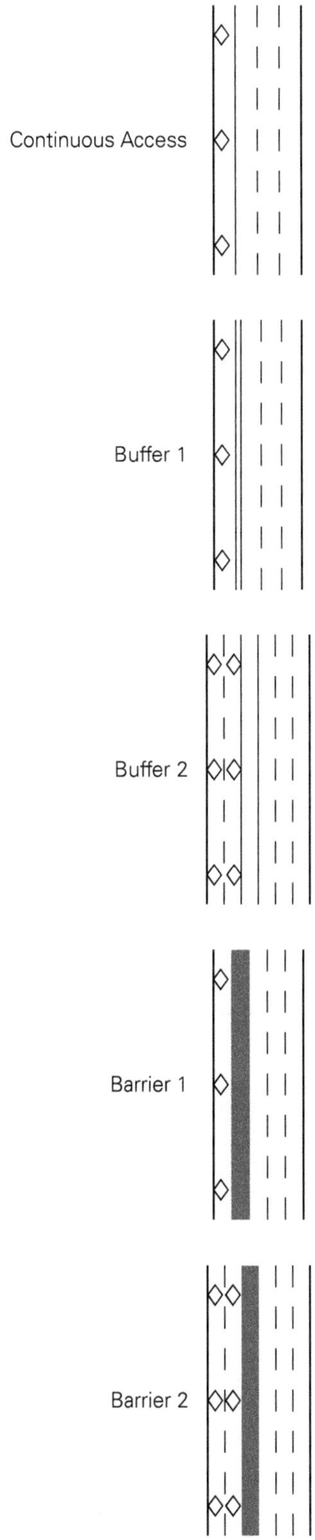

Figure 1.26 *HCM Exhibit 12-11*

FFS (mi/h)	Estimate Lane Capacities for Basic Managed Lane Seqment Type (pc/h/ln) Continuous Access	Buffer 1	Buffer 2	Barrier 1	Barrier 2
75	1,800	1,700	1,850	1,750	2,100
70	1,750	1,650	1,800	1,700	2,050
65	1,700	1,600	1,750	1,650	2,000
60	1,650	1,550	1,700	1,600	1,950
55	1,600	1,500	1,650	1,550	1,900

Figure 1.27 *HCM Exhibit 12-12*

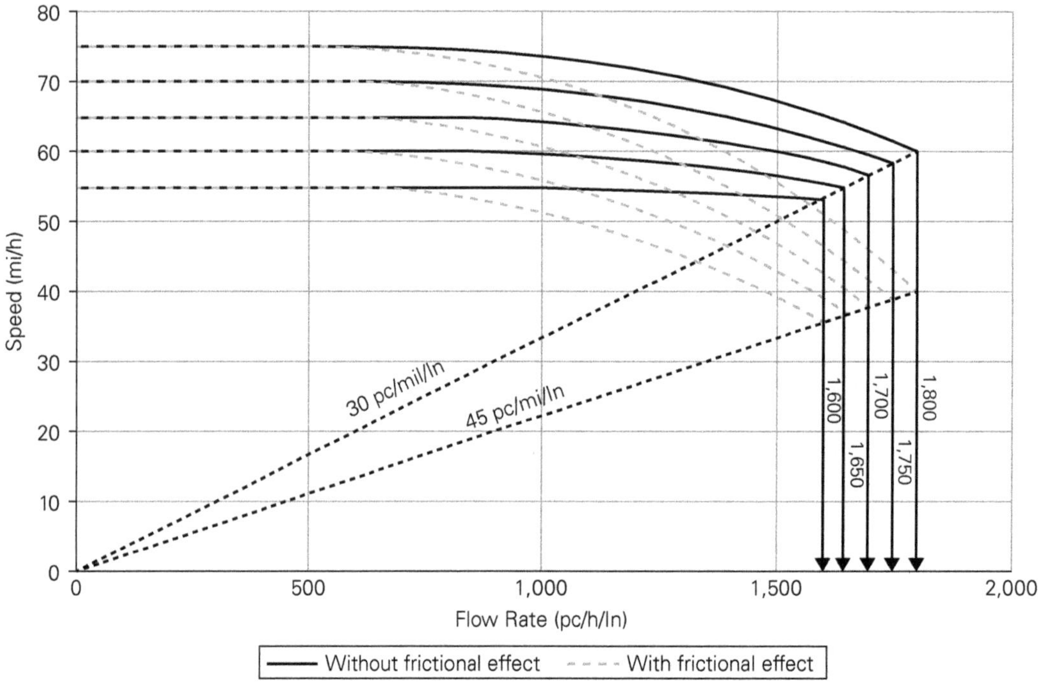

Figure 1.28 *HCM Equation 12-12*

$$S_{ML} = S_1 \qquad v_p \leq BP$$

$$S_{ML} = S_1 - S_2 - I_c \times S_3 \qquad BP < v_p \leq c$$

where

S_{ML} = space mean speed of the basic managed lane segment (mi/h);

S_1 = speed within the linear portion of the speed-flow curve, from Equation 12-15 (mi/h);

S_2 = speed drop within the curvilinear portion of the speed-flow curve, from Equation 12-17 (mi/h);

S_3 = additional speed drop (mi/h) within the curvilinear portion of the speed-flow when the density of the adjacent general purpose lane is more than 35 pc/mi/ln, from Equation 12-19;

I_c = indicator variable, where 1 = presence of densities greater than 35 pc/mi/ln in the adjacent general purpose lane (0 or 1);

BP = breakpoint in the speed-flow curve separating the linear and curvilinear sections (pc/h/ln), from Equation 12-13; and

v_p = 15-min average flow rate (pc/h/ln).

Figure 1.29 *HCM Equation 12-13*

$$BP = [BP_{75} + \lambda_{BP} \times (75 - FFS_{adj})] \times CAF^2$$

where

BP = breakpoint in the speed-flow curve separating the linear and curvilinear sections (pc/h/ln);

BP_{75} = breakpoint for a FFS of 75 mi/h, from Exhibit 12-30 (pc/h/ln);

λ_{BP} = rate of increase in breakpoint per unit decrease in FFS, from Exhibit 12-30 (pc/h/ln);

FFS_{adj} = adjusted free-flow speed (mi/h); and

CAF = capacity adjustment factor (unitless).

Figure 1.30 *HCM Equation 12-14*

$$c_{adj} = CAF \times (c_{75} - \lambda_c \times [75 - FFS_{adj}])$$

where

c_{adj} = adjusted basic managed lane segment capacity (pc/h/ln);

CAF = capacity adjustment factor (unitless);

c_{75} = managed lane capacity for a FFS of 75 mi/h, from Exhibit 12-30 (pc/h/ln);

λ_c = rate of change in capacity per unit change in FFS, from Exhibit 12-30 (pc/h/ln); and

FFS_{adj} = adjusted free-flow speed (mi-h).

Figure 1.31 *HCM Equation 12-15*

$$S_1 = FFS_{adj} - A_1 \times \min(v_p, BP)$$

where A_1 is the speed reduction per unit of flow rate in the linear section

Figure 1.32 *HCM Equation 12-16*

$$A_2 = A_2^{55} + \lambda_{A2} \times [FFS_{adj} - 55]$$

where

A_2 = speed reduction per unit of flow rate in the curvilinear section of the speed-flow curve (mi/h);

A_2^{55} = calibration factor for a FFS of 55 mi/h, from Exhibit 12-30 (mi/h);

λ_{A2} = rate of change in A_2 per unit increase in FFS, from Exhibit 12-30 (mi/h); and

FFS_{adj} = adjusted free-flow speed (mi-h).

Figure 1.33 *HCM Equation 12-17*

$$S_2 = \frac{\left(S_{1,BP} - \frac{c_{adj}}{K_c^{nf}}\right)}{(c_{adj} - BP)^{A_2}}(v_p - BP)^{A_2}$$

where

S_2 = speed drop within the curvilinear portion of the speed-flow curve (mi/h);

$S_{1,BP}$ = speed at the breakpoint of the speed-flow curve, calculated from Equation 12-15 by setting v_p to BP (mi/h);

c_{adj} = adjusted basic managed lane segment capacity (pc/h/ln);

K_c^{nf} = density at capacity, without the frictional effect of the adjacent general purpose lane, from Exhibit 12-30 (pc/mi/ln);

BP = breakpoint in the speed-flow curve separating the linear and curvilinear sections (pc/h/ln);

A_2 = speed reduction per unit of flow rate in the curvilinear section of the speed-flow curve (mi/h); and

v_p = 15-min average flow rate (pc/h/ln).

Figure 1.34 *HCM Equation 12-18*

$I_c = 0$ $K_{GP} \leq 35$ pc/mi//ln
OR segment type is Buffer 2, Barrier 1, or Barrier 2

$I_c = 1$ Otherwise

where K_{GP} is the density of the adjacent general purpose lane (pc/mi/ln).

Figure 1.35 *HCM Equation 12-19*

$$S_3 = \frac{\left(\frac{c_{adj}}{K_c^{nf}}\right) - \left(\frac{c_{adj}}{K_c^{f}}\right)}{(c_{adj} - BP)^2}(v_p - BP)^2$$

where K_c^f is the density at capacity, with the frictional effect

Figure 1.36 *HCM Exhibit 12-30*

Segment Type	BP_{75}	k_{BP}	c_{75}	k_c	A_2^{55}	k_{A2}	A_1	K_c^{nf}	K_c^f
Continuous Access	500	0	1,800	10	2.5	0	0	30	45
Buffer 1	600	0	1,700	10	1.4	0	0.0033	30	42*
Buffer 2	500	10	1,850	10	1.5	0.02	0	45*	NA
Barrier 1	800	0	1,750	10	1.4	0	0.004	35	NA
Barrier 2	700	20	2,100	10	1.3	0.02	0	45	NA

PROBLEM ONE

Determine the LOS for a Basic Freeway Segment with these parameters in the analysis direction: 4000 veh/h with 0.90 PHF with mostly familiar drivers in medium rain; base free-flow speed of 75.4 mi/h; 10% trucks (70% SUT / 30% TT) on a 2% grade for 1 mile; three 11-ft lanes with a 3-ft right-side lateral clearance, and 2 ramps per mile.

1. Compute the free-flow speed by adjusting for lane width, right-side lateral clearance, and ramp density.

Free-Flow Speed: $FFS = 75.4 - 1.9 - 1.2 - 5.8 = 66.5$ mi/h

EQUATION 12-2 / EXHIBITS 12-20 & 12-21

2. Free-flow speed is adjusted for mostly familiar drivers and the medium rain weather event.

Adjustments: $SAF = (0.975)(0.940) = 0.916$

EXHIBITS 11-21 & 26-9

Free-Flow Speed: $FFS_{adj} = 66.5(0.916) = 60.9$

EQUATION 12-5

3. The heavy vehicle adjustment factor is computed for the positive grade with trucks in the traffic stream.

Heavy Vehicles: $E_T = 2.43 \quad f_{HV} = 1 / (1 + 0.10(2.43 - 1)) = 0.875$

EQUATION 12-10 / EXHIBIT 12-28

4. Adjusted flow rate is computed using the heavy vehicle adjustment, peak hour factors with number of lanes.

Flow Rate: $v_p = 4000 / (0.90)(3)(0.875) = 1693$ pc/h/ln

EQUATION 12-9

5. Capacity is computed as a function of free-flow speed, then adjusted for driver familiarity and weather.

Capacity: $c = 2200 + 10(60.9 - 50) = 2309$ pc/h/ln

EXHIBIT 12-6

Adjustments: $CAF = (0.920)(0.968) = 0.891$

EXHIBITS 11-20 & 26-9

Capacity: $c_{adj} = 2309(0.891) = 2057$

EQUATION 12-8

6. The break point can be calculated using the adjusted free-flow speed and capacity adjustment factor.

Break Point: $BP = (1000 + 40(75 - 60.9)) \times (0.891)^2 = 1242$ pc/h/ln

EXHIBIT 12-6

7. This allows speed to be computed from the free-flow speed and flow rate that is beyond the break point.

Speed: $S = 60.9 - ((60.9 - 2057/45)(1693 - 1242)^2) / (2057 - 1242)^2 = 56.3$

EQUATION 12-1

8. Density is calculated from flow rate and speed to determine level of service for the basic freeway segment.

Density: $D = 1693 / 56.3 = 30.1$ pc/mi/ln

EQUATION 12-11

Level of Service: LOS = D (between 26 and 35)

EXHIBIT 12-16

HCS7

These screenshots show how Problem One is coded into *HCS7* in only one screen for this relatively simple procedure, followed by the results as presented in the formatted report. All input data, intermediate computations, and final results are shown in logical order, leading to density and level of service.

Figure 1.37 *HCS7 Basic1 Input*

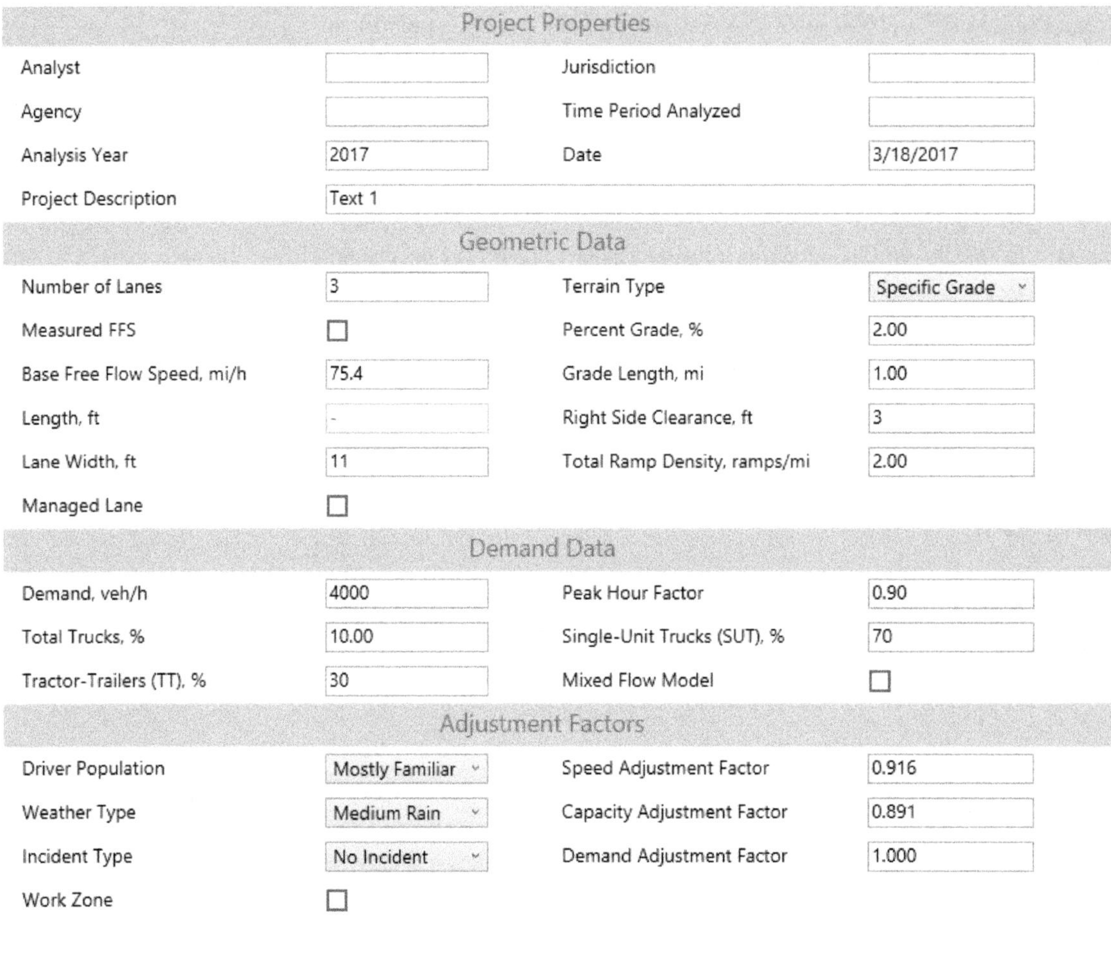

Project Properties

Analyst		Jurisdiction	
Agency		Time Period Analyzed	
Analysis Year	2017	Date	3/18/2017
Project Description	Text 1		

Geometric Data

Number of Lanes	3	Terrain Type	Specific Grade
Measured FFS	☐	Percent Grade, %	2.00
Base Free Flow Speed, mi/h	75.4	Grade Length, mi	1.00
Length, ft	-	Right Side Clearance, ft	3
Lane Width, ft	11	Total Ramp Density, ramps/mi	2.00
Managed Lane	☐		

Demand Data

Demand, veh/h	4000	Peak Hour Factor	0.90
Total Trucks, %	10.00	Single-Unit Trucks (SUT), %	70
Tractor-Trailers (TT), %	30	Mixed Flow Model	☐

Adjustment Factors

Driver Population	Mostly Familiar	Speed Adjustment Factor	0.916
Weather Type	Medium Rain	Capacity Adjustment Factor	0.891
Incident Type	No Incident	Demand Adjustment Factor	1.000
Work Zone	☐		

Figure 1.38 *HCS7 Basic1 Output*

HCS7 Basic Freeway Report			
Project Information			
Analyst		Date	3/18/2017
Agency		Analysis Year	2017
Jurisdiction		Time Period Analyzed	
Project Description	Text 1		
Geometric Data			
Number of Lanes (N), ln	3	Terrain Type	Specific Grade
Segment Length (L), ft	-	Percent Grade, %	2.00
Measured or Base Free-Flow Speed	Base	Grade Length, mi	1.00
Base Free-Flow Speed (BFFS), mi/h	75.4	Total Ramp Density (TRD), ramps/mi	2.00
Lane Width, ft	11	Free-Flow Speed (FFS), mi/h	66.5
Right-Side Lateral Clearance, ft	3		
Adjustment Factors			
Driver Population	Mostly Familiar	Final Speed Adjustment Factor (SAF)	0.916
Weather Type	Medium Rain	Final Capacity Adjustment Factor (CAF)	0.891
Incident Type	No Incident	Demand Adjustment Factor (DAF)	1.000
Demand and Capacity			
Volume (V), veh/h	4000	Heavy Vehicle Adjustment Factor (f_{HV})	0.875
Peak Hour Factor (PHF)	0.90	Flow Rate (v_p), pc/h/ln	1693
Total Trucks, %	10.00	Capacity (c), pc/h/ln	2309
Single-Unit Trucks (SUT), %	70	Adjusted Capacity (c_{adj}), pc/h/ln	2057
Tractor-Trailers (TT), %	30	Volume-to-Capacity Ratio (v/c)	0.82
Passenger Car Equivalent (E_T)	2.430		
Speed and Density			
Lane Width Adjustment (f_{LW})	1.9	Average Speed (S), mi/h	56.3
Right-Side Lateral Clearance Adj. (f_{RLC})	1.2	Density (D), pc/mi/ln	30.1
Total Ramp Density Adjustment	5.8	Level of Service (LOS)	D
Adjusted Free-Flow Speed (FFS_{adj}), mi/h	60.9		

PROBLEM TWO

Expanding on Problem One, one managed lane with continuous access is added to determine the LOS for the managed lane facility with these additional parameters: 1000 veh/h in the managed lane with a base free-flow speed of 65 mi/h and no trucks.

The computations for free-flow speed, flow rate, capacity, and break point are modified for the conditions within the managed lane facility. These values are used in the managed lane process for computing speed through that series of equations. The revised flow rate and speed are then used to compute density and determine level of service for the managed lane, independently of the general purpose lanes analyzed in Problem One.

1. The given free-flow speed is adjusted for mostly familiar drivers and the medium rain weather event.

Adjustments: $SAF = (0.975)(0.940) = 0.916$
EXHIBITS 11-21 & 26-9

Free-Flow Speed: $FFS_{adj} = 65.0(0.916) = 59.5$
EQUATION 12-5

2. Adjusted flow rate is computed using the peak hour factors with no trucks and only one lane.

Flow Rate: $v_p = 1000 / (0.90)(1)(1.000) = 1111$ pc/h/ln
EQUATION 12-9

3. Capacity is computed as a function of free-flow speed, then adjusted for driver familiarity and weather.

Capacity: $c = 1800 - 10\,(75 - 59.5)) = 1645$ pc/h/ln
EQUATION 12-14 / EXHIBIT 12-30

Adjustments: $CAF = (0.920)(0.968) = 0.891$
EXHIBITS 11-5 & 26-9

Capacity: $c_{adj} = 1645(0.891) = 1466$
EQUATION 12-8

4. The break point can be calculated using the adjusted free-flow speed and capacity adjustment factor.

Break Point: $BP = (500 + 0(75 - 59.5)) \times (0.891)^2 = 398$ pc/h/ln

EQUATION 12-13 / EXHIBIT 12-30

5. Speed in the managed lane is computed following the modified procedure.

Speed: $S_1 = 59.5 - (0)(398) = 59.5$

EQUATION 12-15 / EXHIBIT 12-30

$S_2 = ((59.5 - 1466/30)/(1466 - 398)^{2.5})(1111 - 398)^{2.5} = 3.9$ mi/h

EQUATION 12-17 / EXHIBIT 12-30

$S_3 = ((1466/30) - (1466/45))/(1466 - 398)^2)(1111 - 398)^2 = 7.3$ mi/h

EQUATION 12-19 / EXHIBIT 12-30

$S_{ML} = 59.5 - 3.9 - 7.3(0) = 55.6$

EQUATION 12-12

6. Density is calculated from flow rate and speed to determine level of service for the managed lane.

Density: $D_{ML} = 1111 / 55.6 = 20.0$ pc/mi/ln

EQUATION 12-11

Level of Service: $LOS_{ML} = C$ (between 18 and 26)

EXHIBIT 12-16

HCS7

These screens show the managed lane portions of the input screen and results report to illustrate this extension to the analysis that computes density and level of service for the managed lanes, again independently of the general purpose lanes.

Figure 1.39 *HCS7 Basic2 Input*

Geometric Data			
Number of Lanes	3	Terrain Type	Specific Grade
Measured FFS	☐	Percent Grade, %	2.00
Base Free Flow Speed, mi/h	75.4	Grade Length, mi	1.00
Length, ft	-	Right Side Clearance, ft	3
Lane Width, ft	11	Total Ramp Density, ramps/mi	2.00
Managed Lane	☑		
Demand Data			
Demand, veh/h	4000	Peak Hour Factor	0.90
Total Trucks, %	10.00	Single-Unit Trucks (SUT), %	70
Tractor-Trailers (TT), %	30	Mixed Flow Model	☐
Adjustment Factors			
Driver Population	Mostly Familia	Speed Adjustment Factor	0.916
Weather Type	Medium Rain	Capacity Adjustment Factor	0.891
Incident Type	No Incident	Demand Adjustment Factor	1.000
Work Zone	☐		
Managed Lane Geometric Data			
Number of Lanes	1	Terrain Type	Specific Grade
Managed Lane Type	Continuous Ac	Percent Grade, %	2.00
Free Flow Speed, mi/h	65.0	Grade Length, mi	1.00
Managed Lane Demand Data			
Demand, veh/h	1000	Peak Hour Factor	0.90
Total Trucks, %	0.00	Single-Unit Trucks (SUT), %	30
Tractor-Trailers (TT), %	70		
Managed Lane Adjustment Factors			
Driver Population	Mostly Familia	Speed Adjustment Factor	0.916
Demand Adjustment Factor	1.000	Capacity Adjustment Factor	0.891

Figure 1.40 *HCS7 Basic2 Output*

HCS7 Basic Freeway Report			
Project Information			
Analyst		Date	3/18/2017
Agency		Analysis Year	2017
Jurisdiction		Time Period Analyzed	
Project Description	Text 1		
General Purpose Geometric Data			
Number of General Purpose Lanes, ln	3	Terrain Type	Specific Grade
Segment Length (L), ft	-	Percent Grade, %	2.00
Measured or Base Free-Flow Speed	Base	Grade Length, mi	1.00
Base Free-Flow Speed (BFFS), mi/h	75.4	Total Ramp Density (TRD), ramps/mi	2.00
Lane Width, ft	11	Free-Flow Speed (FFS), mi/h	66.5
Right-Side Lateral Clearance, ft	3		
General Purpose Adjustment Factors			
Driver Population	Mostly Familiar	Final Speed Adjustment Factor (SAF)	0.916
Weather Type	Medium Rain	Final Capacity Adjustment Factor (CAF)	0.891
Incident Type	No Incident	Demand Adjustment Factor (DAF)	1.000
General Purpose Demand and Capacity			
Volume (V), veh/h	4000	Heavy Vehicle Adjustment Factor (f_{HV})	0.875
Peak Hour Factor (PHF)	0.90	Flow Rate ($v_{p,GP}$), pc/h/ln	1693
Total Trucks, %	10.00	Capacity (c), pc/h/ln	2309
Single-Unit Trucks (SUT), %	70	Adjusted Capacity (c_{adj}), pc/h/ln	2057
Tractor-Trailers (TT), %	30	Volume-to-Capacity Ratio (v/c)	0.82
Passenger Car Equivalent (E_T)	2.430		
General Purpose Speed and Density			
Lane Width Adjustment (f_{LW})	1.9	Average Speed (S), mi/h	56.3
Right-Side Lateral Clearance Adj. (f_{RLC})	1.2	Density (D_{GP}), pc/mi/ln	30.1
Total Ramp Density Adjustment	5.8	Level of Service (LOS)	D
Adjusted Free-Flow Speed (FFS_{adj}), mi/h	60.9		

Managed Lane Geometric Data			
Managed Lane Type	Continuous Access	Free-Flow Speed (FFS), mi/h	65.0
Number of Managed Lanes, ln	1	Terrain Type	Specific Grade
Managed Lane Length, ft	5280	Percent Grade, %	2.00
Managed Lane Adjustment Factors			
Driver Population	Mostly Familiar	Driver Population CAF	0.968
Weather Type	Medium Rain	Weather Type CAF	0.920
Driver Population SAF	0.968	Final Speed Adjustment Factor (SAF)	0.916
Weather Type SAF	0.940	Final Capacity Adjustment Factor (CAF)	0.891
Demand Adjustment Factor (DAF)	1.000		
Managed Lane Demand and Capacity			
Volume (V_{ML}), veh/h	1000	Heavy Vehicle Adjustment Factor (f_{HV})	1.000
Peak Hour Factor	0.90	Flow Rate ($V_{p,ML}$), pc/h/ln	1111
Total Trucks, %	0.00	Capacity (c), pc/h/ln	1645
Single-Unit Trucks (SUT), %	30	Adjusted Cpacity (c_{adj}), pc/h/ln	1466
Tractor-Trailers (TT), %	70	Volume-to-Capacity Ratio (v/c)	0.76
Passenger Car Equivalent (E_T)	4.867		
Managed Lane Speed and Density			
Breakpoint (BP_{ML})	398	Indicator Variable	0
Speed 1 (S_1), mi/h	59.5	Average Speed (S_{ML}), mi/h	55.6
Speed 2 (S_2), mi/h	3.9	Density (D_{ML}), pc/mi/ln	20.0
Speed 2 (S_3), mi/h	7.3	Level of Service (LOS)	C

Freeway Facilities

Freeway Facilities are analyzed as the combination of multiple freeway segments over multiple time periods to generate facility-wide results requiring the length for each segment be defined to create a continuous facility. Using this methodology, oversaturated conditions and work zones can also be modeled to compute delay and queuing if present due to a bottleneck situation. Procedures are defined for modeling travel time reliability with demand, weather, incidents, work zones, and special events varying for a distribution of results for hundreds of scenarios to produce travel time reliability indices at three different percentiles.

The analysis of a Freeway Facility defined in HCM Chapter 10 involves consolidating multiple freeway segments into a continuous facility to compute facility-wide density and level of service results. For undersaturated conditions, these values will generally match the individual segment analysis results. Except merge and diverge segment densities may differ for freeways with more than two lanes in each direction, since these density results will be computed across all freeway lanes (not just the two adjacent lanes within the segment influence area as required for ramp segment level of service).

Undersaturated Conditions

To analyze the facility, all data required for each included segment as defined in Chapter One must be provided. Additionally, segment lengths to create a continuous facility are required, and the facility should begin and end with basic segments. Ramp segments are assigned a 1500-foot segment length as their influence areas when possible. If restricted distances between ramps don't allow this, an overlap segment can be used. Weaving segment lengths are defined as 500 feet beyond both gores of the ramps. Multiple time periods are accommodated for undersaturated conditions and are required for oversaturated conditions.

Freeway volume entering the initial basic segment is adjusted through the facility with additions from on-ramps and subtractions for off-ramps to generate the volume exiting the system though the last basic segment. All segments within the facility are modeled as appropriate with weaving segments that exceed the maximum length analyzed as basic segments and merges, with lane additions and diverges with lane drops also analyzed as basic segments—with ramp capacity checks in both cases.

Work zones can be modeled to calculate capacity and speed adjustment factors following a procedure defined in HCM Chapter 10, Section 4. Additional data required include type of closure (shoulder or lane); barrier type (cone, drum, or concrete); lateral distance (barrier to travel lane); work zone seed (limit and ratio to normal); area (urban or rural); and time (day or night). The Lane Closure Severity Index (LCSI) is computed as a function of the open ratio (total to open lanes) according to HCM Exhibit 10-15. HCM Equations 10-8 through 10-12 are used to compute the capacity and speed in the work zone for use in comparing with non–work zone capacity and speed to generate the adjustment factors (CAF and SAF).

Figure 1.41 *HCM Page 10-41*

$LCSI$ = Lane closure severity index (described below);

F_{Br} = Indicator variable for barrier type:

= 0 for concrete and hard barrier separation, and

= 1 for cone, plastic drum, or other soft barrier separation;

f_{AT} = Indicator factor for area type:

= 0 for urban areas (i.e., typified by high development densities or concentrations of population), and

= 1 fir rural areas (i.e., areas with widely scattered development and low housing and employment densities);

f_{LAT} = Lateral distance from the edge of travel lane adjacent to the work zone to the barrier, barricades, or cones (0–12 ft);

f_{DN} = Indicator variable for daylight or night:

= 0 for daylight, and

= 1 for night;

f_{Sr} = Speed ratio (decimal); the ratio of non–work zone speed limit (before the work zone was established) to work zone speed limit;

SL_{wz} = Work zone speed limit (mi/h); and

TRD = Total ramp density along the facility (ramps/mi); for isolated segment analyses, ramps should be counted 3 mi upstream and 3 mi downstream of the center of the work zone.

Figure 1.42 *HCM Exhibit 10-15*

Number of Total Lane(s)	Number of Open Lane(s)	Open Ratio	LCSI
3	3	1.00	0.33
2	2	1.00	0.50
4	3	0.75	0.44
3	2	0.67	0.75
4	2	0.50	1.00
2	1	0.50	2.00
3	1	0.33	3.00
4	1	0.25	4.00

Figure 1.43 *HCM Equation 10-8*

$$QDR_{WZ} = 2{,}093 - 154 \times LCSI - 194 \times f_{Br} - 179 \times f_{AT} + 9 \times f_{LAT} - 59 \times f_{DN}$$

Figure 1.44 *HCM Equation 10-9*

$$c_{WZ} = \frac{QDR_{WZ}}{100 - \alpha_{WZ}} \times 100$$

Figure 1.45 *HCM Equation 10-10*

$$FFS_{wz} = 9.95 + 33.49 \times f_{Sr} + 0.53 \times SL_{wz} - 5.60 \times LCSI - 3.84 \times f_{Br} - 1.71 \times f_{DN} - 8.7 \times TRD$$

Figure 1.46 *HCM Equation 10-11*

$$CAF_{wz} = \frac{c_{wz}}{c}$$

where

CAF_{wz} = capacity adjustment factor for a work zone (decimal),

c = basic freeway segment capacity in non-work-zone conditions (pc/h/ln), and

c_{wz} = work zone capacity (pre-breakdown flow rate) (pc/h/ln).

Figure 1.47 *HCM Equation 10-12*

$$SAF_{wz} = \frac{FFS_{wz}}{FFS}$$

where

SAF_{wz} = free-flow-speed adjustment factor for work zone (decimal),

$FFSc$ = freeway free-flow speed in non-work-zone conditions (mi/h), and

FFS_{wz} = work zone free-flow speed (mi/h)

Oversaturated Conditions

This method is the only way to model oversaturated conditions: across multiple segments over multiple time periods to capture the interactive effects of both time and space. The interaction of bottleneck segments upstream (generating queues) and downstream (protecting from flow) are modeled to generate results beyond density to include queuing and delay.

The analysis of oversaturated conditions requires beginning and ending undersaturated in both time and space. This means that the first and last time period must be undersaturated for all segments, and the first and last segments must be undersaturated for all time periods.

When segments have demand-to-capacity (*d/c*) ratios greater than one, the facility is classified as operating in oversaturated conditions—creating a bottleneck. Queues can be expected upstream of these segments and the volume served downstream of these segments is expected to be less than the potential demand. In a freeway queue, there is usually movement as stop-and-go traffic that is different from a standing queue (like at a red light). In a standing queue, passenger cars take up about 25 feet; that includes the length of the vehicle and the distance between vehicles. On a freeway, the distance between the vehicles fluctuates to create a longer average distance than for stopped vehicles. This is why jam density is estimated at 190 pc/mi/ln, not just 5280 feet in a mile divided by 25 feet per vehicle in a queue (which would be about 210 pc/mi/ln).

The d/c ratio for the first bottleneck segment is assumed to be 1.0. If demand increases over time, the unserved demand will be passed forward to the subsequent time period, generating d/c ratios greater than 1.0. This will likely create queuing that requires the speed and density computations to be redone, yielding reduced speeds and higher densities.

Think about your own driving:
You probably have run into stop-and-go congestion on the freeway (where demand exceeds capacity, likely when an on-ramp adds traffic to the freeway) that breaks free at some point beyond the bottleneck (where the restriction has protected the downstream freeway).

Level of Service

Density and level of service results are computed for each segment, then used to generate a facility density as a length and number of lanes weighted average to determine a facility-wide level of service for each time period, following HCM Equation 10-1 and HCM Exhibit 10-6. Density is averaged across time periods, but this value cannot be used for a composite level of service. Level of service is computed for both general purpose and managed lanes independently, while still considering potential friction as defined for Basic Freeway Segments.

Figure 1.48 *HCM Equation 10-1*

$$D_F = \frac{\sum_{i=1}^{n} D_i \times L_i \times N_i}{\sum_{i=1}^{n} L_i \times N_i}$$

where

D_F = average density for the facility in a given 15-min analysis period (pc/mi/ln),

D_i = density for segment i (pc/mi/ln),

L_i = length of segment i (miles),

N_i = number of lanes in segment i, and

n = number of segments in the defined facility.

Figure 1.49 *HCM Exhibit 10-6*

	Freeway Facility Density (pc/mi/ln)	
LOS	**Urban**	**Rural**
A	≤11	≤6
B	>11–18	>6–14
C	>18–26	>14–22
D	>26–35	>22–29
E	>35–45	>29–39
F	>45 or any component segment v_d/c ratio > 1.00	>39 or any component segment v_d/c ratio >1.00

Travel Time Reliability

Most HCM analyses are based on fifteen-minute data from a typical peak period, but there may be a need to look at many more scenarios to cover the variations throughout the year. Modeling (potentially) hundreds of scenarios with varying parameters like demand, weather, incidents, work zones, and special events can generate a distribution of results to produce travel time reliability results like travel time indices and reliability ratings, following HCM Exhibit 11-7.

SCENARIOS

Demand data can normally be obtained from permanent counting stations that generate adjustments by time of day, day of week, and month of year to create the distribution over time. Weather data are provided in a database for over one hundred cities to be able to select the one nearest the analysis facility. Incident data would generally come from crash reports for a given area, but can be in various formats with a variety of details. Work Zone scenarios would be run independently with the work zone parameters in place for inserting during the appropriate time periods. Special Events would also be analyzed separately with the typical increase in demand for inserting for those time periods.

Once the demand, weather, and incident data bases are built and the analyses for any work zones and/or special events are run, a base data set is provided for the typical fifteen-minute analysis for generating the additional scenarios for a specified set of days and times. While the work zone and special event timing is known,

Figure 1.50 *HCM Exhibit 11-7*

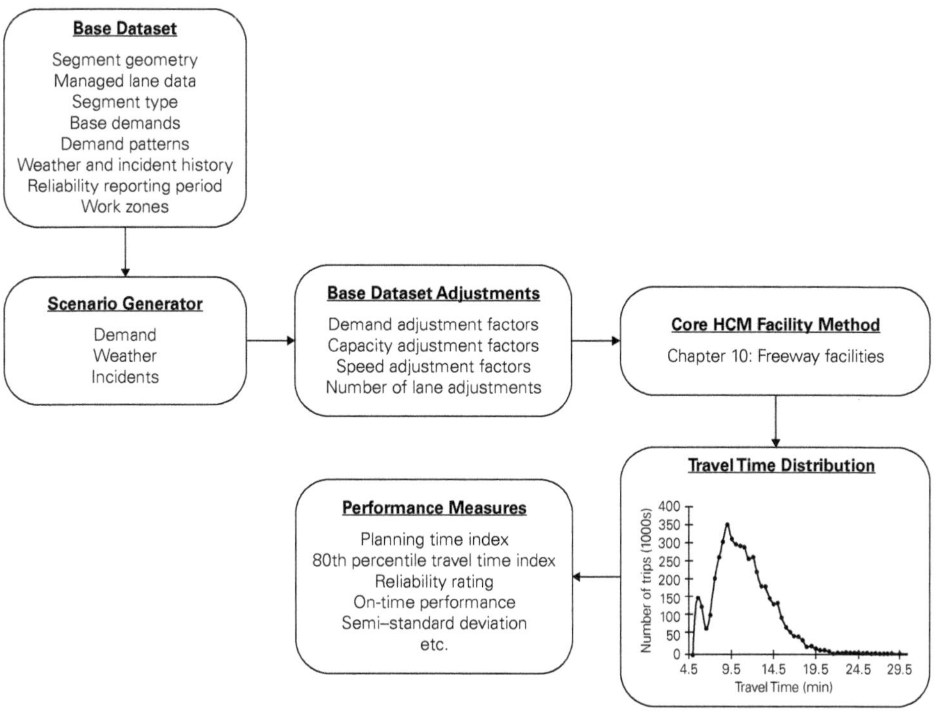

Figure 1.51 *HCM Equation 37-1*

$$AveCap(s) = \frac{CapShldr(s) + CapMFlanes(s) \times MFlanes(s)}{1 + MFlanes(s)}$$

where

$AveCap(s)$ = average capacity per lane for section s (veh/h/ln),

$CapShldr(s)$ = capacity per shoulder lane for section s (veh/h/ln),

$CapMFlanes(s)$ = capacity per mixed-flow lane in section s (veh/h/ln), and

$MFlanes(s)$ = number of mixed-flow lanes in section s (ln).

Figure 1.52 *HCM Equation 37-2*

$$R(t) = \frac{(CM - VM(t))}{NR}$$

subject to

$$MinRate < R(t) < MaxRate$$

$$R(t) > \frac{VR(t) + QR(t-1) - QRS}{NR}$$

where

$R(t)$ = ramp-metering rate for analysis period t (veh/h/ln),

NR = number of metered lanes on ramp (ln),

CM = capacity of downstream section (veh/h),

$VM(t)$ = volume on upstream section for analysis period t (veh/h),

$VR(t)$ = volume on ramp during analysis period t (veh/h),

$QR(t-1)$ = queue on ramp at end of previous analysis period $t-1$ (veh),

QRS = queue storage capacity of ramp (veh),

$MinRate$ = user-defined minimum ramp-metering rate (veh/h/ln) (default value is 240 veh/h/ln), and

$MaxRate$ = user-defined maximum ramp-metering rate (veh/h/ln) (default value is 900 veh/h/ln).

the demand, weather, and incident data must be randomized to populate the individual runs.

RESULTS

Results from all scenario analyses are compiled as a distribution to generate specific performance measures, including travel time, free-flow travel time, travel time index, percentile travel time index, and reliability rating; actual travel time over the entire facility that will vary with the conditions for each scenario. Free-flow travel time is simply the facility length divided by free-flow speed. Travel Time Index (TTI) is the ratio of the actual travel time to the free-flow travel time. Percentile travel time index is provided for the 50th (mean index), 80th (operational), and 95th (planning time index) percentiles. Reliability rating is the percentage of vehicle miles traveled with a travel time index under 1.33 that is considered reliable.

ATDM

Active Transportation Demand Management (ATDM) strategies, including ramp metering; traveler information; managed lanes; and variable speed limits. There are five types of strategies, including demand management (all segments and time periods); weather management (during a severe weather event); incident management (when an incident is present); work zone management (when a work zone is present); and other segment-specific strategies (like hard-shoulder running and ramp metering). While all are discussed, only ramp metering and hard-shoulder running have analytical guidance as shown in HCM Equations 37-1 and 37-2.

PROBLEM ONE

Since the procedures for this analysis do not lend themselves to performing manually, the HCM Example Problem is illustrated here. There are eleven segments as shown in HCM Exhibits 25-43 and 25-44. All pertinent data are shown from HCM pages 25–86, including demands in five time periods in HCM Exhibit 12-45 for an undersaturated flow analysis.

Results for volumes served (HCM Exhibit 25-48), density (HCM Exhibit 25-50), and level of service (HCM Exhibit 25-51) for all segments through all time periods are shown below, with the composite facility density for all time periods from HCM Exhibit 25-52. Notice that Segments 8, 9, 10, and 11 are operating at LOS E in time period three.

Figure 1.53 *HCM Exhibit 25-43*

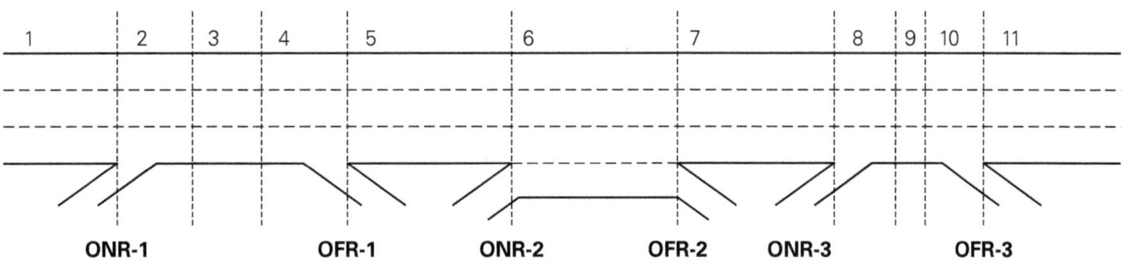

Figure 1.54 *HCM Exhibit 25-44*

Segment No.	1	2	3	4	5	6	7	8	9	10	11
Segment type	B	ONR	B	OFR	B	B or W	B	ONR	R	OFR	B
Segment length (ft)	5,280	1,500	2,280	1,500	5,280	2,640	5,280	1,140	360	1,140	5,280
No. of lanes	3	3	3	3	3	4	3	3	3	3	3

Figure 1.55 *HCM Page 25-86*

SUTs and buses = 1.25% (all movements);

Mainline TTs = 1.00% (all movements);

Driver population = regular commuters;

FFS = 60 mi/h (all mainline segments);

Ramp *FFS* = 40 mi/h (all ramps);

Acceleration lane length = 500 ft (all ramps);

Deceleration lane length = 500 ft (all ramps);

$D_{jam} = 190$ pc/mi/ln;

$c_{IFL} = 2{,}300$ pc/h/ln (for *FFS* = 60 mi/h);

$L_S = 1{,}640$ ft (for Weaving Segment 6);

TRD = 1.0 ramp/mi;

Terrain = level; and

Analysis duration = 75 min (divided into five 15-min intervals).

A queue discharge capacity drop of 7% will be assumed.

Figure 1.56 *HCM Exhibit 25-45*

Time Step (15 min)	Entering Flow Rate (veh/h)	Ramp Flow Rates by Time Period (veh/h)						Exiting Flow Rate (veh/h)
		ONR1	ONR2*	ONR3	OFR1	OFR2	OFR3	
1	4,505	450	540 (50)	450	270	360	270	5,045
2	4,955	540	720 (100)	540	360	360	270	5,765
3	5,225	630	810 (150)	630	270	360	450	6,215
4	4,685	360	360 (80)	450	270	360	270	4,955
5	3,785	180	270 (50)	270	270	180	180	3,875

Figure 1.57 *HCM Exhibit 25-48*

Time Step	Volumes Served (veh/h) by Segment										
	1	2	3	4	5	6	7	8	9	10	11
1	4,505	4,955	4,955	4,955	4,685	5,225	4,865	5,315	5,315	5,315	5,045
2	4,955	5,495	5,495	5,495	5,135	5,855	5,495	6,035	6,035	6,035	5,765
3	5,225	5,855	5,855	5,855	5,585	6,395	6,035	6,665	6,665	6,665	6,215
4	4,685	5,045	5,045	5,045	4,775	5,135	4,775	5,225	5,225	5,225	4,955
5	3,785	3,965	3,965	3,965	3,695	3,965	3,785	4,055	4,055	4,055	3,875

Figure 1.58 *HCM Exhibit 25-50*

Time Step	Density (veh/mi/ln) by Segment										
	1	2	3	4	5	6	7	8	9	10	11
1	25.0	30.6	27.6	29.4	26.0	27.2	27.1	33.2	33.2	31.6	28.1
2	27.6	34.5	31.2	32.8	28.7	31.3	31.2	38.5	38.5	36.1	33.4
3	29.3	37.1	34.1	35.0	31.9	34.6	35.8	43.9	43.9	42.9	37.6
4	26.0	31.3	28.1	30.0	26.5	25.8	26.5	32.5	32.5	31.1	27.6
5	21.0	24.1	22.0	23.5	20.5	18.9	21.0	24.7	24.7	23.9	21.5

Figure 1.59 *HCM Exhibit 25-51*

Time Step	LOS by Segment										
	1	2	3	4	5	6	7	8	9	10	11
1	C	C	D	C	D	C	D	D	D	D	D
2	D	D	D	D	D	D	D	D	E	D	D
3	D	D	D	D	D	D	E	E	E	D	E
4	D	C	D	C	D	C	D	C	D	D	D
5	C	C	C	C	C	B	C	C	C	C	C

Figure 1.60 *HCM Exhibit 25-52*

Time Step	Performance Measures: Space Mean Speed (mi/h)	Performance Measures: Average Density (veh/mi/ln)	LOS
1	57.6	27.5	D
2	56.6	31.3	D
3	55.0	34.8	E
4	57.9	27.5	D
5	58.4	21.4	C
Total	**56.9**	**28.4**	—

PROBLEM TWO

Building on Problem One, all entering and exiting flow rates (HCM Exhibit 25-53) are increased by 11% to generate an oversaturate analysis, yielding demand exceeding capacity in four segments during time period three (HCM Exhibit 25-55). Again, density (HCM Exhibit 25-58) and level of service (HCM Exhibit 25-59) values are shown. Notice that Segments 8, 9, 10, and 11 are operating at LOS F in time period three.

Figure 1.61 *HCM Exhibit 25-53*

Time Step (15 min)	Entering Flow Rate (veh/h)	Ramp Flow Rates by Time Period (veh/h)						Exiting Flow Rate (veh/h)
		ONR1	ONR2*	ONR3	OFR1	OFR2	OFR3	
1	5,001	500	599 (56)	500	300	400	300	5,600
2	5,500	599	799 (111)	599	400	400	300	6,399
3	5,800	699	899 (167)	699	300	400	500	6,899
4	5,200	400	400 (89)	500	300	400	300	5,500
5	4,201	200	200 (56)	300	300	200	200	4,301

Figure 1.62 *HCM Exhibit 25-55*

Time Step	Demand-to-Capacity Ratios by Segment										
	1	2	3	4	5	6	7	8	9	10	11
1	0.74	0.82	0.82	0.82	0.77	0.70	0.80	0.87	0.87	0.87	0.83
2	0.82	0.90	0.90	0.90	0.84	0.78	0.90	0.99	0.99	0.99	0.95
3	0.86	0.96	0.96	0.96	0.92	0.85	0.99	**1.10**	**1.10**	**1.10**	**1.02**
4	0.77	0.83	0.83	0.83	0.79	0.68	0.79	0.86	0.86	0.86	0.82
5	0.62	0.65	0.65	0.65	0.61	0.52	0.62	0.67	0.67	0.67	0.64

Figure 1.63 *HCM Exhibit 25-58*

Time	Density (veh/mi/ln) by Segment										
Step	1	2	3	4	5	6	7	8	9	10	11
1	27.9	34.5	31.3	32.8	29.2	31.0	30.5	37.4	37.4	35.3	32.0
2	31.3	39.0	36.4	36.7	32.8	35.8	36.4	44.2	44.2	43.3	39.6
3	33.7	42.4	40.8	40.8	42.9	64.8	66.4	40.6	40.6	38.3	34.2
4	36.7	39.3	36.3	38.6	33.4	63.9	65.1	40.4	40.4	38.2	34.8
5	23.3	26.9	24.5	26.1	22.8	22.4	31.7	31.9	31.9	30.5	27.3

Figure 1.64 *HCM Exhibit 25-59*

Time	Density-Based LOS by Segment										
Step	1	2	3	4	5	6	7	8	9	10	11
1	D	D	D	D	D	D	D	D	E	D	D
2	D	D	E	D	D	E	E	E	E	D	E
3	D	D	E	D	E	F	F	D	E	D	D
4	E	E	E	E	D	F	F	D	E	D	E
5	C	C	C	C	C	C	D	C	D	C	D
Time	**Demand-Based LOS by Segment**										
Step	**1**	**2**	**3**	**4**	**5**	**6**	**7**	**8**	**9**	**10**	**11**
1											
2											
3								**F**	**F**	**F**	**F**
4											
5											

PROBLEM THREE

To rectify the congestion in Problem Two, the auxiliary lane for the weaving segment (Segment 6) is extended to Segments 7 through 11 (HCM Exhibit 25-61). The additional capacity provided by this lane results in the reduction of all d/c ratios to under 1.0 (HCM Exhibit 25-64) and lowering density values (HCM Exhibit 25-66) for those segments. All LOS F values are eliminated as shown in HCM Exhibit 25-67.

Figure 1.65 *HCM Exhibit 25-61*

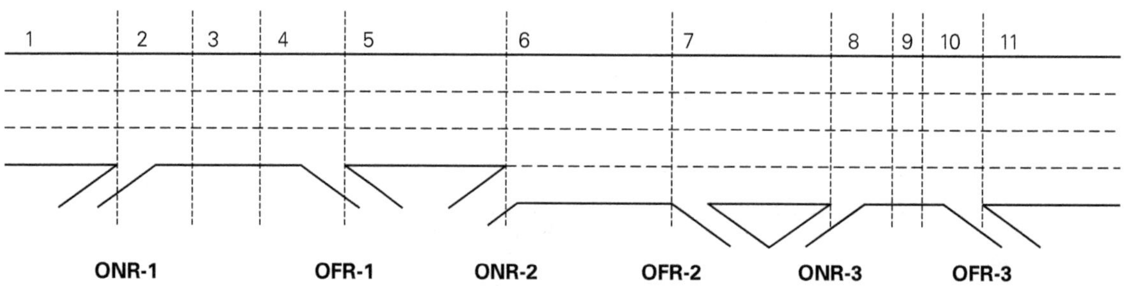

Figure 1.66 *HCM Exhibit 25-64*

Time	Demand-to-Capacity Ratios by Segment										
Step	1	2	3	4	5	6	7	8	9	10	11
1	0.74	0.82	0.82	0.82	0.77	0.70	0.60	0.66	0.66	0.66	0.62
2	0.82	0.90	0.90	0.90	0.84	0.78	0.68	0.74	0.74	0.74	0.71
3	0.86	0.96	0.96	0.96	0.92	0.85	0.74	0.82	0.82	0.82	0.77
4	0.77	0.83	0.83	0.83	0.79	0.68	0.59	0.64	0.64	0.64	0.61
5	0.62	0.65	0.65	0.65	0.61	0.52	0.47	0.50	0.50	0.50	0.48

Figure 1.67 *HCM Exhibit 25-66*

Time	Density (veh/mi/ln) by Segment										
Step	1	2	3	4	5	6	7	8	9	10	11
1	27.9	34.5	31.3	32.8	29.2	28.7	22.5	26.8	26.8	25.4	23.3
2	31.3	39.0	36.4	36.7	32.8	32.5	25.4	30.9	30.9	29.0	26.7
3	33.7	42.4	40.8	40.8	37.4	35.7	28.0	34.5	34.5	32.4	29.0
4	29.2	35.2	32.0	33.4	29.8	28.1	22.1	26.4	26.4	24.9	22.9
5	23.3	26.9	24.5	26.1	22.8	20.6	17.5	20.1	20.1	19.1	17.9

Figure 1.68 *HCM Exhibit 25-67*

Time	LOS for Segment										
Step	1	2	3	4	5	6	7	8	9	10	11
1	D	D	D	D	D	D	C	C	D	C	C
2	D	D	E	D	D	D	C	C	D	C	D
3	D	D	E	D	E	E	D	D	D	D	D
4	D	D	D	D	D	D	C	C	D	C	C
5	C	C	C	C	C	C	B	B	C	B	B

HCS7

These screenshots show how Problems One, Two, and Three are coded into *HCS7*, followed by the results as presented in the formatted report. All input data, intermediate computations, and final results are shown in logical order, leading to density and level of service.

Figure 1.69 *HCS7 Freeways1 Input*

Segments Global Inputs

Input	Checked	Value	Input	Checked	Value
Freeway Lanes	☐	3	Ramp Lanes	☐	1
Freeway FFS, mi/h	☐	75.4	Ramp FFS, mi/h	☐	35.0
Freeway Terrain Type	☐	Level	Ramp Terrain Type	☐	Level
Freeway Peak Hour Factor	☐	0.94	Ramp Peak Hour Factor	☐	0.94
Freeway Total Trucks, %	☐	0.00	Ramp Total Trucks, %	☑	2.25

	Type	Name	Length, ft	Lanes	Demand, veh/h	Volume, veh/h	Segment Analysis Details
1	Basic		5280	3	4505	4505	Details
2	Merge		1500	3	4505	4955	Details
3	Basic		2280	3	4955	4955	Details
4	Diverge		1500	3	4955	4955	Details
5	Basic		5280	3	4685	4685	Details
6	Weaving		2640	4	4865	5275	Details
7	Basic		5280	3	4865	4865	Details
8	Merge		1140	3	4865	5314	Details
9	Overlap		360	3	5315	5315	Details
10	Diverge		1140	3	5315	5315	Details
11	Basic		5280	3	5045	5045	Details

Facility

Type	Basic	Merge	Basic	Diverge	Basic	Weaving	Basic	Merge	Overlap	Diverge	Basic
Length	5280	1500	2280	1500	5280	2640	5280	1140	360	1140	5280
Segment	1	2	3	4	5	6	7	8	9	10	11
Lanes	3	3	3	3	3	4	3	3	3	3	3

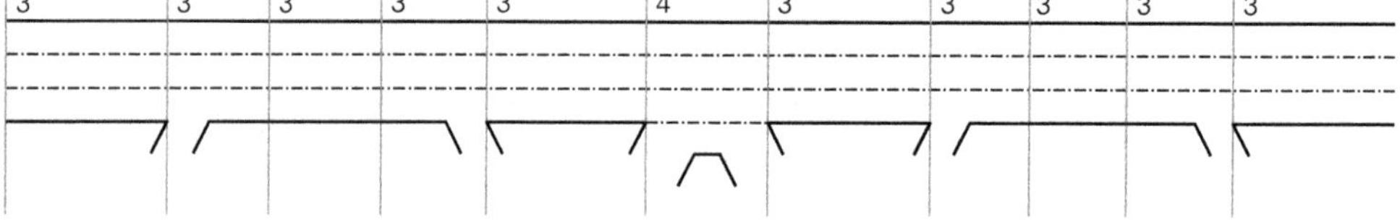

Figure 1.70 *HCS7 Freeways1 Output*

HCS7 Freeway Facilities Report

Project Information

Analyst		Agency	
Jurisdiction		Time Period Analyzed	
Analysis Year	2017	Date	2/10/2017
Project Description	Chapter 25: Example Problem 1		

Facility Global Input

Jam Density, pc/mi/ln	190.0	Density at Capacity, pc/mi/ln	45.0
Queue Discharge Capacity Drop, %	7	Total Segments	11
Total Time Periods	5	Time Period Duration, min	15

Segment Geometric Data

No.	Coded	Analyzed	Name	Length, ft	Lanes
1	Basic	Basic		5280	3
2	Merge	Merge		1500	3
3	Basic	Basic		2280	3
4	Diverge	Diverge		1500	3
5	Basic	Basic		5280	3
6	Weaving	Weaving		2640	4
7	Basic	Basic		5280	3
8	Merge	Merge		1140	3
9	Overlap	Overlap		360	3
10	Diverge	Diverge		1140	3
11	Basic	Basic		5280	3

Facility Time Period Results

T	Speed, mi/h	Density, pc/mi/ln	Density, veh/mi/ln	Travel Time, min	LOS
1	57.6	28.2	27.6	6.2	D
2	56.5	32.2	31.5	6.4	D
3	55.0	35.8	35.0	6.5	E
4	57.8	28.2	27.6	6.2	D
5	58.4	21.9	21.4	6.2	C

Facility Overall Results

Space Mean Speed, mi/h	56.9	Density, veh/mi/ln	28.6

Figure 1.71 *HCS7 Freeways2 Input*

	Type	Name	Length, ft	Lanes	Demand, veh/h	Volume, veh/h	Segment Analysis Details
1	Basic		5280	3	4505	5001	Details
2	Merge		1500	3	5001	5500	Details
3	Basic		2280	3	5451	5451	Details
4	Diverge		1500	3	5451	5451	Details
5	Basic		5280	3	5181	5181	Details
6	Weaving		2640	4	5391	5801	Details
7	Basic		5280	3	5391	5391	Details
8	Merge		1140	3	5391	5890	Details
9	Overlap		360	3	5841	5841	Details
10	Diverge		1140	3	5841	5841	Details
11	Basic		5280	3	5571	5571	Details

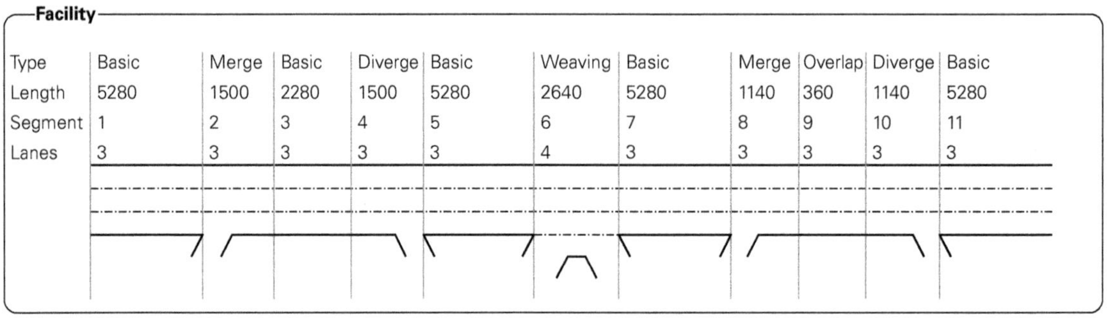

Figure 1.72 *HCS7 Freeways2 Output*

Facility Time Period Results

T	Speed, mi/h	Density, pc/mi/ln	Density, veh/mi/ln	Travel Time, min	LOS
1	56.9	31.6	30.9	6.3	D
2	54.8	36.6	35.8	6.6	E
3	43.3	45.6	44.6	8.3	F
4	45.4	42.5	41.5	7.9	E
5	58.2	25.7	25.1	6.2	C

Facility Overall Results

Space Mean Speed, mi/h	50.4	Density, veh/mi/ln	35.6

Figure 1.73 *HCS7 Freeways3 Input*

	Type	Name	Length, ft	Lanes	Demand, veh/h	Volume, veh/h	Segment Analysis Details
1	Basic		5280	3	4505	5001	Details
2	Merge		1500	3	5001	5500	Details
3	Basic		2280	3	5451	5451	Details
4	Diverge		1500	3	5451	5451	Details
5	Basic		5280	3	5181	5181	Details
6	Weaving		2640	4	5361	5771	Details
7	Basic		5280	4	5361	5361	Details
8	Merge		1140	4	5361	5861	Details
9	Overlap		360	4	5811	5811	Details
10	Diverge		1140	4	5811	5811	Details
11	Basic		5280	4	5541	5541	Details

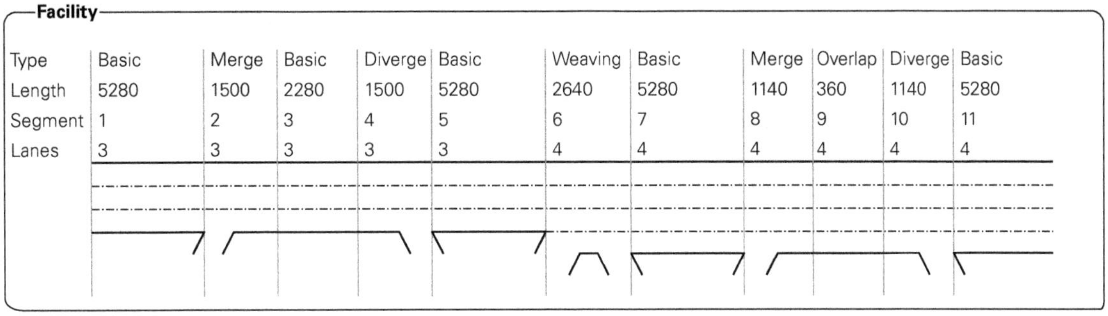

Figure 1.74 *HCS7 Freeways3 Output*

Facility Time Period Results

T	Speed, mi/h	Density, pc/mi/ln	Density, veh/mi/ln	Travel Time, min	LOS
1	58.0	27.2	26.6	6.2	D
2	57.2	30.8	30.2	6.3	D
3	56.1	33.9	33.2	6.4	D
4	57.9	27.4	26.8	6.2	D
5	58.6	21.3	20.9	6.1	C

Facility Overall Results

Space Mean Speed, mi/h	57.4	Density, veh/mi/ln	27.5

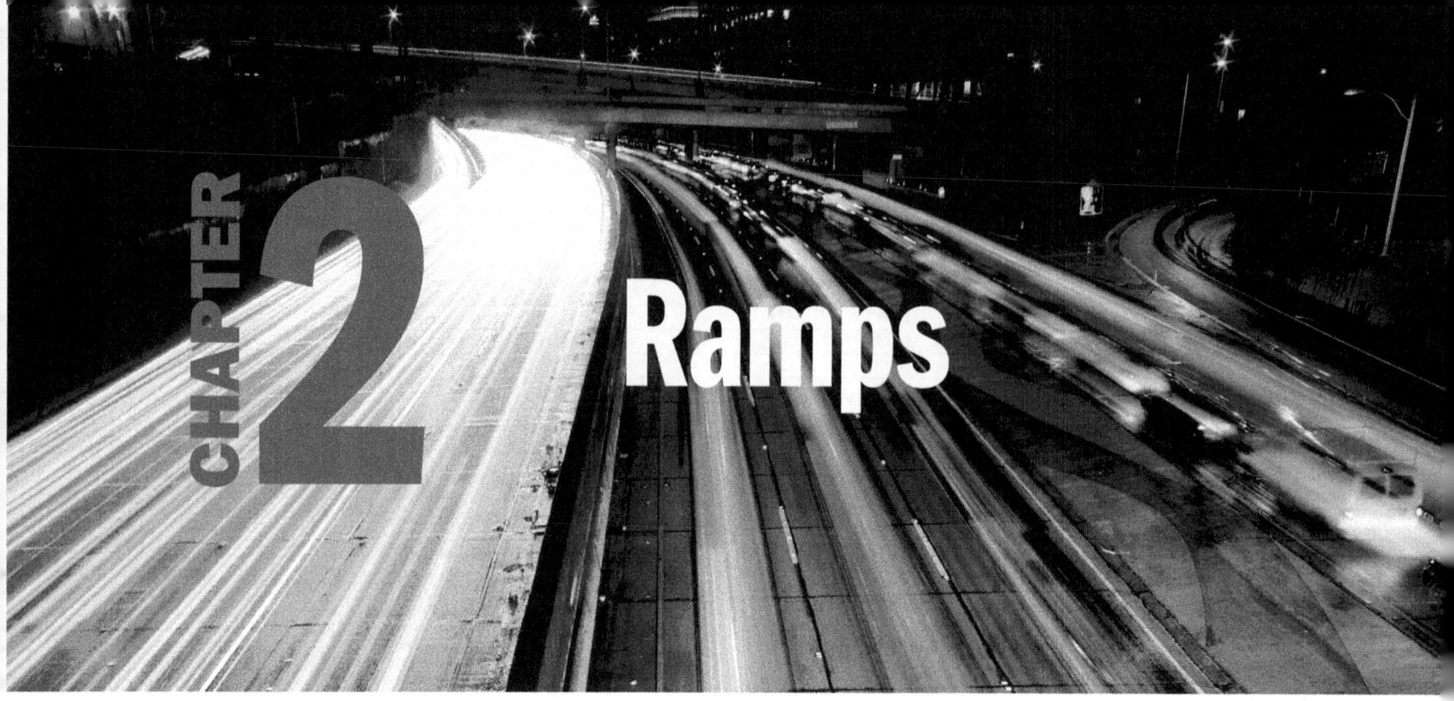

CHAPTER 2 Ramps

Freeways are divided into segments for analyzing one direction at a time, consisting of Basic Freeway Segments, Freeway Weaving Segments, and Freeway Merge and Diverge Segments. A Basic Freeway Segment is a portion of the freeway outside of any ramp influence areas. A Freeway Weaving Segment consists of either one with one-sided (on-ramp followed by an off-ramp with a continuous auxiliary lane between) weaving or two-sided (a left-side on-ramp followed by a right-side off-ramp or a right-side on-ramp followed by a left-side off-ramp) weaving. A Freeway Merge or Diverge Segment models the on-ramp (merge) or off-ramp (diverge) interaction with the adjacent two freeway lanes. This chapter deals with the analysis of individual freeway segments, while Chapter Two defines combining multiple segments into a Freeway Facility to be able to model multiple time periods for analyzing oversaturated conditions as well as travel time reliability. Since the segment procedures do not model oversaturated conditions, any segment analysis resulting in a failing level of service must be modeled within a facility analysis.

Freeway Weaving Segments

The analysis of a Freeway Weaving Segment focuses on lane-changing rates with key checks to ensure capacity is not exceeded by flow rate and that the length of the segment does not exceed the defined maximum. When both are satisfied, lane configuration is defined to compute lane changing rates that are used for calculating speed. As it is for basic segments, speed is used with flow rate to compute density for determining level of service.

Required Data

To compute the flow rate, the demand data required include the prevailing volume (with driver familiarity information), peak hour factor, truck percentage, and the terrain type or specific grade (for which truck data must include the split between tractor trailers and single-unit trucks).

Some basic information about the geometrics is needed, including the number of segment lanes (N) as the cross-section between the ramps and the short length (L_S) that includes delineation if present. To compute lane-changing rates, the lane configuration is defined using the minimum lane changes required to complete a maneuver to get on or get off the freeway. The minimum lane changes required means from the most advantageous lane, usually the rightmost lane on the freeway to leave the freeway (LC_{FR}) and the leftmost lane on the (right-side) ramp to access the freeway (LC_{RF}).

One other parameter used to define lane configuration is the number of weaving lanes (N_{WL}), defined as the total number of lanes from which a maneuver (accessing or leaving the freeway) can be made with zero or one lane change. For the designs covered by the HCM procedures, this value will always be either two or three. To illustrate these three parameters, consider HCM Exhibit 13-5.

This example shows four lanes in the segment (N = 4) with a short length of 1500 feet (L_S = 1500). The outside freeway lane requires one lane change to exit (LC_{FR} = 1), the leftmost ramp lane can enter with no lane changes (LC_{RF} = 0), and the three shaded lanes can make maneuvers with zero or one lane change. (N_{WL} = 3).

Figure 2.1 *HCM Exhibit 13-5*

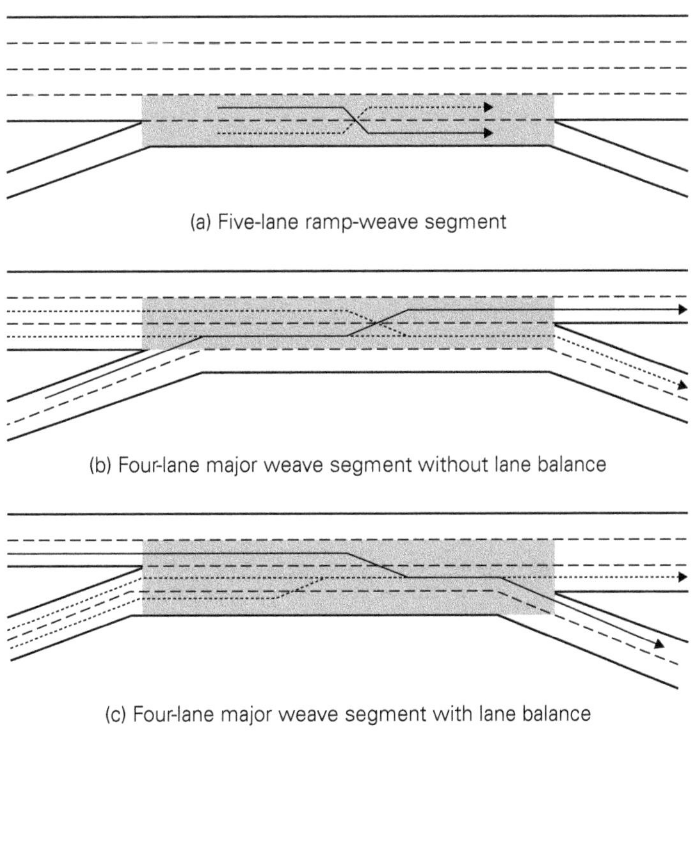

(a) Five-lane ramp-weave segment

(b) Four-lane major weave segment without lane balance

(c) Four-lane major weave segment with lane balance

Flow Rate

As for basic segments, flow rate (v_p) is estimated by modifying the hourly volume in vehicles per hour to adjust for the presence of heavy vehicles and to convert the value to a peak fifteen-minute rate using the peak hour factor, following HCM Equation 13-1. The heavy vehicle adjustment factor (f_{HV}) is determined by the passenger-car equivalents (PCE) as a function of terrain type or specific grade data in combination with the proportion of trucks. This proportion is defined as a split between tractor-trailers (TT) and single unit trucks (SUT) that include buses and recreational vehicles using HCM Equation 12-10. Refer to the Basic Freeway Segments section for this procedure.

There are four flow rates required that include freeway to freeway (v_{FF}), ramp to freeway (v_{RF}), freeway to ramp (v_{FR}), and ramp to ramp (v_{RR}) as shown in HCM Exhibit 13-9.

Beyond the base flow rates, the weaving (v_W) flow rate is computed as the ramp-to-freeway plus the freeway-to-ramp flow rates, and non-weaving (v_{NW}) flow rate is computed as the

Figure 2.2 *HCM Equation 13-1*

$$v_i = \frac{V_i}{PHF \times f_{HV}}$$

where

v_i = flow rate i under ideal conditions (pc/h);

V_i = hourly volume for flow i under prevailing conditions in vehicles per hour (veh/h);

PHF = peak hour factor (decimal);

f_{HV} = adjustment factor for heavy-vehicle presence (decimal); and
the subscript for the type of flow i can take on the following values:

FF = freeway to freeway; FR = freeway to ramp;

RF = ramp to freeway; RR = ramp to ramp;

W = weaving; and NW = non-weaving.

Figure 2.3 *HCM Equation 12-10*

$$f_{HV} = \frac{1}{1 + P_T(E_T - 1)}$$

where

f_{HV} = heavy-vehicle adjustment factor (decimal),

P_T = proportion of SUT and TTs in traffic stream (decimal), and

E_T = passenger-car equivalent of one heavy vehicle in the traffic stream (PCEs).

Figure 2.4 *HCM Exhibit 13-9*

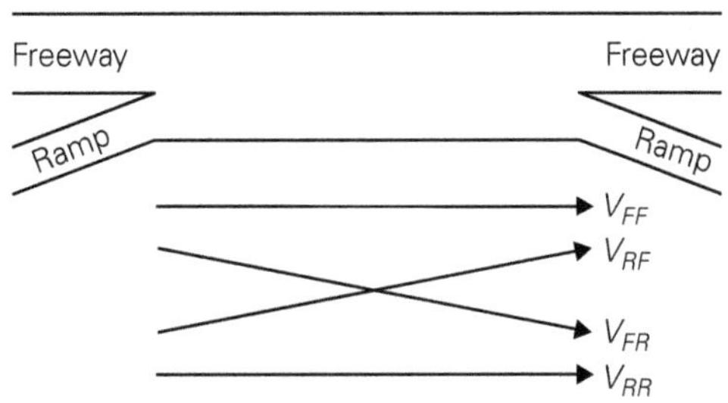

v_{FF} = freeway-to-freeway demand flow rate in the weaving segment in passenger cars per hour (pc/h);

v_{RF} = ramp-to-freeway demand flow rate in the weaving segment (pc/h);

v_{FR} = freeway-to-ramp demand flow rate in the weaving segment (pc/h);

v_{RR} = ramp-to-ramp demand flow rate in the weaving segment (pc/h);

v_W = weaving demand flow rate in the weaving segment (pc/h), $v_W = v_{RF} + v_{FR}$;

v_{NW} = nonweaving demand flow rate in the weaving segment (pc/h), $v_{NW} = v_{FF} + v_{RR}$;

v = total demand flow rate in the weaving segment (pc/h), $v = v_W + v_{NW}$;

VR = volume ratio (decimal), v_W/v;

freeway-to-freeway plus the ramp-to-ramp flow rates. The total flow rate (v) in the segment combines the weaving and non-weaving flow rates to compute the volume ratio (VR) as the weaving flow rate divided by the total flow rate.

Maximum Weaving Length

The underlying purpose of this analysis is to model the level of friction experienced by vehicles accessing the freeway concurrently with those exiting the freeway within the same segment. However, when the distance between the on-ramp and the adjacent off-ramp becomes long enough, the friction can dissipate to the point that the segment behaves like a basic freeway segment. That maximum length (L_{MAX}) within which a segment continues to operate as a freeway weaving segment is computed as a function of the volume ratio (VR) and the number of weaving lanes (N_{WL}) using the following HCM Equation 13-4.

When the short length (L_S) exceeds the maximum length (L_{MAX}), the segment is modeled as a basic freeway segment, but including capacity checks for both ramps (defined in the next section). Care must be taken when this

Figure 2.5 *HCM Equation 13-4*

$$L_{MAX} = [5{,}728(1 + VR)^{1.6}] - [1{,}566 N_{WL}]$$

maximum length exceeds 2800 feet, since the research did not calibrate these results beyond that distance. Especially if this value exceeds 6000 feet, some level of reality check should be made to confirm whether the segment appears to operate with friction between the ramp-to-freeway and freeway-to-ramp vehicles. This can be accomplished using animation from a simulation tool.

Think about your own driving:
Either getting on or getting off the freeway, you feel the friction when you have a limited distance within which to make the maneuver—especially when there are vehicles feeling the same pressure to do the opposite within the same short segment. You have to find the gap and dodge other vehicles in the weaving process that does not exist when the distance is longer and you feel no pressure or friction.

Capacity Checks

As in all freeway segment methodologies, level of service is defined as LOS F if demand exceeds capacity. For freeway weaving segments, capacity of the segment is computed based on density and demand with the lower of the two values compared with the segment flow rate for this check. The capacity values for both ramps are also compared with their respective flow rates as an additional check. If any of the three checks fails, LOS F is dictated.

Capacity as defined by density is computed using the initial capacity of a basic freeway segment (c_{IFL}) for the given free-flow speed, then as a function of the volume ratio (VR), short length (L_S), and number of weaving lanes (N_{WL}) using HCM Equations 13-5 and 13-6.

Capacity as defined by demand is computed using the volume ratio (VR) for either two or three weaving lanes (N_{WL}) using HCM Equations 13-7 and 13-8.

The lower of the resulting capacity values and the previously computed flow rate are

Figure 2.6 *HCM Equation 13-5*

$$c_{IWL} = c_{IFL} - [438.2(1+VR)^{1.6}] + [0.0765L_S] + [119.8N_{WL}]$$

where

c_{IWL} = capacity of the weaving segment under equivalent ideal conditions, per lane (pc/h/ln), and

c_{IFL} = capacity of a basic freeway segment with the same FFS as the weaving segment under equivalent ideal conditions, per lane (pc/h/ln).

Figure 2.7 *HCM Equation 13-6*

$$c_W = c_{IWL} \times N \times f_{HV}$$

Figure 2.8 *HCM Equation 13-7*

$$c_{IW} = \begin{cases} \dfrac{2{,}400}{VR} & \text{for } N_{WL} = 2 \text{ lanes} \\ \dfrac{3{,}500}{VR} & \text{for } N_{WL} = 3 \text{ lanes} \end{cases}$$

Figure 2.9 *HCM Equation 13-8*

$$c_W = c_{IW} \times f_{HV}$$

Figure 2.10 *HCM Equation 13-10*

$$v/c = \frac{v \times f_{HV}}{c_{wa}}$$

converted from passenger cars to vehicles before generating the volume-to-capacity (*v/c*) ratio using HCM Equation 13-10 to determine if the segment is operating at LOS F and a facility analysis is required.

The capacity for each ramp is compared with those defined in the Freeway Merge and Diverge procedures using HCM Exhibit 14-12 as functions of free-flow speeds, which vary by ramp geometric design where long, diamond interchange ramps will have higher free-flow speeds (and therefore higher capacity values) than tight, cloverleaf ramps.

Capacity can be adjusted using the Capacity Adjustment Factor (*CAF*) for the effects of weather, incidents, and/or driver familiarity, following HCM Equation 12-8 with HCM Exhibits 11-20, 11-23, and/or 26-9. For incidents, the adjustment is applied to the open lanes.

Figure 2.11 *HCM Exhibit 14-12*

Ramp FFS S_{FR} (mi/h)	Single-Lane Ramps	Two-Lanes Ramps
>50	2,200	4,400
>40–50	2,100	4,200
>30–40	2,000	4,000
≥20–30	1,900	3,800
<20	1,800	3,600

Figure 2.12 *HCM Equation 12-8*

$$c_{adj} = c \times CAF$$

Figure 2.13 *HCM Exhibit 11-20*

		Capacity Adjustment Factors				
Weather Type	**Weather Event Definition**	**55 mi/h**	**60 mi/h**	**65 mi/h**	**70 mi/h**	**75 mi/h**
Medium rain	> 0.10–0.25 in/h	0.94	0.93	0.92	0.91	0.90
Heavy rain	> 0.25 in/h	0.89	0.88	0.86	0.84	0.82
Light snow	>0.00 – 0.05 in/h	0.97	0.96	0.96	0.95	0.95
Light–medium snow	>0.05 – 0.10 in/h	0.95	0.94	0.92	0.90	0.88
Medium–heavy snow	>0.10 – 0.50 in/h	0.93	0.91	0.90	0.88	0.87
Heavy snow	>0.50 in/h	0.80	0.78	0.76	0.74	0.72
Severe cold	<–4° F	0.93	0.92	0.92	0.91	0.90
Low visibility	0.50 – 0.99 mi	0.90	0.90	0.90	0.90	0.90
Very low visibility	0.25 – 0.49 mi	0.88	0.88	0.88	0.88	0.88
Minimal visibility	< 0.25 mi	0.90	0.90	0.90	0.90	0.90
Non-severe weather	All conditions not listed above	1.00	1.00	1.00	1.00	1.00

Figure 2.14 *HCM Exhibit 11-23*

Directional Lanes	No Incident	Shoulder Closed	1 Lane Closed	2 Lanes Closed	3 Lanes Closed	4 Lanes Closed
2	1.00	0.81	0.70	N/A	N/A	N/A
3	1.00	0.83	0.74	0.51	N/A	N/A
4	1.00	0.85	0.77	0.50	0.52	N/A
5	1.00	0.87	0.81	0.67	0.50	0.50
6	1.00	0.89	0.85	0.75	0.52	0.52
7	1.00	0.91	0.88	0.80	0.63	0.63
8	1.00	0.93	0.89	0.84	0.66	0.66

Figure 2.15 *HCM Exhibit 26-9*

Level of Driver Familiarity	CAF_{pop}
All familiar drivers, regular commuters	1.000
Mostly familiar drivers	0.968
Balanced mix of familiar and unfamiliar drivers	0.939
Mostly unfamiliar drivers	0.898
All or overwhelmingly unfamiliar drivers	0.852

Managed Lanes

When traffic from an on-ramp cross-weaves across traffic in the general purpose lanes to access a managed-lane facility, capacity is reduced for the general purpose lane facility. The same is true for traffic cross-weaving from the managed-lane facility to an off-ramp. This reduction is quantified by HCM Equations 13-24 and 13-25 as functions of the cross-weave demand and the distance between the ramp and the managed-lane access as illustrated in HCM Exhibit 13-12.

Think of your own driving: Similar to general weaving, there is pressure and friction as these vehicles navigate from the on-ramp to the managed lane or from the managed lane to the off-ramp within a limited distance: they get in your way using the general purpose lanes, which reduces capacity.

Figure 2.16 *HCM Exhibit 13-12*

Figure 2.17 *HCM Equation 13-24*

$$CAF = 1 - CRF$$

$$CRF = -0.0897 + 0.0252\ln(CW) - 0.00001453L_{cw-min} + 0.002967N_{GP}$$

where

CRF = capacity reducion factor (decimal),

CAF = capacity adjustment factor (decimal),

CW = cross-weave demand flow rate (pc/h),

L_{cw-min} = cross-weave length (ft), and

N_{GP} = number of general purpose lanes (ln).

Figure 2.18 *HCM Equation 13-25*

$$c_{GPA} = c_{GP} \times CAF$$

where

c_{GPA} = adjusted capacity of the general purpose lanes (veh/h), and

c_{GP} = unadjusted capacity of the general purpose lanes, estimated using basic freeway procedures in Chapter 12 (veh/h).

A managed lane access segment can model the weaving between the managed-lane facility and the general purpose lanes, as well as weaving from direct external access to the managed-lane facility if separated from the general purpose lanes by a barrier as illustrated in HCM Exhibit 13-13. The only difference from the standard procedure is that the capacity check uses the capacity of the basic freeway segment in

Figure 2.19 *HCM Exhibit 13-13*

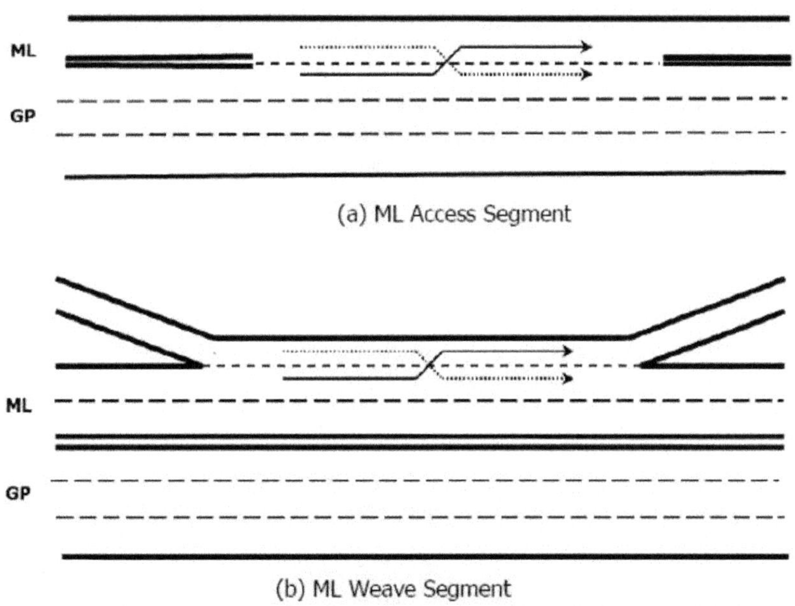

place of the density- and demand-based values defined in the weaving procedures.

Lane-Changing Rates

The minimum number of lane changes required to access and exit the freeway are multiplied by the flow rate making each maneuver to compute the minimum weaving lane-changing rate using HCM Equation 13-2. For two-sided weaving, the ramp-to-ramp demand is multiplied by the number of freeway lanes as the lane changes required for this maneuver using HCM Equation 13-3.

These minimum rates are expanded to include optional weaving using HCM Equation 13-11 for weaving vehicles. For non-weaving vehicles, HCM Equation 13-12 is used to compute an index that is applied to HCM

Figure 2.20 *HCM Equation 13-2*

LC_{RF} = minimum number of lane changes that must be made by one ramp-to-freeway vehicle to execute the desired maneuver successfully (lc), and

LC_{FR} = minimum number of lane changes that must be made by one freeway-to-ramp vehicle to execute the desired maneuver successfully (lc).

LC_{MIN} for one-sided weaving segments is given by Equation 13-2:

$$LC_{MIN} = (LC_{RF} \times v_{RF}) + (LC_{FR} \times v_{FR})$$

Figure 2.21 *HCM Equation 13-3*

LC_{RR} = minimum number of lane changes that must be made by one ramp-to-ramp vehicle to execute the desired maneuver successfully (lc).

LC_{MIN} for two-sided weaving segments is given by Equation 13-3:

$$LC_{MIN} = LC_{RR} \times v_{RR}$$

Figure 2.21a *HCM Equation 13-11*

$$LC_W = LC_{MIN} + 0.39[(L_S - 300)^{0.5} N^2 (1 + ID)^{0.8}]$$

where

LC_W = equivalent hourly rate at which weaving vehicles make lane changes within the weaving segment (lc/h);

LC_{MIN} = minimum equivalent hourly rate at which weaving vehicles must make lane changes within the weaving segment to complete all weaving maneuvers successfully (lc/h);

L_S = length of the weaving segment, using the short length definition (ft) (300 ft is the minimum value);

N = number of lanes within the weaving segment (ln); and

ID = interchange density, the number of interchanges within 3 mi upstream and downstream of the center of the subject weaving segment divided by 6, in interchanges per mile (int/mi).

Equation 13-16 to select among HCM Equations 13-13, 13-14, and 13-15 for computing the non-weaving lane changing rate.

The total lane-changing rate is the sum of the weaving and non-weaving rates using HCM Equation 13-17 that computes LC_{ALL}. This value is applied with the short length in HCM Equation 13-20 to calculate the weaving intensity factor (W).

Figure 2.22 *HCM Equation 13-12*

$$I_{NW} = \frac{L_S \times ID \times v_{NW}}{10{,}000}$$

Figure 2.23 *HCM Equation 13-16*

If $I_{NW} \leq 1{,}300$:	$LC_{NW} = LC_{NW1}$
If $I_{NW} \geq 1{,}950$:	$LC_{NW} = LC_{NW2}$
If $1{,}300 < I_{NW} < 1{,}950$:	$LC_{NW} = LC_{NW3}$
If $LC_{NW1} \geq LC_{NW2}$:	$LC_{NW} = LC_{NW2}$

Figure 2.24 *HCM Equation 13-13*

$$LC_{NW1} = (0.206 v_{NW}) + (0.542 L_S) - (192.6N)$$

Figure 2.25 *HCM Equation 13-14*

$$LC_{NW2} = 2{,}135 + 0.223(v_{NW} - 2{,}000)$$

Figure 2.26 *HCM Equation 13-15*

$$LC_{NW3} = LC_{NW1} + (LC_{NW2} - LC_{NW1})\left(\frac{I_{NW} - 1{,}300}{650}\right)$$

Figure 2.27 *HCM Equation 13-17*

$$LC_{ALL} = LC_W + LC_{NW}$$

Figure 2.28 *HCM Equation 13-20*

$$W = 0.226\left(\frac{LC_{ALL}}{L_S}\right)^{0.789}$$

Speed

The speed (*S*) of all weaving vehicles (S_W) is computed using HCM Equation 13-19 with the free-flow speed and the weaving intensity factor. The speed of all non-weaving vehicles is computed using HCM Equation 13-21. The average speed of all vehicles in the weaving segment is a weighted average of these two values as defined in HCM Equation 13-22.

Free-flow speed can be adjusted using the Speed Adjustment Factor (*SAF*) for the effects of weather and/or driver familiarity, following HCM Equation 12-5 with HCM Exhibits 11-21 and/or 26-9.

Figure 2.29 *HCM Equation 13-19*

$$S_W = 15 + \left(\frac{FFS \times SAF - 15}{1 + W}\right)$$

Figure 2.30 *HCM Equation 13-21*

$$S_{NW} = FFS \times SAF - (0.0072 LC_{MIN}) - \left(0.0048 \frac{v}{N}\right)$$

Figure 2.31 *HCM Equation 13-22*

$$S = \frac{v_W + v_{NW}}{\left(\frac{v_W}{S_W}\right) + \left(\frac{v_{NW}}{S_{NW}}\right)}$$

Figure 2.32 *HCM Equation 12-5*

$$FFS_{adj} = FFS \times SAF$$

Density

The density (*D*) of all vehicles in the weaving segment is the total flow rate converted to a per-lane value divided by the composite speed according to HCM Equation 13-23.

Figure 2.33 *HCM Exhibit 11-21*

Weather Type	Weather Event Definition	Speed Adjustment Factors 55 mi/h	60 mi/h	65 mi/h	70 mi/h	75 mi/h
Medium rain	> 0.10–0.25 in/h	0.96	0.95	0.94	0.93	0.93
Heavy rain	> 0.25 in/h	0.94	0.93	0.93	0.92	0.91
Light snow	>0.00 – 0.05 in/h	0.94	0.92	0.89	0.87	0.84
Light–medium snow	>0.05 – 0.10 in/h	0.92	0.90	0.88	0.86	0.83
Medium–heavy snow	>0.10 – 0.50 in/h	0.90	0.88	0.86	0.84	0.82
Heavy snow	>0.50 in/h	0.88	0.86	0.85	0.83	0.81
Severe cold	<–4° F	0.95	0.95	0.94	0.93	0.92
Low visibility	0.50 – 0.99 mi	0.96	0.95	0.94	0.94	0.93
Very low visibility	0.25 – 0.49 mi	0.95	0.94	0.93	0.92	0.91
Minimal visibility	< 0.25 mi	0.95	0.94	0.93	0.92	0.91
Non-severe weather	All conditions not listed above	1.00	1.00	1.00	1.00	1.00

Figure 2.34 *HCM Exhibit 26-9*

Level of Driver Familiarity	SAF_{pop}
All familiar drivers, regular commuters	1.000
Mostly familiar drivers	0.975
Balanced mix of familiar and unfamiliar drivers	0.950
Mostly unfamiliar drivers	0.913
All or overwhelmingly unfamiliar drivers	0.863

Figure 2.35 *HCM Equation 13-23*

$$D = \frac{(v/N)}{S}$$

Figure 2.36 *HCM Exhibit 13-6*

LOS	Density (pc/mi/ln) Freeway Weaving Segments	Weaving Segments on Multilane Highways or C-D Roads
A	0–10	0–12
B	>10–20	>12–24
C	>20–28	>24–32
D	>28–35	>32–36
E	>35–43	>36–40
F	>43, or demand exceeds capacity	>40, or demand exceeds capacity

LOS

Level of Service (LOS) is defined by the thresholds shown in HCM Exhibit 13-6 for both freeway and collector-distributor (C-D) roads or multilane highways. The scale for freeway weaving segments is different (similar to merge and diverge segments) from the one that defines LOS for basic freeway segments (and freeway facilities). There is also a separate scale for weaving on a multilane highway or collector-distributor roadway with a slightly compressed range.

PROBLEM ONE

Determine the LOS for a Freeway Weaving Segment with the lane configuration, volumes, and short length of 2000 feet shown in the HCM Exhibit 27-2 below, 70 mi/h free-flow speed on the freeway and 45 mi/h free-flow speeds on both ramps, 0.50 interchanges per mile, 0.90 PHF, and no trucks throughout with all familiar drivers, no weather events, or incidents.

Figure 2.37 *HCM Exhibit 27-2*

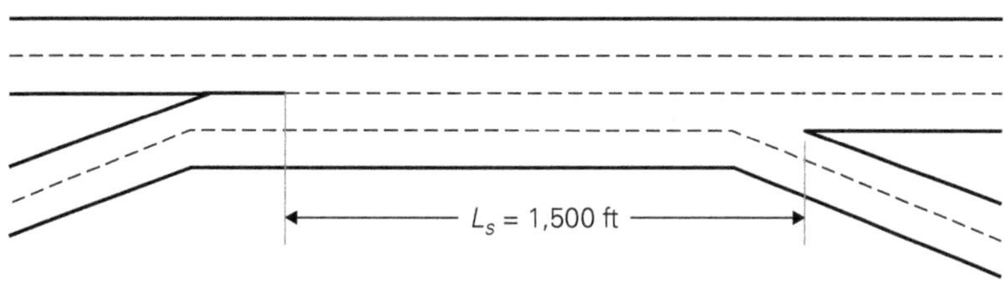

1. Convert the four volumes into flow rates by applying the peak hour factor (only since there are no trucks).

Flow Rate: $v_{FF} = 3000/0.90 = 3333$ pc/h $v_{RF} = 1000/0.90 = 1111$ pc/h

EQUATION 13-1

$v_{RR} = 200/0.90 = 222$ pc/h $v_{FR} = 800/0.90 = 889$ pc/h $v = 5555$ pc/h

2. Compute the weaving and non-weaving flow rates to calculate the volume ratio.

$v_W = 1111 + 889 = 2000$ pc/h $v_{NW} = 3333 + 222 = 3555$ pc/h

EXHIBIT 13-9

Volume Ratio: $VR = 2000/5555 = 0.360$

3. Define the minimum lane changes and number of weaving lanes for lane-changing rates.

Lane Changes: $LC_{RF} = 0$ $LC_{FR} = 1$ $N_{WL} = 3$

EXHIBIT 13-5

4. Compute the maximum weaving length to ensure the segment is operating as a weave with the short length.

Maximum Length: $L_{MAX} = (5728(1 + 0.360)^{1.6}) - (1566)(3) = 4670$ feet

EQUATION 13-4

5. Compute the capacity based on density and demand to compare the lowest with the total flow rate.

Capacity (Demand): $c_W = 3500/0.360 = 9722$ pc/h

EQUATIONS 13-7 & 13-8

Capacity (Density): $c_{IWL} = 2400 - (438.2(1 + 0.360)^{1.6}) + (0.0765)(2000) + (119.8)(3) = 2196$ pc/h/ln
$cW = 2196(4) = 8784$ pc/h > 5554 pc/h

EQUATIONS 13-5 & 13-6

6. Compare both ramp flow rates to those capacities.

Capacity (Ramps): $c_{RF} = 4200$ pc/h > 1111 pc/h $c_{FR} = 4200$ pc/h > 889 pc/h

EXHIBIT 14-12

7. Produce the total lane changing rate for use in calculating the weaving intensity factor.

Weaving Rate: $L_{CW} = (1111)(0) + (889)(1) + 0.39((2000 - 300)^{0.5} (4)^2(1 + 0.5)^{0.8}) = 1245$ lc/h

EQUATIONS 13-2 & 13-11

Non-Weaving Index: $I_{NW} = (2000)(0.5)(3555)/10000 = 356 < 1300$

EQUATION 13-12

Non-Weaving Rate: $LC_{NW1} = (0.206)(3555) + (0.542)(2000) - (192.6)(4) = 1046$ lc/h

EQUATION 13-13

Total Rate: $LC_{ALL} = 1245 + 1046 = 2291$ lc/h

EQUATION 13-17

Intensity Factor: $W = 0.226(2291/2000)^{0.789} = 0.252$

EQUATION 13-20

8. This allows the computation of weaving and non-weaving speeds to generate a composite speed in the segment for use in calculating density to determine level of service for the freeway weaving segment.

Weaving Speed: $S_W = 15 + (70 - 15)/(1 + 0.252) = 58.9$ mi/h

EQUATION 13-19

Non-Weaving Speed: $S_{NW} = 70 - (0.0072)(888) - (0.0048)(5555/4) = 56.9$ mi/h

EQUATION 13-21

Segment Speed: $S = (2000 + 3555)/(2000/58.9) + (3555/56.9) = 57.6$ mi/h

EQUATION 13-22

9. Density is calculated from flow rate and speed to determine level of service for the basic freeway segment.

Density: $D = (5555/4)/57.6 = 24.1$ pc/mi/ln

EQUATION 13-23

Level of Service LOS = C (between 20 and 28)

EXHIBIT 13-6

HCS7

These screenshots show how Problem One is coded into *HCS7*, followed by the results as presented in the formatted report. All input data, intermediate computations, and final results are shown in logical order, leading to density and level of service.

Figure 2.38 *HCS7 Weaving1 Input*

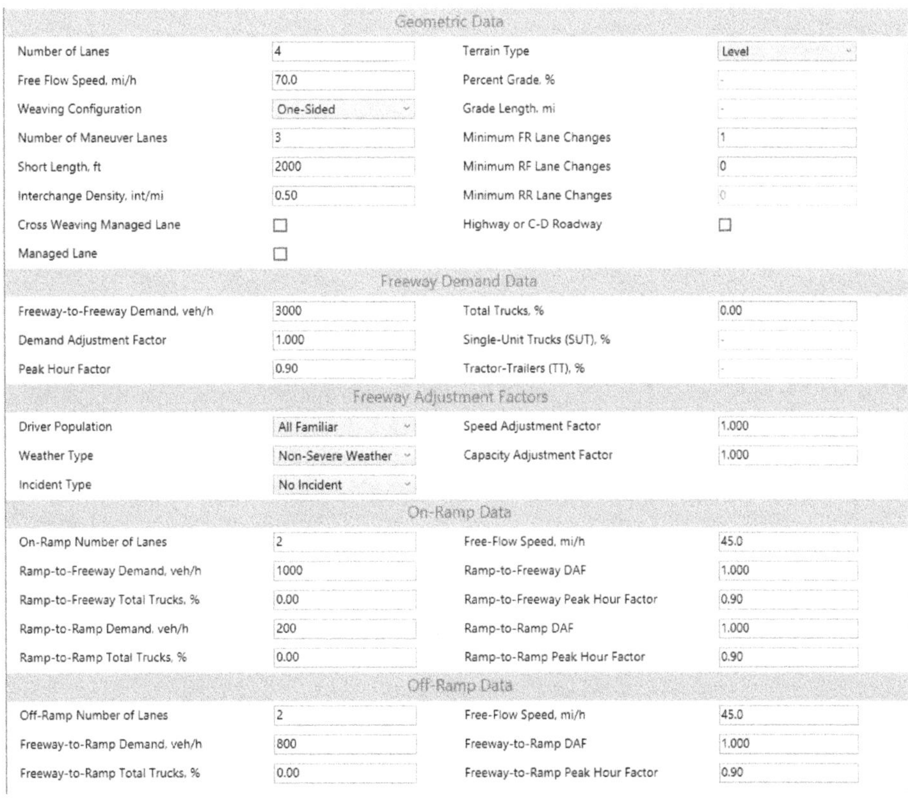

Geometric Data			
Number of Lanes	4	Terrain Type	Level
Free Flow Speed, mi/h	70.0	Percent Grade, %	-
Weaving Configuration	One-Sided	Grade Length, mi	-
Number of Maneuver Lanes	3	Minimum FR Lane Changes	1
Short Length, ft	2000	Minimum RF Lane Changes	0
Interchange Density, int/mi	0.50	Minimum RR Lane Changes	0
Cross Weaving Managed Lane	☐	Highway or C-D Roadway	☐
Managed Lane	☐		

Freeway Demand Data			
Freeway-to-Freeway Demand, veh/h	3000	Total Trucks, %	0.00
Demand Adjustment Factor	1.000	Single-Unit Trucks (SUT), %	-
Peak Hour Factor	0.90	Tractor-Trailers (TT), %	-

Freeway Adjustment Factors			
Driver Population	All Familiar	Speed Adjustment Factor	1.000
Weather Type	Non-Severe Weather	Capacity Adjustment Factor	1.000
Incident Type	No Incident		

On-Ramp Data			
On-Ramp Number of Lanes	2	Free-Flow Speed, mi/h	45.0
Ramp-to-Freeway Demand, veh/h	1000	Ramp-to-Freeway DAF	1.000
Ramp-to-Freeway Total Trucks, %	0.00	Ramp-to-Freeway Peak Hour Factor	0.90
Ramp-to-Ramp Demand, veh/h	200	Ramp-to-Ramp DAF	1.000
Ramp-to-Ramp Total Trucks, %	0.00	Ramp-to-Ramp Peak Hour Factor	0.90

Off-Ramp Data			
Off-Ramp Number of Lanes	2	Free-Flow Speed, mi/h	45.0
Freeway-to-Ramp Demand, veh/h	800	Freeway-to-Ramp DAF	1.000
Freeway-to-Ramp Total Trucks, %	0.00	Freeway-to-Ramp Peak Hour Factor	0.90

Figure 2.39 *HCS7 Weaving1 Output*

HCS7 Freeway Weaving Report

Project Information

Analyst		Date	3/22/2017
Agency		Analysis Year	2017
Jurisdiction		Time Period Analyzed	
Project Description			

Geometric Data

Number of Lanes (N), ln	4	Segment Type	Freeway
Short Length (L_S), ft	2000	Number of Maneuver Lanes (N_{WL}), ln	3
Weaving Configuration	One-Sided	Ramp-to-Freeway Lane Changes (LC_{RF}), lc	0
Terrain Type	Level	Freeway-to-Ramp Lane Changes (LC_{FR}), lc	1
Percent Grade, %	-	Ramp-to-Ramp Lane Changes (LC_{RR}), lc	0
Interchange Density (ID), int/mi	0.50	Cross Weaving Managed Lane	No

Adjustment Factors

Driver Population	All Familiar	Final Speed Adjustment Factor (SAF)	1.000
Weather Type	Non-Severe Weather	Final Capacity Adjustment Factor (CAF)	1.000
Incident Type	No Incident	Demand Adjustment Factor (DAF)	1.000

Demand and Capacity

	FF	RF	RR	FR
Volume (V_i), veh/h	3000	1000	200	800
Peak Hour Factor (PHF)	0.90	0.90	0.90	0.90
Total Trucks, %	0.00	0.00	0.00	0.00
Heavy Vehicle Adjustment Factor (f_{HV})	1.000	1.000	1.000	1.000
Flow Rate (v_i), pc/h	3333	1111	222	889

Weaving Flow Rate (v_W), pc/h	2000	Freeway Max Capacity (c_{IFL}), pc/h/ln	2400
Non-Weaving Flow Rate (v_{NW}), pc/h	3555	Density-Based Capacity (c_{IWL}), pc/h/ln	2196
Total Flow Rate (v), pc/h	5555	Demand Flow-Based Capacity (c_{IW}), pc/h	9722
Volume Ratio (VR)	0.360	Weaving Segment Capacity (c_W), veh/h	8784
Minimum Lane Change Rate (LC_{MIN}), lc/h	889	Adjusted Weaving Area Capacity (c_{WA}), veh/h	8784
Maximum Weaving Length (L_{MAX}), ft	4670	Volume-to-Capacity Ratio (v/c)	0.63

Speed and Density

Non-Weaving Vehicle Index (I_{NW})	356	Average Weaving Speed (S_W), mi/h	58.9
Non-Weaving Lane Change Rate (LC_{NW}), lc/h	1046	Average Non-Weaving Speed (S_{NW}), mi/h	56.9
Weaving Lane Change Rate (LC_W), lc/h	1245	Average Speed (S), mi/h	57.6
Total Lane Change Rate (LC_{All}), lc/h	2291	Density (D), pc/mi/ln	24.1
Weaving Intensity Factor (W)	0.252	Level of Service (LOS)	C

PROBLEM TWO

Building on Problem One to add one continuous access managed lane with an on-ramp cross-weave volume of 200 veh/h with a minimum length of 500 feet (with all other parameters unchanged) to compute the reduced capacity to (only) check for LOS F, since density will not be affected.

1. Convert the cross-weaving volume to a flow rate by applying the peak hour factor (only since there are no trucks).

Flow Rate: $v_{CW} = 200/0.90 = 222$ pc/h

EQUATION 13-1

2. Compute the capacity reduction factor and the capacity adjustment factor using the cross-weave flow rate and the minimum distance to access the managed lanes.

Capacity Reduction: $CRF = -0.0897 + 0.0252\ln(222) - 0.00001453(500) + 0.002867(3) = 0.0481$

$CAF = 1 - 0.0481 = 0.952$

EQUATION 13-24

Capacity: $c_{GPA} = 7200(0.952) = 6854$ pc/mi/ln < 8784 pc/mi/ln (no LOS F)

EQUATION 13-25

HCS7

These screenshots show how Problem Two expands on Problem One adding the cross-weave information, followed by the results showing the reduced capacity as presented in the formatted report. All input data, intermediate computations, and final results are shown in logical order, leading to density and level of service.

Figure 2.40 *HCS7 Weaving2 Input*

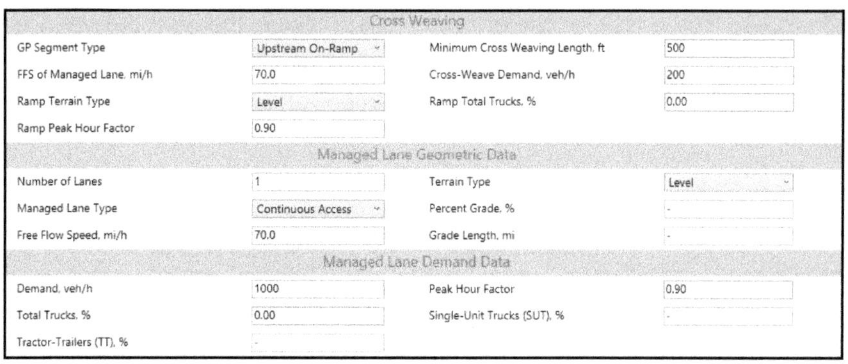

Figure 2.41 *HCS7 Weaving2 Output*

Cross-Weaving Between Ramps and the ML Access Segment		
GP Segment Type	Yes	mi/h
Managed Lane Free-Flow Speed	70.0	
Ramp Terrain Type	Level	
Ramp Percent Total Trucks, %	0.00	%
Cross-Weave Demand Volume	200	veh/h
Peak Hour Factor, PHF	0.90	
Cross-Weave Demand, Flow Rate, CW	222	pc/h
Cross-Weave Length, Lcw-min	500	ft
Number of General Purpose Lanes, NGP	3	ln
Capacity Reduction Factor, CRF	0.0481	
Capacity Adjustment Factor, CAF	0.952	
Unadjusted Capacity of General Purpose Lanes, cGP	7200	pc/h
Adjusted Capacity of General Purpose Lanes, cGPA	6854	pc/h

Merge and Diverge Segments

The analysis of merge and diverge segments models the interaction between the on-ramp or off-ramp and the adjacent two freeway lanes with density within the influence area including only to those lanes and 1500 feet from the gore as illustrated in HCM Exhibit 14-1. This procedure can model one-lane or two-lane ramps on the left or right side of the freeway. Adjacent ramps are considered for possible influence on the analysis ramp for six-lane freeways only (three lanes in each direction).

Required Data

To compute the flow rate, the demand data required include the prevailing volume for the freeway in advance of the ramp and the ramp itself (with driver familiarity information), peak hour factor, truck percentage, and the terrain type or specific grade (for which truck data must include the split between tractor trailers and single-unit trucks). Other data include the number of lanes on the freeway and the ramp,

Figure 2.42 *HCM Exhibit 14-1*

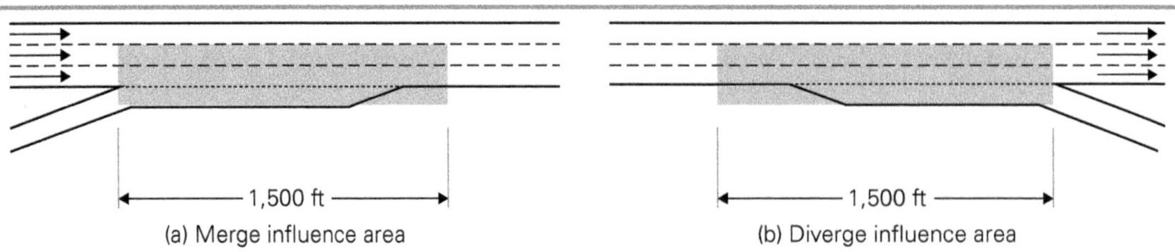

the ramp and freeway free-flow speeds, and the acceleration or deceleration lane length of the ramp. For six-lane freeways, the distance to upstream and/or downstream adjacent ramps is needed with the flow rate of each ramp.

Flow Rate

As for basic segments, flow rate (v_p) is estimated by modifying the hourly volume in vehicles per hour to adjust for the presence of heavy vehicles and to convert the value to a peak fifteen-minute rate using the peak hour factor, following HCM Equation 14-1. The heavy vehicle adjustment factor (f_{HV}) is determined by the passenger-car equivalents (PCE) as a function of terrain type or specific grade data in combination with the proportion of trucks. This proportion is defined as a split between tractor-trailers (TT) and single unit trucks (SUT) that include buses and recreational vehicles using HCM Equation 12-10. Refer to the Basic Freeway Segments section for this procedure.

There are typically two flow rates required that include the freeway (v_F) and the ramp (v_R), but for six-lane freeways it can also include upstream (v_{UP}) or downstream (v_{DN}) ramps.

Lanes 1 and 2

To determine the proportion of the freeway flow rate in the two lanes adjacent to single-lane ramps according to HCM Equations 14-2 and 14-8, equations are provided in HCM Exhibit 14-8 for merges and HCM Exhibit 14-9 for diverges. This value will be 1.00 (100%) for four-lane freeways (two lanes in each direction), or calculated for six- or eight-lane freeways as shown in HCM Equations 14-3 through 14-5 for merges and HCM Equations 14-9 through 14-11 for diverges.

Figure 2.43 *HCM Equation 14-1*

$$v_i = \frac{V_i}{PHF \times f_{HV}}$$

where

v_i = demand flow rate for movement *i* (pc/h),

V_i = demand volume for movement *i* (veh/h),

PHF = peak hour factor (decimal),

f_{HV} = adjustment factor for heavy vehicle presence (decimal).

Figure 2.44 *HCM Equation 12-10*

$$f_{HV} = \frac{1}{1 + P_T(E_T - 1)}$$

where

f_{HV} = heavy-vehicle adjustment factor (decimal),

P_T = proportion of SUT and TTs in traffic stream (decimal), and

E_T = passenger-car equivalent of one heavy vehicle in the traffic stream (PCEs).

Figure 2.45 *HCM Equation 14-2*

$$v_{12} = v_F \times P_{FM}$$

where

v_{12} = flow rate in Lanes 1 and 2 (pc/h),

v_F = total flow rate on freeway immediately upstream of the on-ramp (merge) influence area (pc/h), and

P_{FM} = proportion of freeway vehicles remaining in Lanes 1 and 2 immediately upstream of the on-ramp influence area (decimal).

Figure 2.46 *HCM Equation 14-8*

$$v_{12} = v_R + (v_F - v_R)P_{FD}$$

where

v_{12} = flow rate in Lanes 1 and 2 of the freeway immediately upstream of the deceleration lane (pc/h),

v_R = flow rate on the off-ramp (pc/h), and

P_{FD} = proportion of through freeway traffic remaining in Lanes 1 and 2 immediately upstream of the deceleration lane (decimal).

Figure 2.47 *HCM Exhibit 14-8*

No. of Freeway Lanes[a]	Model(s) for Determining P_{FD}
4	$P_{FM} = 1.000$
6	$P_{FM} = 0.5775 + 0.000028L_A$ $P_{FM} = 0.7289 - 0.0000135(v_F + v_R) - 0.003296S_{FR} + 0.000063L_{UP}$ $P_{FM} = 0.5487 + 0.2628(v_D/L_{DOWN})$
8	For $v_F/S_{FR} \le 72$: $P_{FM} = 0.2178 - 0.000125v_R + 0.01115(L_A/S_{FR})$ For $v_F/S_{FR} > 72$: $P_{FM} = 0.2178 - 0.000125v_R$

SELECTING EQUATIONS FOR P_{FM} FOR SIX-LANE FREEWAYS

Adjacent Upstream Ramp	Subject Ramp	Adjacent Downstream Ramp	Equation(s) Used
None	On	None	Equation 14-3
None	On	On	Equation 14-3
None	On	Off	Equation 14-5 or 14-3
On	On	None	Equation 14-3
Off	On	None	Equation 14-4 or 14-3
On	On	On	Equation 14-3
On	On	Off	Equation 14-5 or 14-3
Off	On	On	Equation 14-4 or 14-3
Off	On	Off	Equation 14-5 or 14-4 or 14-3

Figure 2.48 *HCM Equation 14-3*

$$P_{FM} = 0.5775 + 0.000028L_A$$

Figure 2.49 *HCM Equation 14-4*

$$P_{FM} = 0.7289 - 0.0000135(v_F + v_R) - 0.003296S_{FR} + 0.000063L_{UP}$$

Figure 2.50 *HCM Equation 14-5*

$$P_{FM} = 0.5487 + 0.2628(v_D / L_{DOWN})$$

Figure 2.51 *HCM Exhibit 14-9*

No. of Freeway Lanes[a]	Model(s) for Determining P_{FD}		
4	$P_{FD} = 1.000$		
6	$P_{FD} = 0.760 - 0.000025v_F - 0.000046V_R$		
	$P_{FD} = 0.717 - 0.000039v_F + 0.604(v_U/L_{UP})$		when $v_U/L_{UP} \leq 0.2$[b]
	$P_{FD} = 0.616 - 0.000021v_F + 0.124(v_D/L_{DOWN})$		
8	$P_{FD} = 0.436$		
SELECTING EQUATIONS FOR P_{FD} FOR SIX-LANE FREEWAYS			
Adjacent Upstream Ramp	**Subject Ramp**	**Adjacent Downstream Ramp**	**Equation(s) Used**
None	Off	None	Equation 14-9
None	Off	On	Equation 14-9
None	Off	Off	Equation 14-11 or 14-9
On	Off	None	Equation 14-10 or 14-9
Off	Off	None	Equation 14-9
On	Off	On	Equation 14-10 or 14-9
On	Off	Off	Equation 14-11, 14-10 or 14-9
Off	Off	On	Equation 14-9
Off	Off	Off	Equation 14-11 or 14-9

Figure 2.52 *HCM Equation 14-9*

$$P_{FD} = 0.760 - 0.000025v_F - 0.000046v_R$$

Figure 2.53 *HCM Equation 14-10*

$$P_{FD} = 0.717 - 0.000039v_F + 0.604(v_U/L_{UP})$$

Figure 2.54 *HCM Equation 14-11*

$$P_{FD} = 0.616 - 0.000021v_F + 0.124(v_D/L_{DOWN})$$

For six-lane freeways, adjacent ramps are considered by computing the equilibrium length to determine whether influence on the analysis ramp exists using HCM Equations 14-6 and 14-7 for merges (adjacent upstream off-ramp or downstream off-ramp only) and HCM Equations 14-12 and 14-13 for diverges (adjacent upstream on-ramp or downstream off-ramp only).

Figure 2.55 *HCM Equation 14-6*

$$L_{EQ} = 0.214(v_F + v_R) + 0.444L_A + 52.32S_{FR} - 2{,}403$$

Figure 2.56 *HCM Equation 14-7*

$$L_{EQ} = \frac{v_D}{0.1096 + 0.000107L_A}$$

Figure 2.57 *HCM Equation 14-12*

$$L_{EQ} = \frac{v_U}{0.071 + 0.000023v_F - 0.000076v_R}$$

Figure 2.58 *HCM Equation 14-13*

$$L_{EQ} = \frac{v_D}{1.15 - 0.000032v_F - 0.000369v_R}$$

Lane Balancing

The flow rate in the outer lane or lanes on the freeway is computed from HCM Equations 14-14 and 14-17 to adjust the flow rate in lanes 1 and 2 if the out lanes exceed 2700 pc/ln, following HCM Equations 14-15 and 14-16 for six-lane freeways or HCM Equations 14-18 and 14-19 for eight-lane freeways.

Two-Lane Ramps

For two-lane ramps, the computation of the flow rate in Lanes 1 and 2 uses provided proportions as functions of the number of freeway lanes. For merges, these are 1.00 (four), 0.555 (six), and 0.209 (eight), and for diverges, these are 1.00 (four), 0.450 (six), and 0.260 (eight). Also, the effective acceleration or deceleration lane lengths are computed, following HCM Exhibit 14-16 and HCM Equation 14-25 for merges and HCM Exhibit 14-17 and HCM Equation 14-26 for diverges.

Figure 2.59 *HCM Equation 14-14*

$$v_3 = v_F - v_{12}$$

where

v_3 = flow rate in Lane 3 of the freeway (pc/h/ln),

v_F = flow rate on freeway immediately upstream of the ramp influcence area (pc/h). and

v_{12} = flow rate in Lane 1 and 2 immediately upstream of the ramp influence area (pc/h).

Figure 2.60 *HCM Equation 14-15*

$$v_{12a} = v_F - 2,700$$

Figure 2.61 *HCM Equation 14-16*

$$v_{12a} = \left(\frac{v_F}{1.75}\right)$$

Figure 2.62 *HCM Equation 14-17*

$$v_{av34} = \frac{v_F - v_{12}}{2}$$

Figure 2.63 *HCM Equation 14-18*

$$v_{12a} = v_F - 5,400$$

Figure 2.64 *HCM Equation 14-19*

$$v_{12a} = \left(\frac{v_F}{2.50}\right)$$

Figure 2.65 *HCM Exhibit 14-16*

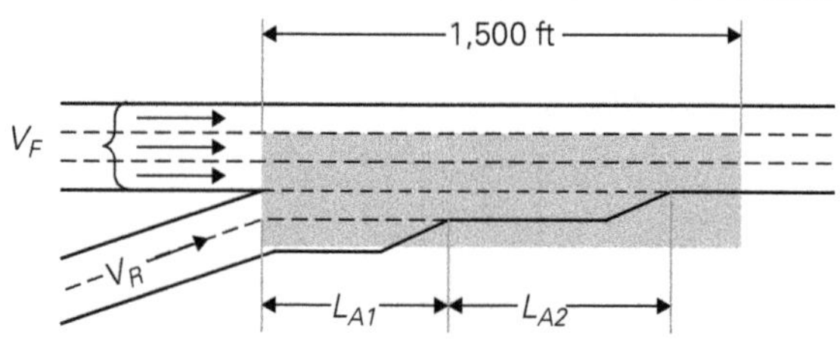

Figure 2.66 *HCM Equation 14-25*

$$L_{Aeff} = 2L_{A1} + L_{A2}$$

Figure 2.67 *HCM Exhibit 14-17*

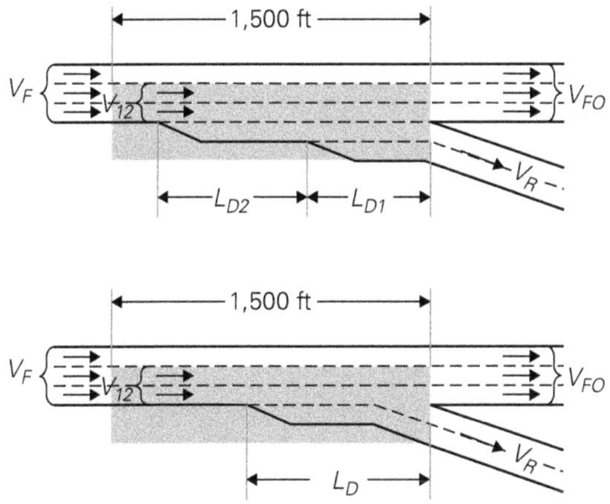

Figure 2.68 *HCM Equation 14-26*

$$L_{Deff} = 2L_{D1} + L_{D2}$$

Left-Side Ramps

Research has shown that left-side ramps capture more through vehicles than right-side ramps with all else being equal. The procedure increases the flow rate computed for right-side ramps when the ramp is on the left side to account for this difference according to HCM Exhibit 14-18.

Think about your own driving:

On a long trip, you will tend to stay away from traffic navigating on- and off-ramps, since you plan to keep going. That means you stay in the leftmost lanes, away from the right-side ramps. When the very occasional left-side ramp appears, you don't bother to move away from it as a one-out-of-twenty-or-thirty event—including you in the two lanes closest to the ramp, even going through.

Ten-Lane Freeways

The procedure only accommodates up to eight-lane freeways specifically, but provides for ten-lane freeways by estimating the flow rate in the fifth lane in HCM Exhibit 14-19 to generate a four-lane effective value using HCM Equation 14-27.

Figure 2.69 *HCM Exhibit 14-18*

Freeway Size	On-Ramps	Off-Ramps
Four-lane	1.00	1.00
Six-lane	1.12	1.05
Eight-lane	1.20	1.10

Figure 2.70 *HCM Exhibit 14-19*

On-Ramps		Off-Ramps	
Approaching Freeway Flow v_F (pc/h)	**Approaching Lane 5 Flow v_5 (pc/h)**	**Approaching Freeway Flow v_F (pc/h)**	**Approaching Lane 5 Flow v_5 (pc/h)**
≥8,500	2,500	≥7,000	0.200 v_F
7,500–8,499	0.285 v_F	5,500–6,999	0.150 v_F
6,500–7,499	0.270 v_F	4,000–5,499	0.100 v_F
5,500–6,499	0.240 v_F	<4,000	0
<5,500	0.220 v_F		

Figure 2.71 *HCM Equation 14-27*

$$v_{F4eff} = v_F - v_5$$

where

v_{F4eff} = effective approaching freeway flow in four lanes (pc/h),

v_F = total approaching freeway flow in five lanes (pc/h), and

v_5 = estimated approaching freeway flow in Lane 5 (pc/h).

Capacity Checks

The flow rate immediately downstream of a merge or upstream of a diverge is compared with the capacity values in HCM Exhibit 14-10. The ramp flow rate is compared with the capacity values in HCM Exhibit 14-12 for one- or two-lane ramps as a function of ramp free-flow speed, which is controlled by geometry. Long diamond ramps will have higher free-flow speeds and tight loop ramps will have lower free-flow speeds. If either check is exceeded, the segment is operating at LOS F. There is a separate table for capacity on multilane highways or collector-distributor roadways with slightly lower values in HCM Exhibit 14-11.

Figure 2.72 *HCM Exhibit 14-10*

FFS (mi/h)	Capacity of Upstream or Downstream Freeway Segment[a] Number of Lanes in One Direction				Maximum Desirable Flow Rate (v_{R12}) Entering Merge Influence Area[b]	Maximum Desirable Flow Rate (v_{12}) Entering Diverge Influence Area[b]
	2	3	4	>4		
≥70	4,800	7,200	9,600	2,400/ln	4,600	4,400
65	4,700	7,050	9,400	2,350/ln	4,600	4,400
60	4,600	6,900	9,200	2,300/ln	4,600	4,400
55	4,500	6,750	9,000	2,250/ln	4,600	4,400

Figure 2.73 *HCM Exhibit 14-12*

Ramp FFS S_{FR} (mi/h)	Single-Lane Ramps	Two-Lane Ramps
>50	2,200	4,400
>40–50	2,100	4,200
>30–40	2,000	4,000
≥20–30	1,900	3,800
<20	1,800	3,600

Figure 2.74 *HCM Exhibit 14-11*

FFS (mi/h)	Capacity of Upstream or Downstream Highway or C-D Segment[a] Number of Lanes in One Direction			Maximum Desirable Flow Rate (v_{R12}) Entering Merge Influence Area[b]	Maximum Desirable Flow Rate (v_{12}) Entering Diverge Influence Area[b]
	2	3	>3		
≥60	4,400	6,600	2,200/ln	4,600	4,400
55	4,200	6,300	2,100/ln	4,600	4,400
50	4,000	6,000	2,000/ln	4,600	4,400
45	3,800	5,700	1,900/ln	4,600	4,400

Capacity can be adjusted using the Capacity Adjustment Factor (*CAF*) for the effects of weather, incidents, and/or driver familiarity, following HCM Equation 12-8 with HCM Exhibits 11-20, 11-23, and/or 26-9. For incidents, the adjustment is applied to the open lanes.

Figure 2.75 *HCM Equation 12-8*

$$c_{adj} = c \times CAF$$

Lane Additions or Drops

Unless part of a weaving segment, merges with lane additions or diverges with lane drops are modeled as basic segments before and after the ramp. All capacity checks (freeway upstream of a diverge, freeway downstream of a merge, and the ramp itself) must still be made.

Figure 2.76 *HCM Exhibit 11-20*

		Capacity Adjustment Factors				
Weather Type	**Weather Event Definition**	**55 mi/h**	**60 mi/h**	**65 mi/h**	**70 mi/h**	**75 mi/h**
Medium rain	> 0.10 – 0.25 in/h	0.94	0.93	0.92	0.91	0.90
Heavy rain	> 0.25 in/h	0.89	0.88	0.86	0.84	0.82
Light snow	>0.00 – 0.05 in/h	0.97	0.96	0.96	0.95	0.95
Light–medium snow	>0.05 – 0.10 in/h	0.95	0.94	0.92	0.90	0.88
Medium–heavy snow	>0.10 – 0.50 in/h	0.93	0.91	0.90	0.88	0.87
Heavy snow	>0.50 in/h	0.80	0.78	0.76	0.74	0.72
Severe cold	<–4° F	0.93	0.92	0.92	0.91	0.90
Low visibility	0.50 – 0.99 mi	0.90	0.90	0.90	0.90	0.90
Very low visibility	0.25 – 0.49 mi	0.88	0.88	0.88	0.88	0.88
Minimal visibility	< 0.25 mi	0.90	0.90	0.90	0.90	0.90
Non-severe weather	All conditions not listed above	1.00	1.00	1.00	1.00	1.00

Figure 2.77 *HCM Exhibit 11-23*

Directional Lanes	No Incident	Shoulder Closed	1 Lane Closed	2 Lanes Closed	3 Lanes Closed	4 Lanes Closed
2	1.00	0.81	0.70	N/A	N/A	N/A
3	1.00	0.83	0.74	0.51	N/A	N/A
4	1.00	0.85	0.77	0.50	0.52	N/A
5	1.00	0.87	0.81	0.67	0.50	0.50
6	1.00	0.89	0.85	0.75	0.52	0.52
7	1.00	0.91	0.88	0.80	0.63	0.63
8	1.00	0.93	0.89	0.84	0.66	0.66

Figure 2.78 *HCM Exhibit 26-9*

Level of Driver Familiarity	CAF_{pop}
All familiar drivers, regular commuters	1.000
Mostly familiar drivers	0.968
Balanced mix of familiar and unfamiliar drivers	0.939
Mostly unfamiliar drivers	0.898
All or overwhelmingly unfamiliar drivers	0.852

Managed Lanes

For an on-ramp or off-ramp accessing a managed lane directly, it can be modeled normally if there are two managed lanes. For situations with only one managed lane, the flow rate in the managed lane is doubled, then modeled as two freeway lanes as shown in HCM Exhibit 14-22.

Major Merge or Diverge

There are no effective models for major merges as illustrated by HCM Exhibit 14-20. For major diverges, HCM Equation 14-28 can approximate the density for situations as illustrated in HCM Exhibit 14-21.

Figure 2.79 *HCM Exhibit 14-22*

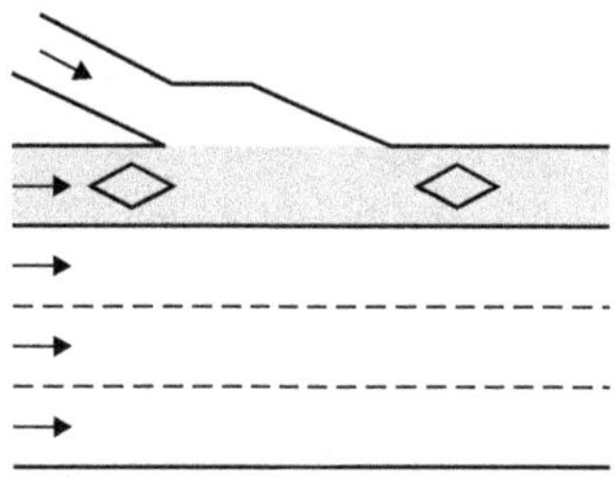
(a) Merge

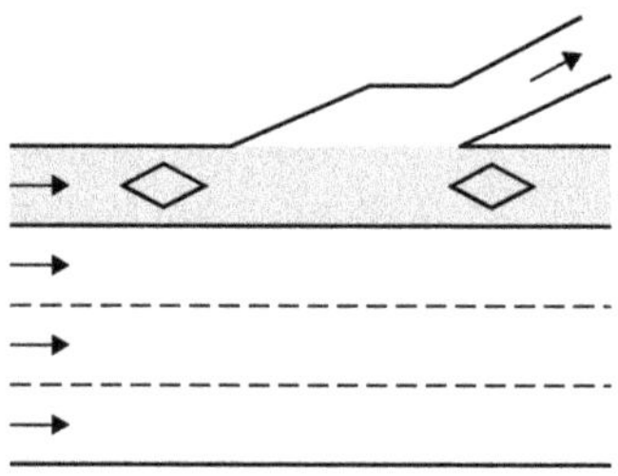
(b) Diverge

Figure 2.80 *HCM Exhibit 14-20*

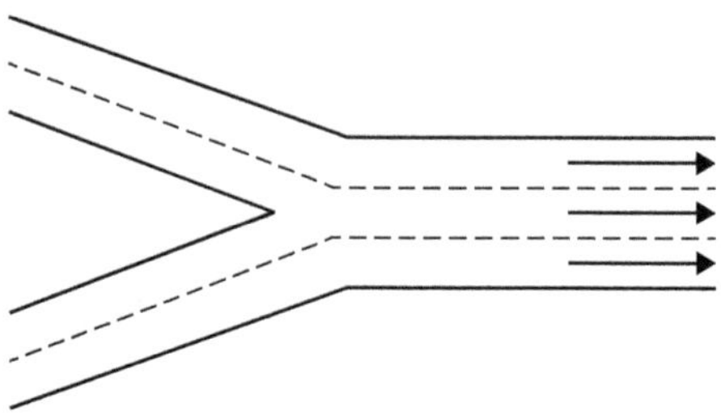
(a) Major merge with one lane dropped

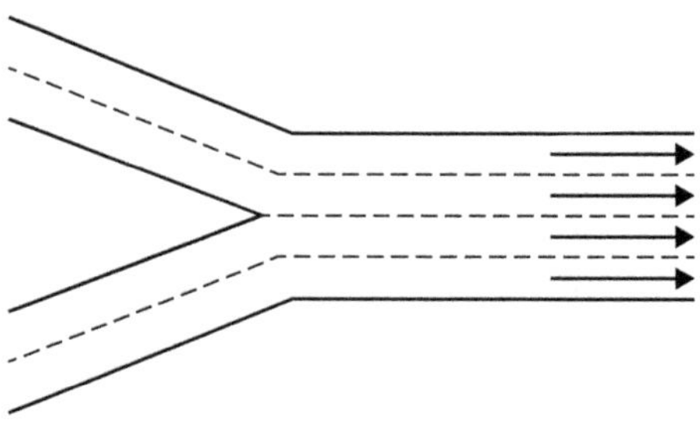
(b) Major merge with no lane dropped

Figure 2.81 *HCM Exhibit 14-21*

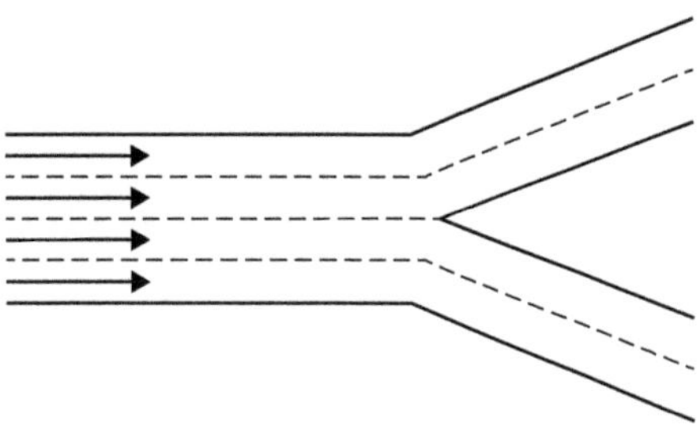
(a) Major diverge area with no lane addition

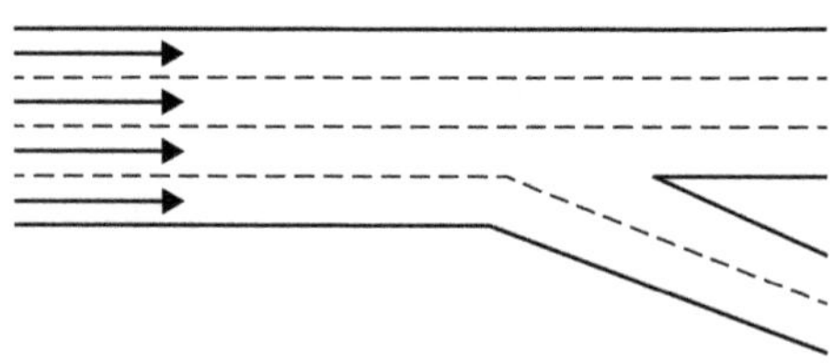
(b) Major diverge area with lane addition

Figure 2.82 *HCM Equation 14-28*

$$D_{MD} = 0.0175\left(\frac{v_F}{N}\right)$$

where

D_{MD} = density in the major diverge influence area (which includes all approaching freeway lanes) (pc/mi/ln),

v_F = demand flow rate immediately upstream of the major diverge influence area (pc/h), and

N = number of lanes approaching the major diverge (ln).

Density

For this procedure, density is computed using regression equations—not flow rate divided by speed as for the other segments. The density equations use the flow rate in lanes 1 and 2 with the acceleration or deceleration lane length as the primary parameters as shown in HCM Equations 14-22 (merge) and 14-23 (diverge). Because the term associated with these lengths is negative, care must be taken to ensure density does not become unreasonably low.

Figure 2.83 *HCM Equation 14-22*

$$D_R = 5.5475 + 0.00734 v_R + 0.0078 v_{12} - 0.00627 L_A$$

Figure 2.84 *HCM Equation 14-23*

$$D_R = 4.252 + 0.0086 v_{12} - 0.009 L_D$$

Figure 2.85 *HCM Exhibit 14-3*

LOS	Density (pc/mi/ln)
A	≤10
B	>10–20
C	>20–28
D	>28–35
E	>35
F	Demand exceeds capacity

LOS

Level of Service is defined by the density value, except for LOS F, which can only be the result if demand exceeds capacity as illustrated in HCM Exhibit 14-3. Notice that the LOS thresholds match those for weaving (except LOS F) and not those for basic.

Speed

Speeds within the ramp influence area and in the outer freeway lanes can be computed following the equations in HCM Exhibits 14-13 (merge) and 14-14 (diverge).

Figure 2.86 *HCM Exhibit 14-13*

Average Speed in	Equation	
Ramp influence area	$S_R = FFS \times SAF - (FFS \times SAF - 42)M_S$ $M_S = 0.321 + 0.00039e^{(v_{R12}/1{,}000)} - 0.002(L_A \times S_{FR} \times SAF/1{,}000)$	
Outer lanes of freeway	$S_O = FFS \times SAF$	$v_{OA} < 500$ pc/h
	$S_O = FFS \times SAF - 0.0036(v_{OA} - 500)$	$500 \le v_{OA} \le 2{,}300$ pc/h
	$S_O = FFS \times SAF - 6.53 - 0.006(v_{OA} - 2{,}300)$	$v_{OA} > 2{,}300$ pc/h

Figure 2.87 *HCM Exhibit 14-14*

Average Speed in	Equation	
Ramp influence area	$S_R = FFS \times SAF - (FFS \times SAF - 42)D_S$	
	$D_S = 0.883 + 0.00009v_R - 0.013S_{FR} \times SAF$	
Outer lanes of freeway	$S_O = 1.097 \times FFS \times SAF$	$v_{OA} < 1{,}000$ pc/h
	$S_O = 1.097 \times FFS \times SAF - 0.0039(v_{OA} - 1{,}000)$	$v_{OA} \geq 1{,}000$ pc/h

Speed consolidated across all freeway lanes can be calculated using the equations in HCM Exhibit 14-15 and HCM Equation 14-24.

Free-flow speed can be adjusted using the Speed Adjustment Factor (*SAF*) for the effects of weather and/or driver familiarity, following HCM Equation 12-5 with HCM Exhibits 11-21 and/or 26-9.

Figure 2.88 *HCM Exhibit 14-15*

Value	Equation
Average flow in outer lanes v_{OA} (pc/h)	$v_{OA} = \frac{v_F - v_{12}}{N_O}$
Average speed for on-ramp (merge) junctions (mi/h)	$S = \frac{v_{R12} + v_{OA}N_O}{\left(\frac{v_{R12}}{S_R}\right) + \left(\frac{v_{OA}N_O}{S_O}\right)}$
Average speed for off-ramp (diverge) junctions (mi/h)	$S = \frac{v_{12} + v_{OA}N_O}{\left(\frac{v_{12}}{S_R}\right) + \left(\frac{v_{OA}N_O}{S_O}\right)}$

Figure 2.89 *HCM Equation 14-24*

$$D = \frac{v}{S}$$

where

D = density including all lanes of the 1,500-ft ramp influence area (pc/mi/ln),

v = total flow rate through the merge or diverge area, all lanes (pc/h/ln), and

S = average speed of all vehicles through the merge or diverge area, all lanes, from Exhibit 14-15 (mi/h).

Figure 2.90 *HCM Equation 12-5*

$$FFS_{adj} = FFS \times SAF$$

Figure 2.91 *HCM Exhibit 11-21*

Weather Type	Weather Event Definition	Speed Adjustment Factors 55 mi/h	60 mi/h	65 mi/h	70 mi/h	75 mi/h
Medium rain	> 0.10–0.25 in/h	0.96	0.95	0.94	0.93	0.93
Heavy rain	> 0.25 in/h	0.94	0.93	0.93	0.92	0.91
Light snow	>0.00 – 0.05 in/h	0.94	0.92	0.89	0.87	0.84
Light–medium snow	>0.05–0.10 in/h	0.92	0.90	0.88	0.86	0.83
Medium–heavy snow	>0.10 –0.50 in/h	0.90	0.88	0.86	0.84	0.82
Heavy snow	>0.50 in/h	0.88	0.86	0.85	0.83	0.81
Severe cold	<–4° F	0.95	0.95	0.94	0.93	0.92
Low visibility	0.50 – 0.99 mi	0.96	0.95	0.94	0.94	0.93
Very low visibility	0.25 – 0.49 mi	0.95	0.94	0.93	0.92	0.91
Minimal visibility	< 0.25 mi	0.95	0.94	0.93	0.92	0.91
Non–severe weather	All conditions not listed above	1.00	1.00	1.00	1.00	1.00

Figure 2.92 *HCM Exhibit 26-9*

Level of Driver Familiarity	SAF_{pop}
All familiar drivers, regular commuters	1.000
Mostly familiar drivers	0.975
Balanced mix of familiar and unfamiliar drivers	0.950
Mostly unfamiliar drivers	0.913
All or overwhelmingly unfamiliar drivers	0.863

PROBLEM ONE

Determine the LOS for a Freeway Merge Segment with three freeway lanes carrying a volume of 3000 veh/h and a free-flow speed of 70 mi/h and one right-side ramp lane with an 800-foot acceleration lane and a volume of 1000 veh/h with a free-flow speed of 35 mi/h. There is an adjacent off-ramp 500 feet upstream with a volume of 500 veh/h. The peak hour factor is 0.90 throughout, and there are no trucks in the traffic stream with all familiar drivers, and no weather events or incidents.

1. Convert the volumes into flow rates by applying the peak hour factor (only since there are no trucks).

Flow Rate: $v_F = 3000/0.90 = 3333$ pc/h $v_R = 1000/0.90 = 1111$ pc/h

EQUATION 14-1

$v_{UP} = 500/0.90 = 556$ pc/h

2. Compute the equilibrium length to determine if the adjacent ramp has influence.

Equilibrium: $L_{EQ} = 0.214(3333{+}1111) + 0.444(800) + 52.32(35) - 2403 = 734$ feet

EQUATION 14-6

3. Compute the proportion of the freeway flow rate that will exist in lanes 1 and 2.

Proportion: $P_{FM} = 0.7289 - 0.0000135(3333 + 1111) + 0.003296(35) + 0.000063(500) = 0.585$

EQUATION 14-4

4. Calculate the flow rate in lanes 1 and 2 using the equation for an upstream ramp having influence.

Lanes 1 & 2: $v_{12} = 3333(0.585) = 1950$ pc/h

EQUATION 14-2

5. Check the outer lane flow rate for any balancing that might be required.

Balancing: $v_3 = 3333 - 1950 = 1383$ pc/h < 2700

EQUATION 14-14

$v_3 = 1.5(1950/2) = 1463$ pc/h > 1383

6. Check the freeway and ramp flow rates against the respective capacity values to ensure no LOS F.

Capacity: 7200 pc/h > 4444

EXHIBIT 14-10

2000 pc/h > 1111

EXHIBIT 14-12

6. Density is calculated for the influence area to determine level of service for the basic freeway segment.

Density: $D = 5.475 + 0.00734(1111) + 0.0078(1950) - 0.00627(800) = 23.8$ pc/mi/ln Equation 14-22

Level of Service: LOS = LOS C (between 20 and 28)

EXHIBIT 14-3

HCS7

These screenshots show how Problem One is coded into *HCS7*, followed by the results as presented in the formatted report. All input data, intermediate computations, and final results are shown in logical order, leading to density and level of service.

Figure 2.93 *HCS7 Merge1 Input*

Geometric Data			
Freeway Lanes	3	Ramp Lanes	1
Freeway FFS, mi/h	70.0	Ramp FFS, mi/h	35.0
Freeway Length, ft	1500	Ramp Side	Right
Freeway Terrain Type	Level	Ramp Terrain Type	Level
Freeway Grade, %	-	Ramp Grade, %	-
Freeway Grade Length, mi	-	Ramp Grade Length, mi	-
Highway or C-D Roadway	☐	Length of First Accel. Lane (LA), ft	800
Managed Lane	☐	Length of Second Accel. Lane (LA2), ft	-
Demand Data			
Freeway Demand, veh/h	3000	Merge Demand, veh/h	1000
Freeway Peak Hour Factor	0.90	Ramp Peak Hour Factor	0.90
Freeway Total Trucks, %	0.00	Ramp Total Trucks, %	0.00
Freeway Single-Unit Trucks (SUT), %	-	Ramp Single-Unit Trucks (SUT), %	-
Freeway Tractor-Trailers (TT), %	-	Ramp Tractor-Trailers (TT), %	-
Adjustment Factors			
Freeway Driver Population	All Familiar	Ramp Driver Population	All Familiar
Freeway Weather Type	Non-Severe Weather	Ramp Weather Type	Non-Severe Weather
Freeway Speed Adjustment Factor	1.000	Ramp Speed Adjustment Factor	1.000
Freeway Capacity Adjustment Factor	1.000	Ramp Capacity Adjustment Factor	1.000
Freeway Demand Adjustment Factor	1.000	Ramp Demand Adjustment Factor	1.000
Incident Type	No Incident		
Adjacent Ramps			
Upstream Ramp	Diverge	Downstream Ramp	No Ramp
Distance to Upstream Ramp, ft	500	Distance to Downstream Ramp, ft	-
Upstream Ramp Terrain	Level	Downstream Ramp Terrain	Level
Upstream Ramp Demand, veh/h	500	Downstream Ramp Demand, veh/h	0
Upstream Ramp PHF	0.90	Downstream Ramp PHF	0.94
Upstream Ramp Trucks, %	0.00	Downstream Ramp Trucks, %	0.00

Figure 2.94 *HCS7 Merge1 Output*

HCS7 Freeway Merge Report			
Project Information			
Analyst		Date	3/23/2017
Agency		Analysis Year	2017
Jurisdiction		Time Period Analyzed	
Project Description			

Geometric Data		
	Freeway	Ramp
Number of Lanes (N)	3	1
Free-Flow Speed (FFS), mi/h	70.0	35.0
Segment Length (L) / Acceleration Length (L_A), ft	1500	800
Terrain Type	Level	Level
Percent Grade, %	-	-
Segment Type / Ramp Side	Freeway	Right
Adjustment Factors		
Driver Population	All Familiar	All Familiar
Weather Type	Non-Severe Weather	Non-Severe Weather
Incident Type	No Incident	-
Final Speed Adjustment Factor (SAF)	1.000	1.000
Final Capacity Adjustment Factor (CAF)	1.000	1.000
Demand Adjustment Factor (DAF)	1.000	1.000
Demand and Capacity		
Volume (V_i), veh/h	3000	1000
Peak Hour Factor (PHF)	0.90	0.90
Total Trucks, %	0.00	0.00
Single-Unit Trucks (SUT), %	-	-
Tractor-Trailers (TT), %	-	-
Heavy Vehicle Adjustment Factor (f_{HV})	1.000	1.000
Flow Rate (v_i), pc/h	3333	1111
Capacity (c), pc/h	7200	2000
Volume-to-Capacity Ratio (v/c)	0.62	0.56

Speed and Density			
Upstream Equilibrium Distance (L_{EQ}), ft	734.4	Density in Ramp Influence Area (D_R), pc/mi/ln	23.9
Distance to Upstream Ramp (L_{UP}), ft	500	Speed Index (M_S)	0.348
Downstream Equilibrium Distance (L_{EQ}), ft	-	Flow Outer Lanes (v_{OA}), pc/h/ln	1383
Distance to Downstream Ramp (L_{DOWN}), ft	-	On-Ramp Influence Area Speed (S_R), mi/h	60.3
Prop. Freeway Vehicles in Lane 1 and 2 (P_{FM})	0.585	Outer Lanes Freeway Speed (S_O), mi/h	66.8
Flow in Lanes 1 and 2 (v_{12}), pc/h	1950	Ramp Junction Speed (S), mi/h	62.2
Flow Entering Ramp-Infl. Area (v_{R12}), pc/h	3061	Average Density (D), pc/mi/ln	23.8
Level of Service (LOS)	C		

PROBLEM TWO

Determine the LOS for a Freeway Diverge Segment including five freeway lanes carrying a volume of 8000 veh/h with a free-flow speed of 70 mi/h and two left-side ramp lanes with 400-foot deceleration lanes (L_{D1} and L_{D2}) carrying a volume of 2000 veh/h with a free-flow speed of 35 mi/h. The peak hour factor is 0.90 throughout, and there are no trucks in the traffic stream with all familiar drivers, and no weather events or incidents.

1. Convert the volumes into flow rates by applying the peak hour factor (only since there are no trucks).

Flow Rate: $v_F = 8000/0.90 = 8889$ pc/h $\quad v_R = 2000/0.90 = 2222$ pc/h

EQUATION 14-1

2. Reduce the flow rate in lane 5 to produce the four-lane rate for further analysis.

Lane 5: $v_5 = 8889(0.20) = 1778$ pc/h

EXHIBIT 14-19

$v_{4eff} = 8889 - 1778 = 7111$ pc/h

EQUATION 14-27

2. Compute the proportion of the freeway flow rate that will exist in lanes 1 and 2.

Proportion: $P_{FD} = 0.260$

Effective Decel: $L_{Deff} = 2(400) + 400 = 1200$ feet

EQUATION 14-26

3. Calculate the flow rate in lanes 1 and 2 with the appropriate modification for a left-side ramp.

Lanes 1 & 2: $v_{12} = 2222 + (7111 - 2222)(0.260) = 3493$ pc/h

EQUATION 14-8

$v_{12} = 3493(1.10) = 3842$ pc/h

EXHIBIT 14-18

4. Check the outer lane flow rate for any balancing that might be required.

Balancing: $v_{av34} = (7111 - 3842)/2 = 1635$ pc/h < 2700
EQUATION 14-14

$v_{av34} = 1.5(3842/2) = 2882$ pc/h < 2700

5. Check the freeway and ramp flow rates against the respective capacity values to ensure no LOS F.

Capacity: 12000 pc/h > 7111
EXHIBIT 14-10

4000 pc/h > 2222
EXHIBIT 14-12

7. Density is calculated for the influence area to determine level of service for the basic freeway segment.

Density: $D = 4.252 + 0.0086(3842) - 0.009(1200) = 26.5$ pc/mi/ln
EQUATION 14-23

Level of Service: LOS = LOS C (between 20 and 28)
EXHIBIT 14-3

HCS7

These screenshots show how Problem Two is coded into *HCS7*, followed by the results as presented in the formatted report. All input data, intermediate computations, and final results are shown in logical order, leading to density and level of service.

Figure 2.95 *HCS7 Diverge2 Input*

Geometric Data			
Freeway Lanes	5	Ramp Lanes	2
Freeway FFS, mi/h	70.0	Ramp FFS, mi/h	35.0
Freeway Length, ft	1500	Ramp Side	Left
Freeway Terrain Type	Level	Ramp Terrain Type	Level
Freeway Grade, %	-	Ramp Grade, %	-
Freeway Grade Length, mi	-	Ramp Grade Length, mi	-
Highway or C-D Roadway	☐	Length of First Decel. Lane (LD1), ft	400
Managed Lane	☐	Length of Second Decel. Lane (LD2), ft	400
Demand Data			
Freeway Demand, veh/h	8000	Diverge Demand, veh/h	2000
Freeway Peak Hour Factor	0.90	Ramp Peak Hour Factor	0.90
Freeway Total Trucks, %	0.00	Ramp Total Trucks, %	0.00
Freeway Single-Unit Trucks (SUT), %	-	Ramp Single-Unit Trucks (SUT), %	-
Freeway Tractor-Trailers (TT), %	-	Ramp Tractor-Trailers (TT), %	-
Adjustment Factors			
Freeway Driver Population	All Familiar	Ramp Driver Population	All Familiar
Freeway Weather Type	Non-Severe Weather	Ramp Weather Type	Non-Severe Weather
Freeway Speed Adjustment Factor	1.000	Ramp Speed Adjustment Factor	1.000
Freeway Capacity Adjustment Factor	1.000	Ramp Capacity Adjustment Factor	1.000
Freeway Demand Adjustment Factor	1.000	Ramp Demand Adjustment Factor	1.000
Incident Type	No Incident		

Figure 2.96 *HCS7 Diverge2 Output*

Geometric Data		
	Freeway	Ramp
Number of Lanes (N)	5	2
Free-Flow Speed (FFS), mi/h	70.0	35.0
Segment Length (L) / Deceleration Length (L_D), ft	1500	1200
Terrain Type	Level	Level
Percent Grade, %	-	-
Segment Type / Ramp Side	Freeway	Left
Adjustment Factors		
Driver Population	All Familiar	All Familiar
Weather Type	Non-Severe Weather	Non-Severe Weather
Incident Type	No Incident	-
Final Speed Adjustment Factor (SAF)	1.000	1.000
Final Capacity Adjustment Factor (CAF)	1.000	1.000
Demand Adjustment Factor (DAF)	1.000	1.000
Demand and Capacity		
Volume (V_i), veh/h	8000	2000
Peak Hour Factor (PHF)	0.90	0.90
Total Trucks, %	0.00	0.00
Single-Unit Trucks (SUT), %	-	-
Tractor-Trailers (TT), %	-	-
Heavy Vehicle Adjustment Factor (f_{HV})	1.000	1.000
Flow Rate (v_i), pc/h	8889	2222
Capacity (c), pc/h	12000	4000
Volume-to-Capacity Ratio (v/c)	0.74	0.56

Speed and Density			
Upstream Equilibrium Distance (L_{EQ}), ft	-	Density in Ramp Influence Area (D_R), pc/mi/ln	26.5
Distance to Upstream Ramp (L_{UP}), ft	-	Speed Index (D_S)	0.628
Downstream Equilibrium Distance (L_{EQ}), ft	-	Flow Outer Lanes (v_{OA}), pc/h/ln	1682
Distance to Downstream Ramp (L_{DOWN}), ft	-	Off-Ramp Influence Area Speed (S_R), mi/h	52.4
Prop. Freeway Vehicles in Lane 3 and 4 (P_{FD})	0.260	Outer Lanes Freeway Speed (S_O), mi/h	74.1
Flow in Lanes 3 and 4 (v_{34}), pc/h	3843	Ramp Junction Speed (S), mi/h	62.8
Flow Entering Ramp-Infl. Area (v_{R34}), pc/h	-	Average Density (D), pc/mi/ln	22.6
Level of Service (LOS)	C		

CHAPTER 3 Highways

Highways are divided into two types: Multilane Highways and Two-Lane Highways for analyzing one direction at a time. A Multilane Highway includes at-grade access points, but operates as uninterrupted flow with signals more than two miles apart. It can be four or six lanes and divided, undivided, or with a two-way left-turn lane (TWLTL), operating at a free-flow speed between 45 and 70 mi/h. A Two-Lane Highway also can include access points, but operates as an uninterrupted facility defined as one of three classes. Each class dictates a different service measure for determining level of service. Procedures are defined for determining bicycle level of service as well.

Multilane Highways

The analysis of a Multilane Highway as defined in HCM Chapter 12 involves estimating the free-flow speed and the flow rate for computing the travel speed, which is used to calculate density for determining level of service. This is very similar to the procedures for Basic Freeway Segments, which is the reason these methods are defined in the same chapter (HCM Chapter 12).

Required Data

To compute the free-flow speed, the geometric data required include number of lanes, lane width, total lateral clearance, median type, and access points per mile. If the free-flow speed is measured in the field, it is assumed to be equal to the adjusted free-flow speed (since these geometric attributes would have influenced the field-measured average).

To compute the flow rate, the demand data required include the prevailing volume (with driver familiarity information), peak hour factor, truck percentage, and the terrain type or specific grade (for which truck data must include the proportions of tractor trailers and single-unit trucks).

Free-Flow Speed

The process for computing adjusted free-flow speed (*FFS*) begins with a base free-flow speed (*BFFS*) estimated as 5 mi/h over the speed limit (7 mi/h over if the limit is 45 mi/h). This value can be changed with justification like field data to defend using a locally derived speed. Adjustments are made if the average lane width is below 12 feet or if the highway is undivided. The total lateral clearance is computed by adding the left- and right-side clearances with a 6-foot maximum for either side (total maximum of 12 feet). For undivided highways, the left-side lateral clearance is assumed to be 6 feet, since the median factor adjusts for this (preventing double counting the effect). (HCM Equation 12-3 and HCM Exhibits 12-20 and 12-22 through 12-24.)

Figure 3.1 *HCM Equation 12-3*

$$FFS = BFFS - f_{LW} - f_{TLC} - f_M - f_A$$

where

FFS = free-flow speed of the multilane highway segment (mi/h);

$BFFS$ = base FFS for the multilane highway segment (mi/h);

f_{LW} = adjustment for lane width, from Exhibit 12-20 (mi/h);

F_{TLC} = adjustment for total lateral clearance, from Exhibit 12-22 (mi/h);

f_M = adjustment for median type, from Exhibit 12-23 (mi/h); and

f_A = adjustment for access-point density, from Exhibit 12-24 (mi/h).

Figure 3.2 *HCM Exhibit 12-20*

Average Lane Width (ft)	Reduction in FFS, f_{LW} (mi/h)
≥12	0.0
≥11–12	1.9
≥10–11	6.6

Figure 3.3 *HCM Exhibit 12-22*

Four-Lane Highways		Six-Lane Highways	
TLC (ft)	Reduction in FFS (mi/h)	TLC (ft)	Reduction in FFS (mi/h)
12	0.0	12	0.0
10	0.4	10	0.4
8	0.9	8	0.9
6	1.3	6	1.3
4	1.8	4	1.7
2	3.6	2	2.8
0	5.4	0	3.9

Figure 3.4 *HCM Exhibit 12-23*

Median Type	Reduction in FFS, f_M (mi/h)
Undivided	1.6
TWLTL	0.0
Divided	0.0

Figure 3.5 *HCM Exhibit 12-24*

Four-Lane Highways		Six-Lane Highways	
TLC (ft)	Reduction in FFS (mi/h)	TLC (ft)	Reduction in FFS (mi/h)
12	0.0	12	0.0
10	0.4	10	0.4
8	0.9	8	0.9
6	1.3	6	1.3
4	1.8	4	1.7
2	3.6	2	2.8
0	5.4	0	3.9

These adjustments reflect how drivers react to geometric restrictions with slower speeds. Field-measured free-flow speed is recommended if there are geometric situations that are not being modeled.

Think about your own driving:
You feel the friction of opposing traffic separated only by two yellow lines and/or access points with other vehicles turning into and pulling out of side streets and driveways. Your speed will be higher on a divided highway with fewer access points.

Free-flow speed can be adjusted using the Speed Adjustment Factor (*SAF*) for the effects of driver familiarity, following HCM Equation 12-5 with HCM Exhibit 26-9.

Figure 3.6 *HCM Equation 12-5*

$$FFS_{adj} = FFS \times SAF$$

Flow Rate

Flow rate (v_p) is estimated by modifying the hourly volume in vehicles per hour to adjust for the presence of heavy vehicles and to convert the value to a peak fifteen-minute rate using the peak hour factor, following HCM Equation 12-9. The heavy vehicle adjustment factor (f_{HV}) is determined by the passenger-car equivalents (*PCE*) as a function of terrain type or specific grade data in combination with the proportion of trucks. This proportion is defined as a split between tractor-trailers (TT) and single-unit trucks (SUT) that include buses and recreational vehicles. HCM Equation 12-10 is used with passenger-car equivalent values from HCM Exhibits 12-25, 12-26, 12-27, and 12-28 to generate the final flow rate value.

Interpolation and extrapolation are utilized for values not specifically listed in the provided tables. If the truck percentage is high along a long, steep upgrade, the mixed-flow model should be used in place of this equation and

Figure 3.7 *HCM Exhibit 26-9*

Level of Driver Familiarity	SAF_{pop}
All familiar drivers, regular commuters	1.000
Mostly familiar drivers	0.975
Balanced mix of familiar and unfamiliar drivers	0.950
Mostly unfamiliar drivers	0.913
All or overwhelmingly unfamiliar drivers	0.863

Figure 3.8 *HCM Equation 12-9*

$$v_p = \frac{V}{PHF \times N \times f_{HV}}$$

where

v_p = demand flow rate under equivalent base conditions (pc/h/ln),

V = demand volume under prevailing conditions (veh/h),

PHF = peak hour factor (decimal),

N = number of lanes in analysis direction (ln), and

f_{HV} = adjustment factor for presence of heavy vehicles (decimal)

Figure 3.9 *HCM Equation 12-10*

$$f_{HV} = \frac{1}{1 + P_T(E_T - 1)}$$

where

f_{HV} = heavy-vehicle adjustment factor (decimal),

P_T = proportion of SUT and TTs in traffic stream (decimal), and

E_T = passenger-car equivalent of one heavy vehicle in the traffic stream (PCEs).

Figure 3.10 *HCM Exhibit 12-25*

Passenger Car Equivalent	Terrain Type	
	Level	**Rolling**
E_T	2.0	3.0

these tables. HCM Exhibit 26-4 illustrates the modified speed-flow curves that are required to model these situations.

Capacity

Capacity (c) is also defined as a function of free-flow speed in the equation from HCM Exhibit 12-6 between 1900 pc/h/ln at 45 mi/h and 2300 pc/h/ln at 70 mi/h.

Capacity can be adjusted using the Capacity Adjustment Factor (CAF) for the effects of driver familiarity, following HCM Equation 12-8 with HCM Exhibit 26-9.

> ***Think about your own driving:***
> You have all been in areas or times with a significant number of visitors who are not familiar with the area. They are looking for street name signs or landmarks and sometimes following an automated navigation system. Their headways are longer and speeds slower, affecting both capacity and speed.

Figure 3.11 *HCM Exhibit 12-26*

%	Length	Percentage of Trucks and Buses (%)								
Grade	**(mi)**	**2%**	**4%**	**5%**	**6%**	**8%**	**10%**	**15%**	**20%**	**>25%**
−2	0.125	2.62	2.37	2.30	2.24	2.17	2.12	2.04	1.99	1.97
	0.375	2.62	2.37	2.30	2.24	2.17	2.12	2.04	1.99	1.97
	0.625	2.62	2.37	2.30	2.24	2.17	2.12	2.04	1.99	1.97
	0.875	2.62	2.37	2.30	2.24	2.17	2.12	2.04	1.99	1.97
	1.25	2.62	2.37	2.30	2.24	2.17	2.12	2.04	1.99	1.97
	1.5	2.62	2.37	2.30	2.24	2.17	2.12	2.04	1.99	1.97
0	0.125	2.62	2.37	2.30	2.24	2.17	2.12	2.04	1.99	1.97
	0.375	2.62	2.37	2.30	2.24	2.17	2.12	2.04	1.99	1.97
	0.625	2.62	2.37	2.30	2.24	2.17	2.12	2.04	1.99	1.97
	0.875	2.62	2.37	2.30	2.24	2.17	2.12	2.04	1.99	1.97
	1.25	2.62	2.37	2.30	2.24	2.17	2.12	2.04	1.99	1.97
	1.5	2.62	2.37	2.30	2.24	2.17	2.12	2.04	1.99	1.97
2	0.125	2.62	2.37	2.30	2.24	2.17	2.12	2.04	1.99	1.97
	0.375	3.76	2.96	2.78	2.65	2.48	2.38	2.22	2.14	2.09
	0.625	4.47	3.33	3.08	2.91	2.68	2.54	2.34	2.23	2.17
	0.875	4.80	3.50	3.22	3.03	2.77	2.61	2.39	2.28	2.21
	1.25	5.00	3.60	3.30	3.09	2.83	2.66	2.42	2.30	2.23
	1.5	5.04	3.62	3.32	3.11	2.84	2.67	2.43	2.31	2.23
2.5	0.125	2.62	2.37	2.30	2.24	2.17	2.12	2.04	1.99	1.97
	0.375	4.11	3.14	2.93	2.78	2.58	2.46	2.28	2.19	2.13
	0.625	5.04	3.62	3.32	3.11	2.84	2.67	2.43	2.31	2.23
	0.875	5.48	3.85	3.51	3.27	2.96	2.77	2.50	2.36	2.28
	1.25	5.73	3.98	3.61	3.36	3.03	2.83	2.54	2.40	2.31
	1.5	5.80	4.02	3.64	3.38	3.05	2.84	2.55	2.41	2.32
3.5	0.125	2.62	2.37	2.30	2.24	2.17	2.12	2.04	1.99	1.97
	0.375	4.88	3.54	3.25	3.05	2.80	2.63	2.41	2.29	2.22
	0.625	6.34	4.30	3.87	3.58	3.20	2.97	2.64	2.48	2.38
	0.875	7.03	4.66	4.16	3.83	3.39	3.12	2.76	2.57	2.46
	1.25	7.44	4.87	4.33	3.97	3.50	3.22	2.82	2.62	2.50
	1.5	7.53	4.92	4.38	4.01	3.53	3.24	2.84	2.63	2.51
4.5	0.125	2.62	2.37	2.30	2.24	2.17	2.12	2.04	1.99	1.97
	0.375	5.80	4.02	3.64	3.38	3.05	2.84	2.55	2.41	2.32
	0.625	7.90	5.11	4.53	4.14	3.63	3.32	2.90	2.68	2.55
	0.875	8.91	5.64	4.96	4.50	3.92	3.56	3.07	2.82	2.67
	1	9.19	5.78	5.08	4.60	3.99	3.62	3.11	2.85	2.70
5.5	0.125	2.62	2.37	2.30	2.24	2.17	2.12	2.04	1.99	1.97
	0.375	6.87	4.58	4.10	3.77	3.35	3.09	2.73	2.55	2.44
	0.625	9.78	6.09	5.33	4.82	4.16	3.76	3.21	2.93	2.77
	0.875	11.20	6.83	5.94	5.33	4.56	4.09	3.45	3.12	2.93
	1	11.60	7.04	6.11	5.47	4.67	4.18	3.51	3.17	2.97
6	0.125	2.62	2.37	2.30	2.24	2.17	2.12	2.04	1.99	1.97
	0.375	7.48	4.90	4.36	3.99	3.52	3.23	2.83	2.63	2.51
	0.625	10.87	6.66	5.79	5.21	4.46	4.01	3.39	3.08	2.89
	0.875	12.54	7.54	6.51	5.81	4.94	4.40	3.67	3.30	3.08
	1	13.02	7.78	6.71	5.99	5.07	4.51	3.75	3.37	3.14

Figure 3.12 *HCM Exhibit 12-27*

% Grade	Length (mi)	Percentage of Trucks and Buses (%) 2%	4%	5%	6%	8%	10%	15%	20%	>25%
−2	0.125	2.67	2.38	2.31	2.25	2.16	2.11	2.02	1.97	1.93
	0.375	2.67	2.38	2.31	2.25	2.16	2.11	2.02	1.97	1.93
	0.625	2.67	2.38	2.31	2.25	2.16	2.11	2.02	1.97	1.93
	0.875	2.67	2.38	2.31	2.25	2.16	2.11	2.02	1.97	1.93
	1.25	2.67	2.38	2.31	2.25	2.16	2.11	2.02	1.97	1.93
	1.5	2.67	2.38	2.31	2.25	2.16	2.11	2.02	1.97	1.93
0	0.125	2.67	2.38	2.31	2.25	2.16	2.11	2.02	1.97	1.93
	0.375	2.67	2.38	2.31	2.25	2.16	2.11	2.02	1.97	1.93
	0.625	2.67	2.38	2.31	2.25	2.16	2.11	2.02	1.97	1.93
	0.875	2.67	2.38	2.31	2.25	2.16	2.11	2.02	1.97	1.93
	1.25	2.67	2.38	2.31	2.25	2.16	2.11	2.02	1.97	1.93
	1.5	2.67	2.38	2.31	2.25	2.16	2.11	2.02	1.97	1.93
2	0.125	2.67	2.38	2.31	2.25	2.16	2.11	2.02	1.97	1.93
	0.375	3.76	2.95	2.77	2.64	2.47	2.36	2.20	2.11	2.06
	0.625	4.32	3.24	3.01	2.84	2.63	2.49	2.29	2.19	2.12
	0.875	4.57	3.37	3.11	2.93	2.70	2.55	2.33	2.22	2.15
	1.25	4.71	3.45	3.17	2.99	2.74	2.58	2.36	2.24	2.17
	1.5	4.74	3.47	3.19	3.00	2.75	2.59	2.36	2.24	2.17
2.5	0.125	2.67	2.38	2.31	2.25	2.16	2.11	2.02	1.97	1.93
	0.375	4.10	3.13	2.92	2.77	2.57	2.44	2.26	2.16	2.10
	0.625	4.84	3.52	3.23	3.03	2.77	2.61	2.38	2.26	2.18
	0.875	5.17	3.69	3.37	3.15	2.87	2.69	2.43	2.30	2.22
	1.25	5.36	3.79	3.45	3.22	2.92	2.73	2.47	2.33	2.24
	1.5	5.40	3.81	3.47	3.24	2.93	2.74	2.47	2.33	2.25
3.5	0.125	2.67	2.38	2.31	2.25	2.16	2.11	2.02	1.97	1.93
	0.375	4.89	3.54	3.25	3.05	2.79	2.62	2.39	2.26	2.19
	0.625	6.05	4.15	3.75	3.47	3.11	2.89	2.58	2.42	2.32
	0.875	6.58	4.43	3.97	3.66	3.26	3.01	2.67	2.49	2.39
	1.25	6.88	4.58	4.10	3.77	3.35	3.09	2.72	2.53	2.42
	1.5	6.95	4.62	4.13	3.80	3.37	3.10	2.73	2.54	2.43
4.5	0.125	2.67	2.38	2.31	2.25	2.16	2.11	2.02	1.97	1.93
	0.375	5.83	4.03	3.65	3.39	3.05	2.84	2.55	2.39	2.30
	0.625	7.53	4.92	4.38	4.01	3.53	3.24	2.83	2.62	2.50
	0.875	8.32	5.34	4.72	4.29	3.75	3.42	2.97	2.73	2.59
	1	8.53	5.45	4.81	4.37	3.81	3.47	3.00	2.76	2.62
5.5	0.125	2.67	2.38	2.31	2.25	2.16	2.11	2.02	1.97	1.93
	0.375	6.97	4.63	4.14	3.81	3.38	3.11	2.74	2.55	2.43
	0.625	9.37	5.89	5.16	4.68	4.05	3.67	3.14	2.88	2.72
	0.875	10.49	6.48	5.65	5.09	4.37	3.93	3.34	3.03	2.85
	1	10.80	6.64	5.78	5.20	4.46	4.01	3.39	3.08	2.89
6	0.125	2.67	2.38	2.31	2.25	2.16	2.11	2.02	1.97	1.93
	0.375	7.64	4.98	4.43	4.05	3.56	3.26	2.85	2.64	2.51
	0.625	10.45	6.45	5.63	5.07	4.36	3.92	3.33	3.03	2.85
	0.875	11.78	7.16	6.20	5.56	4.74	4.24	3.56	3.22	3.01
	1	12.15	7.35	6.36	5.69	4.85	4.33	3.62	3.27	3.05

Figure 3.13 *HCM Exhibit 12-28*

% Grade	Length (mi)	2%	4%	5%	6%	8%	10%	15%	20%	>25%
		Percentage of Trucks and Buses (%)								
−2	0.125	2.39	2.18	2.12	2.07	2.01	1.96	1.89	1.85	1.83
	0.375	2.39	2.18	2.12	2.07	2.01	1.96	1.89	1.85	1.83
	0.625	2.39	2.18	2.12	2.07	2.01	1.96	1.89	1.85	1.83
	0.875	2.39	2.18	2.12	2.07	2.01	1.96	1.89	1.85	1.83
	1.25	2.39	2.18	2.12	2.07	2.01	1.96	1.89	1.85	1.83
	1.5	2.39	2.18	2.12	2.07	2.01	1.96	1.89	1.85	1.83
0	0.125	2.39	2.18	2.12	2.07	2.01	1.96	1.89	1.85	1.83
	0.375	2.39	2.18	2.12	2.07	2.01	1.96	1.89	1.85	1.83
	0.625	2.39	2.18	2.12	2.07	2.01	1.96	1.89	1.85	1.83
	0.875	2.39	2.18	2.12	2.07	2.01	1.96	1.89	1.85	1.83
	1.25	2.39	2.18	2.12	2.07	2.01	1.96	1.89	1.85	1.83
	1.5	2.39	2.18	2.12	2.07	2.01	1.96	1.89	1.85	1.83
2	0.125	2.67	2.32	2.23	2.17	2.08	2.03	1.94	1.89	1.86
	0.375	3.63	2.82	2.64	2.52	2.35	2.25	2.10	2.02	1.97
	0.625	4.12	3.08	2.85	2.69	2.49	2.36	2.18	2.08	2.02
	0.875	4.37	3.21	2.96	2.78	2.56	2.42	2.22	2.11	2.05
	1.25	4.53	3.29	3.02	2.84	2.60	2.45	2.24	2.13	2.07
	1.5	4.58	3.31	3.04	2.86	2.61	2.46	2.25	2.14	2.07
2.5	0.125	2.75	2.36	2.27	2.20	2.11	2.04	1.95	1.90	1.87
	0.375	4.01	3.02	2.80	2.65	2.46	2.33	2.16	2.06	2.01
	0.625	4.66	3.35	3.08	2.88	2.64	2.48	2.26	2.15	2.08
	0.875	4.99	3.52	3.21	3.00	2.73	2.56	2.32	2.19	2.12
	1.25	5.20	3.64	3.30	3.08	2.79	2.60	2.35	2.22	2.14
	1.5	5.26	3.67	3.33	3.10	2.80	2.62	2.36	2.23	2.15
3.5	0.125	2.93	2.45	2.34	2.26	2.16	2.09	1.98	1.92	1.89
	0.375	4.86	3.46	3.16	2.96	2.69	2.53	2.30	2.18	2.10
	0.625	5.88	3.99	3.59	3.32	2.98	2.76	2.46	2.31	2.22
	0.875	6.40	4.26	3.81	3.51	3.12	2.88	2.55	2.38	2.28
	1.25	6.74	4.43	3.96	3.63	3.21	2.96	2.60	2.42	2.32
	1.5	6.83	4.48	3.99	3.66	3.24	2.98	2.62	2.44	2.33
4.5	0.125	3.13	2.56	2.43	2.34	2.21	2.13	2.01	1.95	1.91
	0.375	5.88	3.99	3.59	3.32	2.98	2.76	2.46	2.31	2.22
	0.625	7.35	4.75	4.22	3.85	3.39	3.10	2.71	2.51	2.39
	0.875	8.11	5.15	4.54	4.13	3.60	3.27	2.83	2.61	2.47
	1	8.33	5.27	4.63	4.21	3.66	3.33	2.87	2.64	2.50
5.5	0.125	3.37	2.69	2.53	2.42	2.28	2.19	2.05	1.98	1.94
	0.375	7.09	4.62	4.11	3.76	3.31	3.04	2.66	2.47	2.36
	0.625	9.13	5.68	4.97	4.49	3.88	3.51	3.00	2.74	2.59
	0.875	10.21	6.24	5.43	4.88	4.18	3.76	3.18	2.89	2.71
	1	10.52	6.41	5.57	5.00	4.27	3.83	3.24	2.93	2.75
6	0.125	3.51	2.76	2.59	2.47	2.32	2.22	2.08	2.00	1.95
	0.375	7.78	4.98	4.40	4.01	3.51	3.20	2.78	2.56	2.44
	0.625	10.17	6.23	5.42	4.87	4.17	3.75	3.18	2.88	2.71
	0.875	11.43	6.88	5.95	5.32	4.53	4.04	3.39	3.06	2.86
	1	11.81	7.08	6.11	5.46	4.64	4.13	3.45	3.11	2.90

Figure 3.14 *HCM Exhibit 26-4*

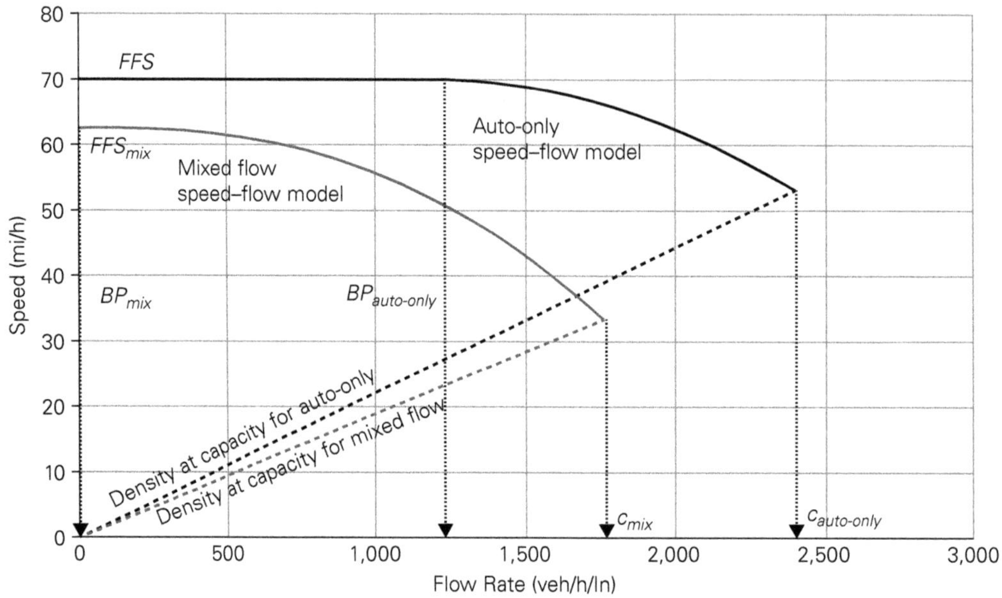

Figure 3.15 *HCM Exhibit 12-6*

$$c = 1{,}900 + 20(FFS - 45)$$
$$c \le 2{,}300$$
$$45 \le FFS \le 70$$

Figure 3.16 *HCM Equation 12-8*

$$c_{adj} = c \times CAF$$

Figure 3.17 *HCM Exhibit 26-9*

Level of Driver Familiarity	CAF_{pop}
All familiar drivers, regular commuters	1.000
Mostly familiar drivers	0.968
Balanced mix of familiar and unfamiliar drivers	0.939
Mostly unfamiliar drivers	0.898
All or overwhelmingly unfamiliar drivers	0.852

Break Point

Speed is unaffected by flow rate until the flow rate exceeds the break point that is defined as 1400 pc/h for all free-flow speeds in HCM Exhibit 12-6. When the flow rate is lower than the break point, speed is equal to free-flow speed.

Figure 3.18 *HCM Exhibit 12-6*

$$BP_{adj} = 1{,}400$$

> ***Think about your own driving:***
> Speeds are generally unaffected by traffic until it reaches a certain level—where other drivers are in front slowing you down, or behind pushing you to speed up, and on either side preventing you from getting out of their way.

Speed

Speed (S) is computed using the adjusted flow rate and free-flow speed values following the speed-flow curves defined by the HCM Equation 12-1 and HCM Exhibit 12-8.

Speed can be adjusted using the Speed Adjustment Factor (SAF) for the effects of driver familiarity, following HCM Equation 12-5 with HCM Exhibit 26-9.

Figure 3.19 *HCM Equation 12-1*

$$S = FFS_{adj} \qquad v_p \leq BP$$

$$S = FFS_{adj} - \frac{\left(FFS_{adj} - \frac{c_{adj}}{D_c}\right)(v_p - BP)^a}{(c_{adj} - BP)^a} \qquad BP < v_p \leq c$$

Figure 3.20 *HCM Exhibit 12-8*

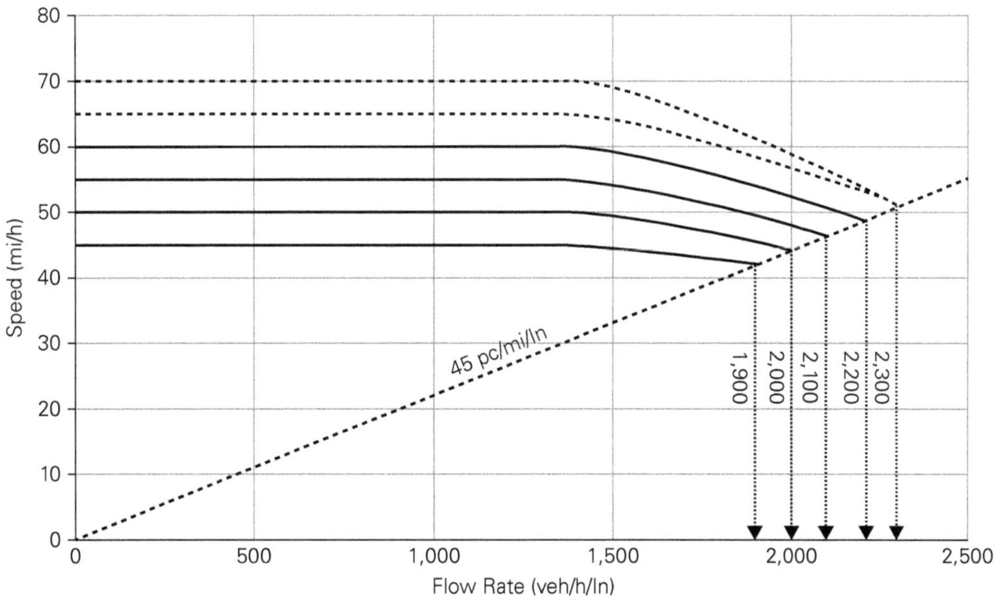

Figure 3.21 *HCM Equation 12-5*

$$FFS_{adj} = FFS \times SAF$$

Figure 3.22 *HCM Exhibit 26-9*

Level of Driver Familiarity	*SAFpop*
All familiar drivers, regular commuters	1.000
Mostly familiar drivers	0.975
Balanced mix of familiar and unfamiliar drivers	0.950
Mostly unfamiliar drivers	0.913
All or overwhelmingly unfamiliar drivers	0.863

Density

Density (D) becomes simply adjusted flow rate divided by speed as shown in HCM Equation 12-11, following the speed-flow relationship illustrated by HCM Exhibit 12-17.

LOS

Level of Service (LOS) is defined by the thresholds shown in HCM Exhibit 12-5 and illustrated within HCM Exhibit 12-17 with LOS F further dictated when demand exceeds capacity.

Figure 3.23 *HCM Equation 12-11*

$$D = \frac{v_p}{S}$$

where

D = density (pc/mi/ln),

v_p = demand flow rate (pc/h/ln), and

S = mean speed of traffic stream under base conditions (mi/h).

Figure 3.24 *HCM Exhibit 12-17*

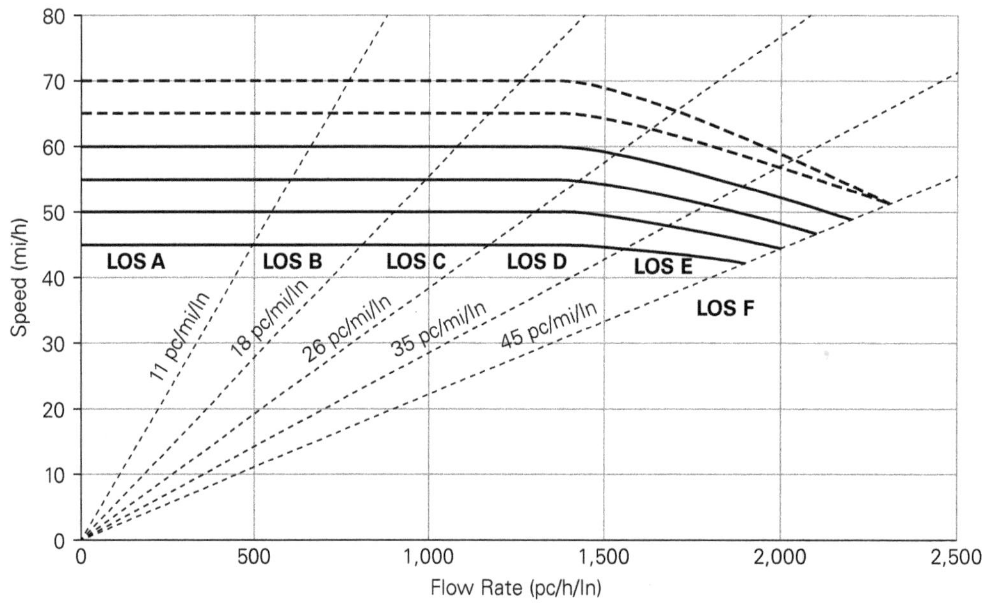

Figure 3.25 *HCM Exhibit 12-15*

LOS	Density (pc/mi/ln)
A	≤11
B	>11–18
C	>18–26
D	>26–35
E	>35–45
F	Demand exceeds capacity OR density >45

PROBLEM ONE

Determine the LOS for a Multilane Highway with these parameters in the analysis direction: 3000 veh/h with 0.92 PHF with mostly unfamiliar drivers; 5% trucks (70% SUT / 30% TT) on a 1% grade for 1 mile; two 11-ft lanes, undivided; 50 mi/h speed limit; 30-ft right-side lateral clearance; and 5 access points per mile.

1. Compute the free-flow speed by adjusting for lane width, right-side lateral clearance, and ramp density.

Free-Flow Speed: $FFS = 55.0 - 1.9 - 0.0 - 1.6 - 1.3 = 50.3$ mi/h

EQUATION 12-3 / EXHIBITS 12-20, 12-22, 12-23, & 12-24

2. Free-flow speed is adjusted for mostly familiar drivers.

Free-Flow Speed: $FFS_{adj} = 50.3(0.913) = 45.9$ mi/h

EXHIBIT 26-9 & EQUATION 12-5

3. The heavy vehicle adjustment factor is computed for the positive grade with trucks in the traffic stream.

Heavy Vehicles: $E_T = 2.55$ $f_{HV} = 1 / (1 + 0.05(2.55 - 1)) = 0.928$

EQUATION 12-10 / EXHIBIT 12-28

4. Adjusted flow rate is computed using the heavy vehicle adjustment and peak hour factors with number of lanes.

Flow Rate: $v_p = 3000 / (0.92)(2)(0.928) = 1757$ pc/h/ln

EQUATION 12-9

5. Capacity is computed as a function of free-flow speed, then adjusted for driver familiarity.

Capacity: $c = 1900 + 20(45.9 - 45) = 1918$ pc/h/ln

EXHIBIT 12-6

Capacity: $c_{adj} = 1918(0.898) = 1722$ pc/h

EQUATION 12-8 / EXHIBIT 26-9

6. The break point is 1400 pc/h.

Break Point: $BP = 1400$ pc/h/ln

EXHIBIT 12-6

7. This allows speed to be computed from the free-flow speed and flow rate that is beyond the break point.

Speed: $S = 45.9 - ((45.9 - 1722/45)(1757 - 1400)^2) / (1722 - 1400)^2 = 36.5$ mi/h

EQUATION 12-1

8. Density is calculated from flow rate and speed to determine level of service for the basic freeway segment.

Density: $D = 1757 / 36.5 = 48.1$ pc/mi/ln

EQUATION 12-11

Level of Service: LOS = F (over 45.0 with flow rate of 1757 exceeding capacity of 1722)

EXHIBIT 12-17

HCS7

These screenshots show how Problem One is coded into *HCS7* in only one screen for this relatively simple procedure, followed by the results as presented in the formatted report. All input data, intermediate computations, and final results are shown in logical order, leading to density and level of service.

Figure 3.26 *HCS7 Multilane1 Input*

Direction 1	
Number of Lanes	2
Measured FFS	☐
Base Free Flow Speed, mi/h	55.0
Median Type	Undivided
Lane Width, ft	11
Right Side Clearance, ft	6
Median (Left) Side Clearance, ft	6
Access Point Density, points/mi	5.0
Terrain Type	Specific Grade
Percent Grade, %	1.00
Grade Length, mi	1.00
Demand, veh/h	3000
Peak Hour Factor	0.92
Total Trucks, %	5.00
Single-Unit Trucks (SUT), %	70
Tractor-Trailers (TT), %	30
Mixed Flow Model	☐
Driver Population	Mostly Unfamiliar
Capacity Adjustment Factor	0.898
Speed Adjustment Factor	0.913

Figure 3.27 *HCS7 Multilane1 Output*

Direction 1 Geometric Data			
Direction 1 Description			
Number of Lanes (N), ln	2	Terrain Type	Specific Grade
Segment Length (L), ft	-	Percent Grade, %	1.00
Measured or Base Free-Flow Speed	Base	Grade Length, mi	1.00
Base Free-Flow Speed (BFFS), mi/h	55.0	Right-Side Lateral Clearance (LC_R), ft	6
Lane Width, ft	11	Left-Side Lateral Clearance (LC_L), ft	6
Median Type	Undivided	Total Lateral Clearance (TLC), ft	12.00
Access Point Density, pts/mi	5.0	Free-Flow Speed (FFS), mi/h	50.3
Direction 1 Adjustment Factors			
Driver Population	Mostly Unfamiliar	Final Speed Adjustment Factor (SAF)	0.913
Driver Population SAF	0.913	Final Capacity Adjustment Factor (CAF)	0.898
Driver Population CAF	0.898		
Direction 1 Demand and Capacity			
Volume (V), veh/h	3000	Heavy Vehicle Adjustment Factor (f_{HV})	0.928
Peak Hour Factor (PHF)	0.92	Flow Rate (v_p), pc/h/ln	1757
Total Trucks, %	5.00	Capacity (c), pc/h/ln	1918
Single-Unit Trucks (SUT), %	70	Adjusted Capacity (c_{adj}), pc/h/ln	1722
Tractor-Trailers (TT), %	30	Volume-to-Capacity Ratio (v/c)	1.02
Direction 1 Speed and Density			
Lane Width Adjustment (f_{LW})	1.9	Average Speed (S), mi/h	-
Total Lateral Clearance Adj. (f_{TLC})	0.0	Density (D), pc/mi/ln	-
Median Type Adjustment (f_M)	1.6	Level of Service (LOS)	F
Access Point Density Adjustment (f_A)	1.3		

PROBLEM TWO

Building on Problem One, compute the LOS for the other direction with the same lanes and traffic, but with no access points.

1. Compute the free-flow speed by adjusting for lane width, right-side lateral clearance, and ramp density.

Free-Flow Speed: $FFS = 55.0 - 1.9 - 0.0 - 1.6 - 0.0 = 51.5$ mi/h
EQUATION 12-3 / EXHIBITS 12-20, 12-22, 12-23, & 12-24

2. Free-flow speed is adjusted for mostly familiar drivers.

Free-Flow Speed: $FFS_{adj} = 51.5(0.913) = 47.0$ mi/h
EXHIBIT 26-9 & EQUATION 12-5

3. The heavy vehicle adjustment factor is computed for the negative grade with trucks in the traffic stream.

Heavy Vehicles: $E_T = 2.12$ $f_{HV} = 1 / (1 + 0.05(2.12 - 1)) = 0.947$
EQUATION 12-10 / EXHIBIT 12-28

4. Adjusted flow rate is computed using the heavy vehicle adjustment, peak hour factor with number of lanes.

Flow Rate: $v_p = 3000 / (0.92)(2)(0.947) = 1722$ pc/h/ln
EQUATION 12-9

5. Capacity is computed as a function of free-flow speed, then adjusted for driver familiarity.

Capacity: $c = 1900 + 20(47.0 - 45) = 1940$ pc/h/ln
EXHIBIT 12-6

Capacity: $c_{adj} = 1940(0.898) = 1742$ pc/h
EQUATION 12-8 / EXHIBIT 26-9

6. The break point is 1400 pc/h.

Break Point: $BP = 1400$ pc/h/ln
EXHIBIT 12-6

7. This allows speed to be computed from the free-flow speed and flow rate that is beyond the break point.

Speed: $S = 47.0 - ((47.0 - 1742/45)(1722 - 1400)^2) / (1742 - 1400)^2 = 39.7$

EQUATION 12-1

8. Density is calculated from flow rate and speed to determine level of service for the basic freeway segment.

Density: $D = 1722 / 39.7 = 43.4$ pc/mi/ln

EQUATION 12-11

Level of Service: LOS = E (between 35 and 45)

EXHIBIT 12-17

HCS7

These screenshots show how Problem Two is coded into *HCS7* in only one screen for this relatively simple procedure, followed by the results as presented in the formatted report. Notice that both directions can be coded together, but the reports are presented separately. All input data, intermediate computations, and final results are shown in logical order, leading to density and level of service.

Figure 3.28 *HCS7 Multilane2 Input*

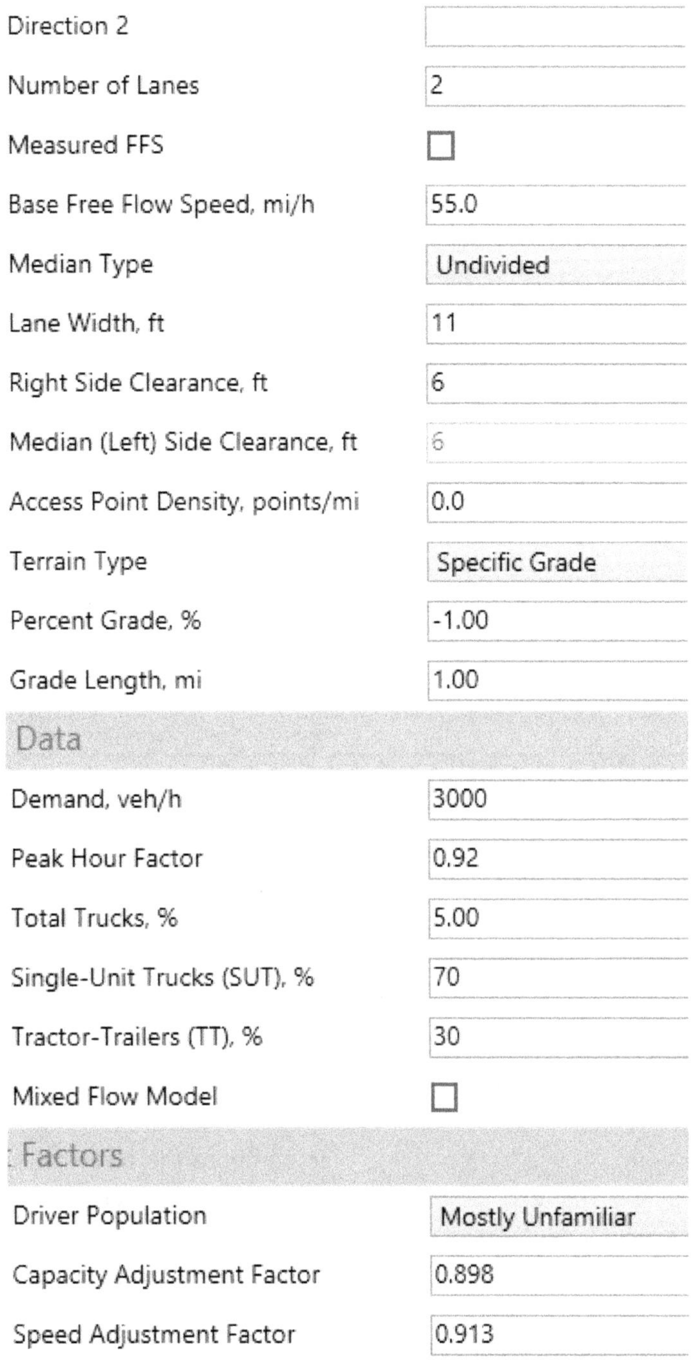

Figure 3.29 *HCS7 Multilane2 Output*

Direction 2 Geometric Data			
Direction 2 Description			
Number of Lanes (N), ln	2	Terrain Type	Specific Grade
Segment Length (L), ft	-	Percent Grade, %	-1.00
Measured or Base Free-Flow Speed	Base	Grade Length, mi	1.00
Base Free-Flow Speed (BFFS), mi/h	55.0	Right-Side Lateral Clearance (LC_R), ft	6
Lane Width, ft	11	Left-Side Lateral Clearance (LC_L), ft	6
Median Type	Undivided	Total Lateral Clearance (TLC), ft	12.00
Access Point Density, pts/mi	0.0	Free-Flow Speed (FFS), mi/h	51.5
Direction 2 Adjustment Factors			
Driver Population	Mostly Unfamiliar	Final Speed Adjustment Factor (SAF)	0.913
Driver Population SAF	0.913	Final Capacity Adjustment Factor (CAF)	0.898
Driver Population CAF	0.898		
Direction 2 Demand and Capacity			
Volume (V), veh/h	3000	Heavy Vehicle Adjustment Factor (f_{HV})	0.947
Peak Hour Factor (PHF)	0.92	Flow Rate (v_P), pc/h/ln	1722
Total Trucks, %	5.00	Capacity (c), pc/h/ln	1940
Single-Unit Trucks (SUT), %	70	Adjusted Capacity (c_{adj}), pc/h/ln	1742
Tractor-Trailers (TT), %	30	Volume-to-Capacity Ratio (v/c)	0.99
Direction 2 Speed and Density			
Lane Width Adjustment (f_{LW})	1.9	Average Speed (S), mi/h	39.7
Total Lateral Clearance Adj. (f_{TLC})	0.0	Density (D), pc/mi/ln	43.4
Median Type Adjustment (f_M)	1.6	Level of Service (LOS)	E
Access Point Density Adjustment (f_A)	0.0		

Two-Lane Highways

The analysis of a Two-Lane Highway generates three different service measures: Average Travel Speed (*ATS*), Percent Time-Spent-Following (*PTSF*), and Percent Free-Flow Speed (*PFFS*). Passing lanes can be modeled to quantify the reduction in percent time spent following and the increase in average travel speed, using the relative lengths of four defined regions in the analysis segment.

Required Data

To compute the free-flow speed, the lane and shoulder width and access points per mile (in both directions) are needed. In this procedure, if free-flow speed is measured, it is still adjusted for flow rate and heavy vehicles. The percent no-passing zones for the total segment being analyzed is also required.

To compute the flow rate, the demand data required include the prevailing volume (for the analysis and opposing directions), peak hour factor, heavy vehicle percentage (for both trucks and buses and recreational vehicles), and the terrain type or specific grade.

Classes

The service measure for this procedure is based on its class, which is defined by the predominant users. Class I highways serve mostly commuting traffic traveling 10 or 15 miles for which speed is important; both *ATS* and *PTSF* are computed in these analyses. Class II highways are used mostly for access traveling much shorter distances to school, work, or the store—only *PTSF* is computed for these analyses. Class III highways are for moderately developed areas in contrast with the rural environment this procedure has always assumed—only *PFFS* is computed for these analyses.

Free-Flow Speed

For field-measured free-flow speeds, HCM Equation 15-1 is used to adjust this value for the presence of traffic (especially heavy vehicles) during data collection. Unlike Multilane Highways and Freeway Segments, low volumes can still affect free-flow speed, since there is only one lane in each direction.

To compute free-flow speed, the base is estimated at 10 mi/h over the posted speed limit, then adjusted for lane and shoulder width and access point density, following HCM Equation 15-2 and using HCM Exhibits 15-7 and 15-8.

Figure 3.30 *HCM Equation 15-1*

$$FFS = S_{FM} + 0.00776\left(\frac{V}{F_{HV,ATS}}\right)$$

where

FFS = free-flow speed (mi/h);

S_{FM} = mean speed of sample (v>200 veh/h) (mi/h);

v = total demand flow rate, both directions, during period of speed measurements (veh/h); and

$f_{HV,ATS}$ = heavy vehicle adjustment factor for ATS, from Equation 15-4 or Equation 15-5.

Figure 3.31 *HCM Equation 15-2*

$$FFS = BFFS - f_{LS} - F_A$$

where

FFS = free-flow speed (mi/h),

$BFFS$ = base free-flow speed (mi/h),

F_{LS} = adjustment for lane and shoulder width (mi/h), and

f_A = adjustment for access-point density (mi/h).

Figure 3.32 *HCM Exhibit 15-7*

Lane Width (ft)	Shoulder Width (ft) ≥0 <2	≥2 <4	≥4 <6	≥6
≥9 <10	6.4	4.8	3.5	2.2
≥10 <11	5.3	3.7	2.4	1.1
≥ 11 <12	4.7	3.0	1.7	0.4
≥12	4.2	2.6	1.3	0.0

Figure 3.33 *HCM Exhibit 15-8*

Access Points per Mile (Both Sides)	Reduction in FFS (mi/h)
0	0.0
10	2.5
20	5.0
30	7.5
40	10.0

ATS Flow Rate

To compute flow rate for *ATS*, the volume is divided by the peak hour factor and adjustments for grade and heavy vehicles according to HCM Equation 15-3. The grade adjustment factor (f_g) is found in HCM Exhibits 15-9 and 15-10. The heavy vehicle adjustment

Figure 3.34 *HCM Equation 15-3*

$$v_{i,ATS} = \frac{V_i}{PHF \times f_{g,ATS} \times f_{HV,ATS}}$$

where

$v_{i,ATS}$ = demand flow rate i for ATS estimation (pc/h);

i = "d" (analysis direction) or "o" (opposing direction);

V_i = demand volume for direction i (veh/h);

$f_{g,ATS}$ = grade adjustment factor, from Exhibit 15-9 or Exhibit 15-10; and

$f_{HV,ATS}$ = heavy vehicle adjustment factor, from Equation 15-4 or Equation 15-5.

Figure 3.35 *HCM Exhibit 15-9*

One-Direction Demand Flow Rate, v_{vph} (veh/h)	Adjustment Factor: Level Terrain and Specific Downgrades	Rolling Terrain
≤100	1.00	0.67
200	1.00	0.75
300	1.00	0.83
400	1.00	0.90
500	1.00	0.95
600	1.00	0.97
700	1.00	0.98
800	1.00	0.99
≥900	1.00	1.00

Figure 3.36 *HCM Exhibit 15-10*

Grade (%)	Grade Length (mi)	Directional Demand Flow Rate, v_{vph} (veh/h)								
		≤100	200	300	400	500	600	700	800	≥900
≥3 <3.5	0.25	0.78	0.84	0.87	0.91	1.00	1.00	1.00	1.00	1.00
	0.50	0.75	0.83	0.86	0.90	1.00	1.00	1.00	1.00	1.00
	0.75	0.73	0.81	0.85	0.89	1.00	1.00	1.00	1.00	1.00
	1.00	0.73	0.79	0.83	0.88	1.00	1.00	1.00	1.00	1.00
	1.50	0.73	0.79	0.83	0.87	0.99	0.99	1.00	1.00	1.00
	2.00	0.73	0.79	0.82	0.86	0.98	0.98	0.99	1.00	1.00
	3.00	0.73	0.78	0.82	0.85	0.95	0.96	0.96	0.97	0.98
	≥4.00	0.73	0.78	0.81	0.85	0.94	0.94	0.95	0.95	0.96
≥3.5 <4.5	0.25	0.75	0.83	0.86	0.90	1.00	1.00	1.00	1.00	1.00
	0.50	0.72	0.80	0.84	0.88	1.00	1.00	1.00	1.00	1.00
	0.75	0.67	0.77	0.81	0.86	1.00	1.00	1.00	1.00	1.00
	1.00	0.65	0.73	0.77	0.81	0.94	0.95	0.97	1.00	1.00
	1.50	0.63	0.72	0.76	0.80	0.93	0.95	0.96	1.00	1.00
	2.00	0.62	0.70	0.74	0.79	0.93	0.94	0.96	1.00	1.00
	3.00	0.61	0.69	0.74	0.78	0.92	0.93	0.94	0.98	1.00
	≥4.00	0.61	0.69	0.73	0.78	0.91	0.91	0.92	0.96	1.00
≥4.5 <5.5	0.25	0.71	0.79	0.83	0.88	1.00	1.00	1.00	1.00	1.00
	0.50	0.60	0.70	0.74	0.79	0.94	0.95	0.97	1.00	1.00
	0.75	0.55	0.65	0.70	0.75	0.91	0.93	0.95	1.00	1.00
	1.00	0.54	0.64	0.69	0.74	0.91	0.93	0.95	1.00	1.00
	1.50	0.52	0.62	0.67	0.72	0.88	0.90	0.93	1.00	1.00
	2.00	0.51	0.61	0.66	0.71	0.87	0.89	0.92	0.99	1.00
	3.00	0.51	0.61	0.65	0.70	0.86	0.88	0.91	0.98	0.99
	≥4.00	0.51	0.60	0.65	0.69	0.84	0.86	0.88	0.95	0.97
≥5.5 <6.5	0.25	0.57	0.68	0.72	0.77	0.93	0.94	0.96	1.00	1.00
	0.50	0.52	0.62	0.66	0.71	0.87	0.90	0.92	1.00	1.00
	0.75	0.49	0.57	0.62	0.68	0.85	0.88	0.90	1.00	1.00
	1.00	0.46	0.56	0.60	0.65	0.82	0.85	0.88	1.00	1.00
	1.50	0.44	0.54	0.59	0.64	0.81	0.84	0.87	0.98	1.00
	2.00	0.43	0.53	0.58	0.63	0.81	0.83	0.86	0.97	0.99
	3.00	0.41	0.51	0.56	0.61	0.79	0.82	0.85	0.97	0.99
	≥4.00	0.40	0.50	0.55	0.61	0.79	0.82	0.85	0.97	0.99
≥6.5	0.25	0.54	0.64	0.68	0.73	0.88	0.90	0.92	1.00	1.00
	0.50	0.43	0.53	0.57	0.62	0.79	0.82	0.85	0.98	1.00
	0.75	0.39	0.49	0.54	0.59	0.77	0.80	0.83	0.96	1.00
	1.00	0.37	0.45	0.50	0.54	0.74	0.77	0.81	0.96	1.00
	1.50	0.35	0.45	0.49	0.54	0.71	0.75	0.79	0.96	1.00
	2.00	0.34	0.44	0.48	0.53	0.71	0.74	0.78	0.94	0.99
	3.00	0.34	0.44	0.48	0.53	0.70	0.73	0.77	0.93	0.98
	≥4.00	0.33	0.43	0.47	0.52	0.70	0.73	0.77	0.91	0.95

factor (f_{HV}) is computed using HCM Equation 15-4 with passenger-car equivalents found in HCM Exhibits 15-11, 15-12, and 15-13. For trucks operating at crawl speed (geared low to maintain speed on a steep downgrade), the heavy vehicle adjustment factor is computed using HCM Equation 15-5 with passenger-car equivalents found in HCM Exhibit 15-14. Flow rate is computed for both the analysis direction (v_d) and the opposing direction (v_o) for use in the computations of ATS ($v_{,ATS}$).

Figure 3.37 *HCM Equation 15-4*

$$f_{HV,ATS} = \frac{1}{1 + P_T(E_T - 1) + P_R(E_R - 1)}$$

where

$f_{HV,ATS}$ = heavy vehicle adjustment factor for ATS estimation,

P_T = proportion of trucks in the traffic stream (decimal),

P_R = proportion of RVs in the traffic stream (decimal),

E_T = passenger car equivalent for trucks from Exhibit 15-11 or Exhibit 15-12, and

E_R = passenger car equivalent for RVs from Exhibit 15-11 or Exhibit 15-13.

Figure 3.38 *HCM Exhibit 15-11*

Vehicle Type	Directional Demand Flow Rate, vvph (veh/h)	Level Terrain and Specific Downgrades	Rolling Terrain
Trucks, E_T	≤100	1.9	2.7
	200	1.5	2.3
	300	1.4	2.1
	400	1.3	2.0
	500	1.2	1.8
	600	1.1	1.7
	700	1.1	1.6
	800	1.1	1.4
	≥900	1.0	1.3
RVs, E_R	All flows	1.0	1.1

Figure 3.39 *HCM Exhibit 15-12*

Grade (%)	Grade Length (mi)	Directional Demand Flow Rate, v_{vph} (veh/h)								
		≤100	200	300	400	500	600	700	800	≥900
≥3 <3.5	0.25	2.6	2.4	2.3	2.2	1.8	1.8	1.7	1.3	1.1
	0.50	3.7	3.4	3.3	3.2	2.7	2.6	2.6	1.9	1.6
	0.75	4.6	4.4	4.3	4.2	3.7	3.6	3.4	2.4	1.9
	1.00	5.2	5.0	4.9	4.9	4.4	4.2	4.1	3.0	2.3
	1.50	6.2	6.0	5.9	5.8	5.3	5.0	4.8	3.6	2.9
	2.00	7.3	6.9	6.7	6.5	5.7	5.5	5.3	4.1	3.5
	3.00	8.4	8.0	7.7	7.5	6.5	6.2	6.0	4.6	3.9
	≥4.00	9.4	8.8	8.6	8.3	7.2	6.9	6.6	4.8	3.7
≥3.5 <4.5	0.25	3.8	3.4	3.2	3.0	2.3	2.2	2.2	1.7	1.5
	0.50	5.5	5.3	5.1	5.0	4.4	4.2	4.0	2.8	2.2
	0.75	6.5	6.4	6.5	6.5	6.3	5.9	5.6	3.6	2.6
	1.00	7.9	7.6	7.4	7.3	6.7	6.6	6.4	5.3	4.7
	1.50	9.6	9.2	9.0	8.9	8.1	7.9	7.7	6.5	5.9
	2.00	10.3	10.1	10.0	9.9	9.4	9.1	8.9	7.4	6.7
	3.00	11.4	11.3	11.2	11.2	10.7	10.3	10.0	8.0	7.0
	≥4.00	12.4	12.2	12.2	12.1	11.5	11.2	10.8	8.6	7.5
≥4.5 <5.5	0.25	4.4	4.0	3.7	3.5	2.7	2.7	2.7	2.6	2.5
	0.50	6.0	6.0	6.0	6.0	5.9	5.7	5.6	4.6	4.2
	0.75	7.5	7.5	7.5	7.5	7.5	7.5	7.5	7.5	7.5
	1.00	9.2	9.2	9.1	9.1	9.0	9.0	9.0	8.9	8.8
	1.50	10.6	10.6	10.6	10.6	10.5	10.4	10.4	10.2	10.1
	2.00	11.8	11.8	11.8	11.8	11.6	11.6	11.5	11.1	10.9
	3.00	13.7	13.7	13.6	13.6	13.3	13.1	13.0	11.9	11.3
	≥4.00	15.3	15.3	15.2	15.2	14.6	14.2	13.8	11.3	10.0
≥5.5 <6.5	0.25	4.8	4.6	4.5	4.4	4.0	3.9	3.8	3.2	2.9
	0.50	7.2	7.2	7.2	7.2	7.2	7.2	7.2	7.2	7.2
	0.75	9.1	9.1	9.1	9.1	9.1	9.1	9.1	9.1	9.1
	1.00	10.3	10.3	10.3	10.3	10.3	10.3	10.3	10.2	10.1
	1.50	11.9	11.9	11.9	11.9	11.8	11.8	11.8	11.7	11.6
	2.00	12.8	12.8	12.8	12.8	12.7	12.7	12.7	12.6	12.5
	3.00	14.4	14.4	14.4	14.4	14.3	14.3	14.3	14.2	14.1
	≥4.00	15.4	15.4	15.3	15.3	15.2	15.1	15.1	14.9	14.8
≥6.5	0.25	5.1	5.1	5.0	5.0	4.8	4.7	4.7	4.5	4.4
	0.50	7.8	7.8	7.8	7.8	7.8	7.8	7.8	7.8	7.8
	0.75	9.8	9.8	9.8	9.8	9.8	9.8	9.8	9.8	9.8
	1.00	10.4	10.4	10.4	10.4	10.4	10.4	10.4	10.3	10.2
	1.50	12.0	12.0	12.0	12.0	11.9	11.9	11.9	11.8	11.7
	2.00	12.9	12.9	12.9	12.9	12.8	12.8	12.8	12.7	12.6
	3.00	14.5	14.5	14.5	14.5	14.4	14.4	14.4	14.3	14.2
	≥4.00	15.4	15.4	15.4	15.4	15.3	15.3	15.3	15.2	15.1

Figure 3.40 *HCM Exhibit 15-13*

Grade (%)	Grade Length (mi)	Directional Demand Flow Rate, v_{vph} (veh/h)								
		≤100	200	300	400	500	600	700	800	≥900
≥3 <3.5	≤0.25	1.1	1.1	1.1	1.0	1.0	1.0	1.0	1.0	1.0
	>0.25 ≤0.75	1.2	1.2	1.1	1.1	1.0	1.0	1.0	1.0	1.0
	>0.75 ≤1.25	1.3	1.2	1.2	1.1	1.0	1.0	1.0	1.0	1.0
	>1.25 ≤2.25	1.4	1.3	1.2	1.1	1.0	1.0	1.0	1.0	1.0
	>2.25	1.5	1.4	1.3	1.2	1.0	1.0	1.0	1.0	1.0
≥3.5 <4.5	≤0.75	1.3	1.2	1.2	1.1	1.0	1.0	1.0	1.0	1.0
	>0.75 ≤3.50	1.4	1.3	1.2	1.1	1.0	1.0	1.0	1.0	1.0
	>3.50	1.5	1.4	1.3	1.2	1.0	1.0	1.0	1.0	1.0
≥4.5 <5.5	≤2.50	1.5	1.4	1.3	1.2	1.0	1.0	1.0	1.0	1.0
	>2.50	1.6	1.5	1.4	1.2	1.0	1.0	1.0	1.0	1.0
≥5.5 <6.5	≤0.75	1.5	1.4	1.3	1.1	1.0	1.0	1.0	1.0	1.0
	>0.75 ≤2.50	1.6	1.5	1.4	1.2	1.0	1.0	1.0	1.0	1.0
	>2.50 ≤3.50	1.6	1.5	1.4	1.3	1.2	1.1	1.0	1.0	1.0
	>3.50	1.6	1.6	1.6	1.5	1.5	1.4	1.3	1.2	1.1
≥6.5	≤2.50	1.6	1.5	1.4	1.2	1.0	1.0	1.0	1.0	1.0
	>2.50 ≤3.50	1.6	1.5	1.4	1.2	1.3	1.3	1.3	1.3	1.3
	>3.50	1.6	1.6	1.6	1.5	1.5	1.5	1.4	1.4	1.4

Figure 3.41 *HCM Equation 15-5*

$$f_{HV,ATS} = \frac{1}{1 + P_{TC} \times P_T (E_{TC} - 1) + (1 - P_{TC}) \times P_T \times (E_T - 1) + P_R (E_R - 1)}$$

where

P_{TC} = proportion of trucks operating at crawl speed (decimal); and

E_{TC} = passenger car equivalent for trucks operating at crawl speed, from Exhibit 15-14.

Figure 3.42 *HCM Exhibit 15-14*

Difference Between FFS and Truck Crawl Speed (mi/h)	Directional Demand Flow Rate, v_{vph} (veh/h)								
	≤100	200	300	400	500	600	700	800	≥900
≤15	4.7	4.1	3.6	3.1	2.6	2.1	1.6	1.0	1.0
20	9.9	8.7	7.8	6.7	5.8	4.9	4.0	2.7	1.0
25	15.1	13.5	12.0	10.4	9.0	7.7	6.4	5.1	3.8
30	22.0	19.8	17.5	15.6	13.1	11.6	9.2	6.1	4.1
35	29.0	26.0	23.1	20.1	17.3	14.6	11.9	9.2	6.5
≥40	35.9	32.3	28.6	24.9	21.4	18.1	14.7	11.3	7.9

Think about your own driving:

Most of us have followed a big truck going very slowly down a long, steep grade and building a very long queue of vehicles caught behind. The very large passenger-car equivalents in the crawl speed adjustment table can reach over thirty-five passenger cars as equivalent to the one truck.

Average Travel Speed

ATS is calculated using HCM Equation 15-6 that requires both analysis ($v_{d,ATS}$) and opposing ($v_{o,ATS}$) direction flow rates with an adjustment for no-passing zones ($f_{np,ATS}$) found in HCM Exhibit 15-15.

PTSF Flow Rate

To compute flow rate for *PTSF*, the volume is divided by the peak hour factor and adjustments for grade and heavy vehicles according to HCM Equation 15-7. The grade adjustment factor (f_g) is found in HCM Exhibits 15-16 and 15-17. The heavy vehicle adjustment factor (f_{HV}) is computed using HCM Equation 15-8 with passenger-car equivalents found in HCM Exhibits 15-18 and 15-19. Flow rate is computed for both the analysis direction (v_d) and the opposing direction (v_o) for use in the computations of *PTSF* ($v_{,PTSF}$).

Percent Time-Spent-Following

Base *PTSF* is computed using HCM Equation 15-10 with coefficients from HCM Exhibit 15-20 and flow rates for the analysis ($v_{d,PTSF}$) and opposing ($v_{o,PTSF}$) directions. Then, the final *PTSF* uses the passing zone adjustment ($f_{np,\ PTSF}$) from HCM Exhibit 15-21 in HCM Equation 15-9.

Think about your own driving:

On a two-lane highway with few or no passing zones, one car (or more likely truck) can force you to go much slower than you would normally—and you have to follow them for a long time. So, your trip takes longer and is more frustrating than it would have been without the hindrance. As a result your following time goes up and the level of service gets worse to reflect the loss in comfort and convenience.

Figure 3.43 *HCM Equation 15-6*

$$ATS_d = FFS - 0.00776(v_{d,ATS} + v_{o,ATS}) - f_{np,ATS}$$

where

ATS_d = average travel speed in the analysis direction (mi/h);

FFS = free-flow speed (mi/h);

$v_{d,ATS}$ = demand flow rate for ATS determination in the analysis direction (pc/h);

$v_{o,ATS}$ = demand flow rate for ATS determination in the opposing direction (pc/h); and

$f_{np,ATS}$ = adjustment factor for ATS determination for the percentage of no-passing zones in the analysis direction, from Exhibit 15-15.

Figure 3.44 *HCM Exhibit 15-15*

Opposing Demand Flow Rate, v_o (pc/h)	Percent No-Passing Zones				
	≤ 20%	40%	60%	80%	100%
	FFS ≥ 65 mi/h				
≤100	1.1	2.2	2.8	3.0	3.1
200	2.2	3.3	3.9	4.0	4.2
400	1.6	2.3	2.7	2.8	2.9
600	1.4	1.5	1.7	1.9	2.0
800	0.7	1.0	1.2	1.4	1.5
1,000	0.6	0.8	1.1	1.1	1.2
1,200	0.6	0.8	0.9	1.0	1.1
1,400	0.6	0.7	0.9	0.9	0.9
≥1,600	0.6	0.7	0.7	0.7	0.8
	FFS = 60 mi/h				
≤100	0.7	1.7	2.5	2.8	2.9
200	1.9	2.9	3.7	4.0	4.2
400	1.4	2.0	2.5	2.7	3.9
600	1.1	1.3	1.6	1.9	2.0
800	0.6	0.9	1.1	1.3	1.4
1,000	0.6	0.7	0.9	1.1	1.2
1,200	0.5	0.7	0.9	0.9	1.1
1,400	0.5	0.6	0.8	0.8	0.9
≥1,600	0.5	0.6	0.7	0.7	0.7
	FFS = 55 mi/h				
≤100	0.5	1.2	2.2	2.6	2.7
200	1.5	2.4	3.5	3.9	4.1
400	1.3	1.9	2.4	2.7	2.8
600	0.9	1.1	1.6	1.8	1.9
800	0.5	0.7	1.1	1.2	1.4
1,000	0.5	0.6	0.8	0.9	1.1
1,200	0.5	0.6	0.7	0.9	1.0
1,400	0.5	0.6	0.7	0.7	0.9
≥1,600	0.5	0.6	0.6	0.6	0.7
	FFS = 50 mi/h				
≤100	0.2	0.7	1.9	2.4	2.5
200	1.2	2.0	3.3	3.9	4.0
400	1.1	1.6	2.2	2.6	2.7
600	0.6	0.9	1.4	1.7	1.9
800	0.4	0.6	0.9	1.2	1.3
1,000	0.4	0.4	0.7	0.9	1.1
1,200	0.4	0.4	0.7	0.8	1.0
1,400	0.4	0.4	0.6	0.7	0.8
≥1,600	0.4	0.4	0.5	0.5	0.5
	FFS ≤ 45 mi/h				
≤100	0.1	0.4	1.7	2.2	2.4
200	0.9	1.6	3.1	3.8	4.0
400	0.9	0.5	2.0	2.5	2.7
600	0.4	0.3	1.3	1.7	1.8
800	0.3	0.3	0.8	1.1	1.2
1,000	0.3	0.3	0.6	0.8	1.1
1,200	0.3	0.3	0.6	0.7	1.0
1,400	0.3	0.3	0.6	0.6	0.7
≥1,600	0.3	0.3	0.4	0.4	0.6

Figure 3.45 *HCM Equation 15-7*

$$v_{i,PTSF} = \frac{V_i}{PHF \times f_{g,PTSF} \times f_{HV,PTSF}}$$

$v_{i,PTSF}$ = demand flow rate i for determination of PTSF (pc/h);

i = "d" (analysis direction) or "o" (opposing direction);

$f_{g,PTSF}$ = grade adjustment factor for PTSF determination, from Exhibit 15-16 or Exhibit 15-17; and

$f_{HV,PTSF}$ = heavy vehicle adjustment factor for PTSF determination, from Exhibit 15-18 or Exhibit 15-19.

Figure 3.46 *HCM Exhibit 15-16*

Directional Demand Flow Rate, v_{vph} (veh/h)	Level Terrain and Specific Downgrades	Rolling Terrain
≤100	1.00	0.73
200	1.00	0.80
300	1.00	0.85
400	1.00	0.90
500	1.00	0.96
600	1.00	0.97
700	1.00	0.99
800	1.00	1.00
≥900	1.00	1.00

Figure 3.47 *HCM Exhibit 15-17*

Grade (%)	Grade Length (mi)	Directional Demand Flow Rate, v_{vph} (veh/h)								
		≤100	200	300	400	500	600	700	800	≥900
	0.25	1.00	0.99	0.97	0.96	0.92	0.92	0.92	0.92	0.92
	0.50	1.00	0.99	0.98	0.97	0.93	0.93	0.93	0.93	0.93
	0.75	1.00	0.99	0.98	0.97	0.93	0.93	0.93	0.93	0.93
≥3	1.00	1.00	0.99	0.98	0.97	0.93	0.93	0.93	0.93	0.93
<3.5	1.50	1.00	0.99	0.98	0.97	0.94	0.94	0.94	0.94	0.94
	2.00	1.00	0.99	0.98	0.98	0.95	0.95	0.95	0.95	0.95
	3.00	1.00	1.00	0.99	0.99	0.97	0.97	0.97	0.96	0.96
	≥4.00	1.00	1.00	1.00	1.00	1.00	0.99	0.99	0.97	0.97
	0.25	1.00	0.99	0.98	0.97	0.94	0.93	0.93	0.92	0.92
	0.50	1.00	1.00	0.99	0.99	0.97	0.97	0.97	0.96	0.95
	0.75	1.00	1.00	0.99	0.99	0.97	0.97	0.97	0.96	0.96
≥3.5	1.00	1.00	1.00	0.99	0.99	0.97	0.97	0.97	0.97	0.97
<4.5	1.50	1.00	1.00	0.99	0.99	0.97	0.97	0.97	0.97	0.97
	2.00	1.00	1.00	0.99	0.99	0.98	0.98	0.98	0.98	0.98
	3.00	1.00	1.00	1.00	1.00	1.00	1.00	1.00	1.00	1.00
	≥4.00	1.00	1.00	1.00	1.00	1.00	1.00	1.00	1.00	1.00
≥4.5	0.25	1.00	1.00	1.00	1.00	1.00	0.99	0.99	0.97	0.97
<5.5	≥0.50	1.00	1.00	1.00	1.00	1.00	1.00	1.00	1.00	1.00
≥5.5	All	1.00	1.00	1.00	1.00	1.00	1.00	1.00	1.00	1.00

Figure 3.48 *HCM Equation 15-8*

$$f_{HV,PTSF} = \frac{1}{1 + P_T(E_T - 1) + P_R(E_R - 1)}$$

$f_{g,PTSF}$ = grade adjustment factor for PTSF determination

P_T = proportion of trucks in the traffic stream (decimal),

P_R = proportion of RVs in the traffic stream (decimal),

E_T = passenger car equivalent for trucks from Exhibit 15-11 or Exhibit 15-12, and

E_R = passenger car equivalent for RVs from Exhibit 15-11 or Exhibit 15-13.

Figure 3.49 *HCM Exhibit 15-18*

Vehicle Type	Directional Demand Flow Rate, v_{vph} (veh/h)	Level and Specific Downgrade	Rolling
Trucks, E_T	≤100	1.1	1.9
	200	1.1	1.8
	300	1.1	1.7
	400	1.1	1.6
	500	1.0	1.4
	600	1.0	1.2
	700	1.0	1.0
	800	1.0	1.0
	≥900	1.0	1.0
RVs, E_R	All	1.0	1.0

Figure 3.50 *HCM Exhibit 15-19*

Grade (%)	Grade Length (mi)	Directional Demand Flow Rate, v_{vph} (veh/h)								
		≤100	200	300	400	500	600	700	800	≥900
Passenger Car Equivalents for Trucks (E_T)										
≥3 <3.5	≤2.00	1.0	1.0	1.0	1.0	1.0	1.0	1.0	1.0	1.0
	3.00	1.5	1.3	1.3	1.2	1.0	1.0	1.0	1.0	1.0
	≥4.00	1.6	1.4	1.3	1.3	1.0	1.0	1.0	1.0	1.0
≥3.5 <4.5	≤1.00	1.0	1.0	1.0	1.0	1.0	1.0	1.0	1.0	1.0
	1.50	1.1	1.1	1.0	1.0	1.0	1.0	1.0	1.0	1.0
	2.00	1.6	1.3	1.0	1.0	1.0	1.0	1.0	1.0	1.0
	3.00	1.8	1.4	1.1	1.2	1.2	1.2	1.2	1.2	1.2
	≥4.00	2.1	1.9	1.8	1.7	1.4	1.4	1.4	1.4	1.4
≥4.5 <5.5	≤1.00	1.0	1.0	1.0	1.0	1.0	1.0	1.0	1.0	1.0
	1.50	1.1	1.1	1.1	1.2	1.2	1.2	1.2	1.2	1.2
	2.00	1.7	1.6	1.6	1.6	1.5	1.4	1.4	1.3	1.3
	3.00	2.4	2.2	2.2	2.1	1.9	1.8	1.8	1.7	1.7
	≥4.00	3.5	3.1	2.9	2.7	2.1	2.0	2.0	1.8	1.8
≥5.5 <6.5	≤0.75	1.0	1.0	1.0	1.0	1.0	1.0	1.0	1.0	1.0
	1.00	1.0	1.0	1.1	1.1	1.2	1.2	1.2	1.2	1.2
	1.50	1.5	1.5	1.5	1.6	1.6	1.6	1.6	1.6	1.6
	2.00	1.9	1.9	1.9	1.9	1.9	1.9	1.9	1.8	1.8
	3.00	3.4	3.2	3.0	2.9	2.4	2.3	2.3	1.9	1.9
	≥4.00	4.5	4.1	3.9	3.7	2.9	2.7	2.6	2.0	2.0
≥6.5	≤0.50	1.0	1.0	1.0	1.0	1.0	1.0	1.0	1.0	1.0
	0.75	1.0	1.0	1.0	1.0	1.1	1.1	1.1	1.0	1.0
	1.00	1.3	1.3	1.3	1.4	1.4	1.5	1.5	1.4	1.4
	1.50	2.1	2.1	2.1	2.1	2.0	2.0	2.0	2.0	2.0
	2.00	2.9	2.8	2.7	2.7	2.4	2.4	2.3	2.3	2.3
	3.00	4.2	3.9	3.7	3.6	3.0	2.8	2.7	2.2	2.2
	≥4.00	5.0	4.6	4.4	4.2	3.3	3.1	2.9	2.7	2.5
Passenger Car Equivalents for RVs (E_R)										
All	All	1.0	1.0	1.0	1.0	1.0	1.0	1.0	1.0	1.0

Figure 3.51 *HCM Equation 15-10*

$$BPTSF_d = 100[1 - \exp(av_d^b)]$$

Figure 3.52 *HCM Exhibit 15-20*

Opposing Demand Flow Rate, v_o (pc/h)	Coefficient *a*	Coefficient *b*
≤200	−0.0014	0.973
400	−0.0022	0.923
600	−0.0033	0.870
800	−0.0045	0.833
1,000	−0.0049	0.829
1,200	−0.0054	0.825
1,400	−0.0058	0.821
≥1,600	−0.0062	0.817

Figure 3.53 *HCM Exhibit 15-21*

Total Two-Way Flow Rate, $v = v_d + v_o$(pc/h)	Percent No-Passing Zones 0	20	40	60	80	100
	Directional Split = 50/50					
≤200	9.0	29.2	43.4	49.4	51.0	52.6
400	16.2	41.0	54.2	61.6	63.8	65.8
600	15.8	38.2	47.8	53.2	55.2	56.8
800	15.8	33.8	40.4	44.0	44.8	46.6
1,400	12.8	20.0	23.8	26.2	27.4	28.6
2,000	10.0	13.6	15.8	17.4	18.2	18.8
2,600	5.5	7.7	8.7	9.5	10.1	10.3
3,200	3.3	4.7	5.1	5.5	5.7	6.1
	Directional Split = 60/40					
≤200	11.0	30.6	41.0	51.2	52.3	53.5
400	14.6	36.1	44.8	53.4	55.0	56.3
600	14.8	36.9	44.0	51.1	52.8	54.6
800	13.6	28.2	33.4	38.6	39.9	41.3
1,400	11.8	18.9	22.1	25.4	26.4	27.3
2,000	9.1	13.5	15.6	16.0	16.8	17.3
2,600	5.9	7.7	8.6	9.6	10.0	10.3
	Directional Split = 70/30					
≤200	9.9	28.1	38.0	47.8	48.5	49.0
400	10.6	30.3	38.6	46.7	47.7	48.8
600	10.9	30.9	37.5	43.9	45.4	47.0
800	10.3	23.6	28.4	33.3	34.5	35.5
1,400	8.0	14.6	17.7	20.8	21.6	22.3
2,000	7.3	9.7	11.7	13.3	14.0	14.5
	Directional Split = 80/20					
≤200	8.9	27.1	37.1	47.0	47.4	47.9
400	6.6	26.1	34.5	42.7	43.5	44.1
600	4.0	24.5	31.3	38.1	39.1	40.0
800	3.8	18.5	23.5	28.4	29.1	29.9
1,400	3.5	10.3	13.3	16.3	16.9	32.2
2,000	3.5	7.0	8.5	10.1	10.4	10.7
	Directional Split = 90/10					
≤200	4.6	24.1	33.6	43.1	43.4	43.6
400	0.0	20.2	28.3	36.3	36.7	37.0
600	−3.1	16.8	23.5	30.1	30.6	31.1
800	−2.8	10.5	15.2	19.9	20.3	20.8
1,400	−1.2	5.5	8.3	11.0	11.5	11.9

Figure 3.54 *HCM Equation 15-9*

$$PTSF_d = BPTSF_d + f_{np,PTSF}\left(\frac{v_{d,PTSF}}{v_{d,PTSF} + v_{o,PTSF}}\right)$$

where

$PTSF_d$ = percent time-spent-following in the analysis direction (decimal);

$BPTSF_d$ = base percent time-spent-following in the analysis direction, from Equation 15-10;

$f_{np,PTSF}$ = adjustment to PTSF for the percentage of no-passing zones in the analysis segment, from Exhibit 15-21;

$v_{d,PTSF}$ = demand flow rate in the analysis direction for estimation of PTSF (pc/h); and

$v_{o,PTSF}$ = demand flow rate in the opposing direction for estimation of PTSF (pc/h).

Passing Zones

The effects of adding a passing lane within the analysis segment can be modeled to increase *ATS* and lower *PTSF* values accordingly. Four regions are defined to provide the lengths required to make these adjustments as illustrated in HCM Exhibit 15-25. The length downstream of the passing lane still including the effects is shown in HCM Exhibit 15-23 as a function of analysis flow rate for *PTSF* and always 1.7 mi for *ATS* (regardless of the flow rate). The adjusted *PTSF* is computed, following HCM Equation 15-20 with the passing lane adjustment factor (f_{np}) from HCM Exhibit 15-26. The adjusted *ATS* is computed, following HCM Equation 15-22 with the passing lane adjustment factor (f_{np}) from HCM Exhibit 15-28.

If the segment is truncated (ends before the end of the downstream effective length), HCM Equation 15-21 is used instead for *PTSF*. Similarly, if the segment is truncated (ends before the end of the downstream effective length), HCM Equation 15-23 is used instead for *ATS*. If the segment uses a specific grade, the passing lane is considered a climbing lane, and the adjustment factor comes from HCM Exhibit 15-29 with all region lengths set to zero (except for the passing lane).

Think about your own driving:
Building on the previous comment, a passing lane allows you to (finally) get in front of the slow truck and no longer have to go slower than you prefer. Your time following goes down, and level of service improves to reflect this restoration of comfort and convenience.

Figure 3.55 *HCM Exhibit 15-25*

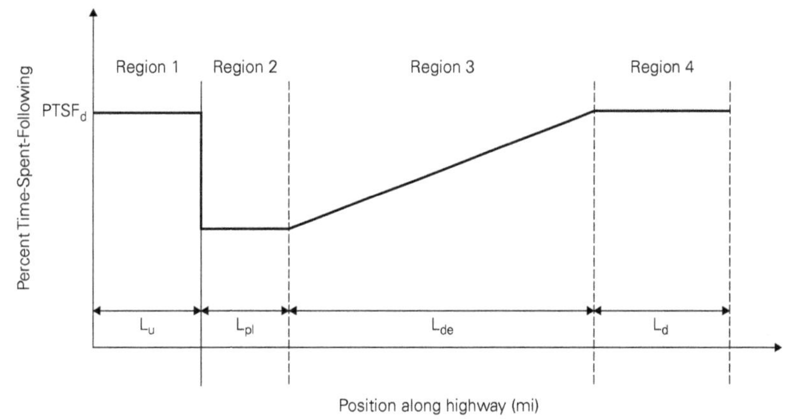

Figure 3.56 *HCM Exhibit 15-23*

Directional Demand Flow Rate, v_d(pc/h)	Downstream Length of Roadway Affected, L_{de} (mi) PTSF	ATS
≤200	13.0	1.7
300	11.6	1.7
400	8.1	1.7
500	7.3	1.7
600	6.5	1.7
700	5.7	1.7
800	5.0	1.7
900	4.3	1.7
≥1,000	3.6	1.7

Figure 3.57 *HCM Equation 15-20*

$$PTSF_{pl} = \frac{PTSF_d \left[L_u + L_d + f_{pl,PTSF} L_{pl} + \left(\frac{1+f_{pl,PTSF}}{2} \right) L_{de} \right]}{L_t}$$

where

$PTSF_{pl}$ = percent time-spent-following for segment as affected by the presence of a passing lane (decimal); and

$f_{pl,PTSF}$ = adjustment factor for the impact of a passing lane on percent time-spent-following, from Exhibit 15-26.

All other variables are as previously defined.

Figure 3.58 *HCM Exhibit 15-26*

Directional Demand Flow Rate, v_d (pc/h)	$f_{pl,PTSF}$
≤100	0.51
200	0.59
300	0.60
400	0.61
500	0.61
600	0.61
700	0.62
800	0.62
≥900	0.62

Figure 3.59 *HCM Equation 15-22*

$$ATS_{pl} = \frac{ATS_d L_t}{L_u + L_d + \left(\frac{L_{pl}}{f_{pl,ATS}} \right) + \left(\frac{2L_{de}}{1+f_{pl,ATS}} \right)}$$

Figure 3.60 *HCM Exhibit 15-28*

Directional Demand Flow Rate, v_d (pc/h)	$f_{pl,ATS}$
≤100	1.08
200	1.09
300	1.10
400	1.10
500	1.10
600	1.11
700	1.11
800	1.11
≥900	1.11

Figure 3.61 *HCM Equation 15-21*

$$PTSF_{pl} = \frac{PTSF_d \left[L_u + f_{pl,PTSF} L_{pl} + f_{pl,PTSF} L'_{de} + \left(\frac{1+f_{pl,PTSF}}{2} \right) \left(\frac{L'^{2}_{de}}{L_{de}} \right) \right]}{L_t}$$

Figure 3.62 *HCM Equation 15-23*

$$ATS_{pl} = \frac{ATS_d L_t}{L_u + \frac{L_{pl}}{f_{pl,ATS}} + \frac{2L'_{de}}{\left[1 + f_{pl,ATS} + (f_{pl,ATS} - 1)\left(\frac{L_{de} - L'_{de}}{L_{de}}\right)\right]}}$$

Figure 3.63 *HCM Exhibit 15-29*

Directional Demand Flow Rate, v_d(pc/h)	f_{pl} ATS	PTSF
0–300	1.02	0.20
>300–600	10.7	0.21
>600	1.14	0.23

Capacity

Capacity is computed using 1700 veh/h as the base (assuming a PHF of 1.0), then modified by the appropriate grade and heavy vehicle adjustment, following HCM Equation 15-12 for *ATS* and HCM Equation 15-13 for *PTSF*.

Level of Service

LOS is determined from HCM Exhibit 15-13 using the appropriate service measure for a given class. The worst LOS between *ATS* and *PTSF* is used for Class I. Class III requires *PFFS* be computed using HCM Equation 15-11. LOS F cannot be achieved from the service measure and only occurs when a flow rate exceeds capacity.

Figure 3.64 *HCM Equation 15-12*

$$c_{d,ATS} = 1{,}700 f_{g,ATS} f_{HV,ATS}$$

Figure 3.65 *HCM Equation 15-13*

$$c_{d,PTSF} = 1{,}700 f_{g,PTSF} f_{HV,PTSF}$$

Figure 3.66 *HCM Exhibit 15-3*

LOS	Class I Highways ATS (mi/h)	Class I Highways PTSF (%)	Class II Highways PTSF (%)	Class III Highways PFFS (%)
A	>55	≤35	≤40	>91.7
B	>50–55	>35–50	>40–55	>83.3–91.7
C	>45–50	>50–65	>55–70	>75.0–83.3
D	>40–45	>65–80	>70–85	>66.7–75.0
E	≤40	>80	>85	≤66.7
F	Demand exceeds capacity			

Figure 3.67 *HCM Equation 15-11*

$$PFFS = \frac{ATS_d}{FFS}$$

Bicycles

The procedure uses vehicular volume (including percentage heavy vehicles) and speed information with the width of the outside through lane and its pavement condition to compute the index for determining level of service.

REQUIRED DATA

The outside through lane and shoulder widths, including pavement condition, using the FHWA five-point scale are needed. The posted speed limit and vehicle volume with the percentage heavy vehicles and percentage on-highway parking are also required.

PROCEDURE

The vehicle directional flow rate in the outside lane is computed using HCM Equation 15-24. The effective width of the outside lane is calculated, following HCM Equations 15-25 through 15-29. The effective speed factor is defined in HCM Equation 15-30.

Figure 3.68 *HCM Equation 15-24*

$$v_{OL} = \frac{V}{PHF \times N}$$

where

v_{OL} = directional demand flow rate in the outside lane (veh/h),

V = hourly directional volume (veh/h),

PHF = peak hour factor, and

N = number of directional lanes (=1 for two-lane highways).

Figure 3.69 *HCM Equation 15-25*

If W_s is greater than or equal to 8 ft:

$$W_e = W_v + W_s - (\%OHP \times 10\text{ ft})$$

Figure 3.70 *HCM Equation 15-26*

If W_s is greater than or equal to 4 ft and less than 8 ft:

$$W_e = W_v + W_s - 2 \times (\%OHP[2\text{ ft} + W_s])$$

Figure 3.71 *HCM Equation 15-27*

If W_s is less than 4 ft:

$$W_e = W_v + (\%OHP[2\text{ ft} + W_s])$$

Figure 3.72 *HCM Equation 15-28*

with, if V is greater than 160 veh/h:

$$W_v = W_{OL} + W_s$$

Figure 3.73 *HCM Equation 15-29*

$$W_v = (W_{OL} + W_s) \times (2 - 0.005V)$$

where

W_v = effective width as a function of traffic volume (ft),
W_{OL} = outside lane width (ft),
W_s = paved shoulder width (ft),
V = hourly directional volume per lane (veh/h),
W_e = average effective width of the outside through lane (ft), and
$\%OHP$ = percentage of segment with occupied on-highway parking (decimal).

Figure 3.74 *HCM Equation 15-30*

$$S_t = 1.1199 \ln(S_p - 20) + 0.8103$$

where

S_t = effective speed factor, and
S_p = posted speed limit (mi/h).

BICYCLE LOS

These are combined in HCM Equation 15-31 along with the pavement rating and heavy vehicle percentage to generate the index score used in HCM Exhibit 15-4.

Figure 3.75 *HCM Equation 15-31*

$$BLOS = 0.507 \ln(v_{OL}) + 0.1999S_t(1 + 10.38HV)^2 + 7.066(1/P)^2 - 0.005(W_e)^2 + 0.760$$

where

$BLOS$ = bicycle level of service score;
v_{OL} = directional demand flow rate in the outside lane (veh/h);
S_t = effective speed factor;
HV = percentage of heavy vehicles (decimal); if $V < 200$ veh/h, then HV should be limited to a maximum of 50%;
P = FHWA's 5-point pavement surface condition rating; and
W_e = average effective width of the outside through lane (ft).

Figure 3.76 *HCM Exhibit 15-4*

LOS	BLOS Score
A	≤1.5
B	>1.5–2.5
C	>2.5–3.5
D	>3.5–4.5
E	>4.5–5.5
F	>5.5

PROBLEM ONE

Determine the LOS for a 10-mile Class II Two-Lane Highway with 1000 veh/h in each direction, peak hour factor of 0.90, 11-foot lanes, 4-foot shoulders, 50 access points, 5% trucks, and 2% RVs, on rolling terrain with a posted speed limit of 50 mi/h and 100% no-passing zones.

1. Compute the free-flow speed with a base of 60 mi/h (50 mi/h limit) for five access points per mile (25 in 5 miles).

Free-Flow Speed: $FFS = 60.0 - 1.7 - 1.3 = 57.0$ mi/h
EQUATION 15-2 / EXHIBITS 15-7 & 15-8

2. The grade adjustment factor for *ATS* is computed for rolling terrain.

Grade: Flow = 1000/0.90 = 1111 veh/h $f_{g,ATS} = 1.00$
EXHIBIT 15-9

3. The heavy vehicle adjustment factor for *ATS* is computed for rolling terrain with trucks and RVs.

Heavy Vehicles: $E_T = 1.3$ $E_R = 1.1$
EXHIBIT 15-11

$f_{HV} = 1 / ((1 + 0.05(1.3 - 1) + 0.02(1.1 - 1)) = 0.983$
EQUATION 15-4

4. Adjusted flow rate for *ATS* is computed using grade and heavy vehicle adjustments with the peak hour factor.

Flow Rate: $v_d = v_o = 1000 / (0.90)(1.00)(0.983) = 1130$ pc/h/ln
EQUATION 15-3

5. Average travel speed is calculated with the free-flow speed, flow rates, and the no-passing zone adjustment.

Average Travel Speed: $ATS = 57.0 - 0.0076(1130 + 1130) - 1.1 = 38.4$ mi/h
EQUATION 15-6 / EXHIBIT 15-15

6. The grade adjustment factor for *PTSF* is computed for rolling terrain.

Grade: Flow = 1000/0.90 = 1111 veh/h $f_{g,PTSF} = 1.00$

EXHIBIT 15-16

7. The heavy vehicle adjustment factor for *PTSF* is computed for rolling terrain with trucks and RVs.

Heavy Vehicles: $E_T = 1.0$ $E_R = 1.0$

EXHIBIT 15-18

$f_{HV} = 1 / ((1 + 0.05(1.0 - 1) + 0.02(1.0 - 1)) = 1.000$

EQUATION 15-80

8. Adjusted flow rate for *PTSF* is computed using grade and heavy vehicle adjustments with peak hour factor.

Flow Rate: $v_d = v_o = 1000 / (0.90)(1.00)(1.000) = 1111$ pc/h/ln

EQUATION 15-7

9. Base percent time-spent-following is calculated using the flow rates and provided coefficients.

Base PTSF: $BPTSF = 100(1 - \exp(-0.0051)(1111)^{0.827}) = 82.0\%$

EQUATION 15-10 / EXHIBIT 15-20

10. Percent time-spent-following is calculated from the Base with the no-passing zone adjustment flow rates.

Percent Time-Spent-Following: $PTSF = 82.0 + 15.7(1111/2222) = 89.8\%$

EQUATION 15-9 / EXHIBIT 15-21

11. Capacity is computed to compare with the flow rate for the LOS F possibility.

Capacity: $1111 < 1700$ $v/c = 1111/1700 = 0.65$ Not LOS F

EQUATION 15-18

12. Level of service is determined from the *PTSF* for Class II highways.

Level of Service: LOS = E ($PTSF > 85\%$)

EXHIBIT 15-3

HCS7

These screenshots show how Problem One is coded into *HCS7*, followed by the results as presented in the formatted report. All input data, intermediate computations, and final results are shown in logical order leading to average travel speed, percent time-spent-following, and level of service.

Figure 3.77 *HCS7 TwoLane1 Input*

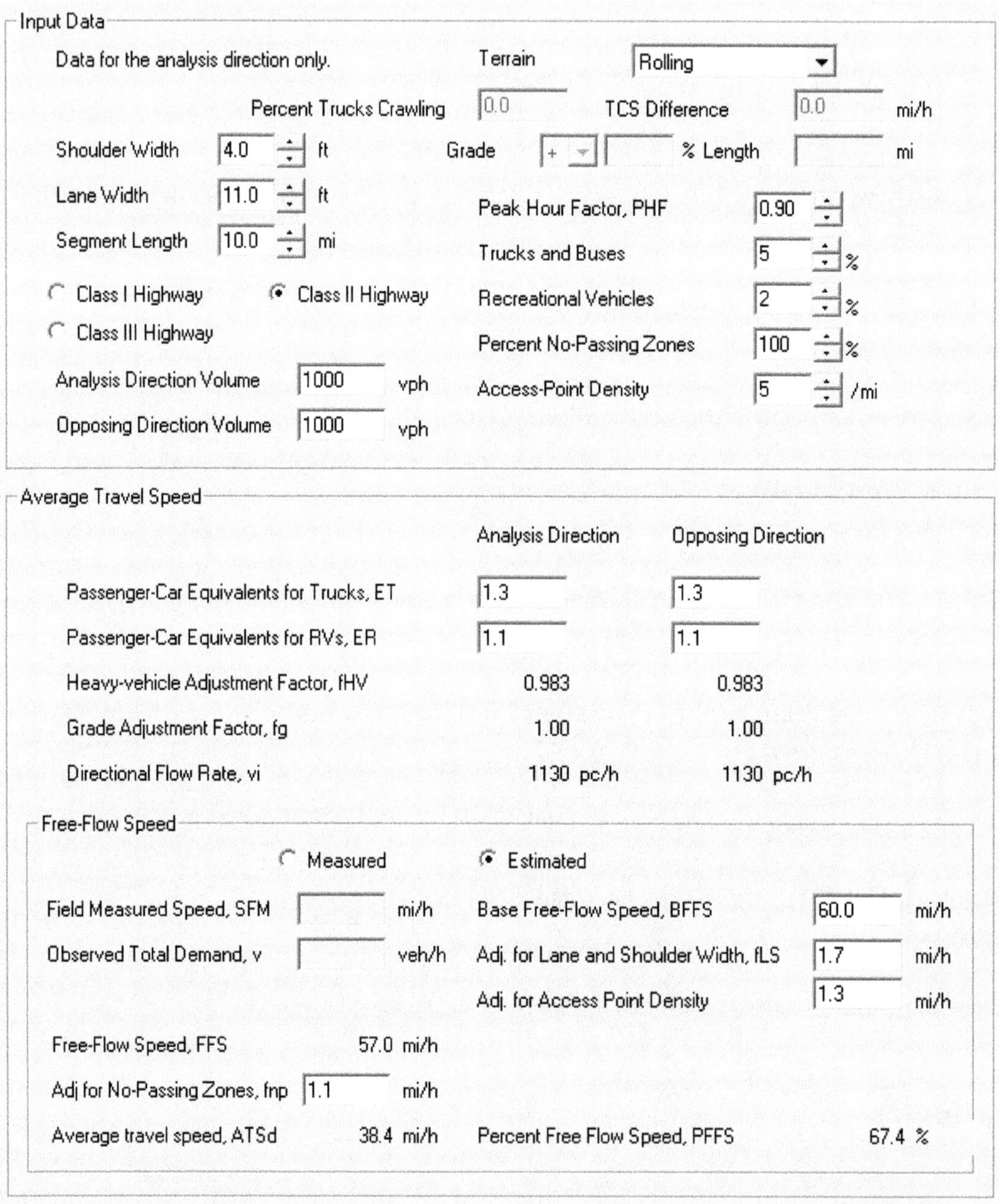

Figure 3.78 *HCS7 TwoLane1 Output*

Average Travel Speed		
	Analysis Direction (d)	Opposing Direction (o)
Passenger-car equivalents for trucks, E_T (Exhibit 15-11 or 15-12)	1.3	1.3
Passenger-car equivalents for RVs, E_R (Exhibit 15-11 or 15-13)	1.1	1.1
Heavy-vehicle adjustment factor, $f_{HV,ATS}$=1/ (1+ P_T(E_T-1)+P_R(E_R-1))	0.983	0.983
Grade adjustment factor[1], $f_{g,ATS}$ (Exhibit 15-9)	1.00	1.00
Demand flow rate[2], v_i (pc/h) v_i=V_i / (PHF* $f_{g,ATS}$ * $f_{HV,ATS}$)	1130	1130
Free-Flow Speed from Field Measurement	**Estimated Free-Flow Speed**	
Mean speed of sample[3], S_{FM} Total demand flow rate, both directions, v Free-flow speed, FFS=S_{FM}+0.00776(v/ $f_{HV,ATS}$) Adj. for no-passing zones, $f_{np,ATS}$ (Exhibit 15-15) 1.1 mi/h	Base free-flow speed[4], BFFS 60.0 mi/h Adj. for lane and shoulder width,[4] f_{LS}(Exhibit 15-7) 1.7 mi/h Adj. for access points[4], f_A (Exhibit 15-8) 1.3 mi/h Free-flow speed, FFS (FSS=BFFS-f_{LS}-f_A) 57.0 mi/h Average travel speed, ATS_d=FFS-0.00776($v_{d,ATS}$ + $v_{o,ATS}$) - $f_{np,ATS}$ 38.4 mi/h Percent free flow speed, PFFS 67.4 %	
Percent Time-Spent-Following		
	Analysis Direction (d)	Opposing Direction (o)
Passenger-car equivalents for trucks, E_T(Exhibit 15-18 or 15-19)	1.0	1.0
Passenger-car equivalents for RVs, E_R (Exhibit 15-18 or 15-19)	1.0	1.0
Heavy-vehicle adjustment factor, f_{HV}=1/ (1+ P_T(E_T-1)+P_R(E_R-1))	1.000	1.000
Grade adjustment factor[1], $f_{g,PTSF}$ (Exhibit 15-16 or Ex 15-17)	1.00	1.00
Directional flow rate[2], v_i(pc/h) v_i=V_i/(PHF*$f_{HV,PTSF}$* $f_{g,PTSF}$)	1111	1111
Base percent time-spent-following[4], $BPTSF_d$(%)=100(1-$e^{av_d^b}$)	82.0	
Adj. for no-passing zone, $f_{np,PTSF}$ (Exhibit 15-21)	15.7	
Percent time-spent-following, $PTSF_d$(%)=$BPTSF_d$+$f_{np,PTSF}$ *($v_{d,PTSF}$ / $v_{d,PTSF}$ + $v_{o,PTSF}$)	89.8	
Level of Service and Other Performance Measures		
Level of service, LOS (Exhibit 15-3)	E	
Volume to capacity ratio, v/c	0.65	
Capacity, $C_{d,ATS}$ (Equation 15-12) veh/h	1671	
Capacity, $C_{d,PTSF}$ (Equation 15-13) veh/h	1700	
Percent Free-Flow Speed $PFFS_d$(Equation 15-11 - Class III only)	67.4	

PROBLEM TWO

Building on Problem One, compute the LOS after adding a 2-mile passing lane 1 mile from the beginning of the segment with all other conditions unchanged.

1. The effective length downstream of the passing lane is determined from the flow rate.

Downstream Effective Length: $L_{de} = 3.6$ mi

EXHIBIT 15-23

2. The lengths of the four regions within the analysis segment are noted.

Regions: $L_u 1 = 1$ mi $L_{pl} = 2$ mi $L_{de} = 3.6$ mi $L_d = 10 - (1 + 2 + 3.6) = 3.4$ mi

EXHIBIT 15-25 / EQUATION 15-19

3. Percent time-spent-following is adjusted to use the four region lengths and the passing lane adjustment.

Percent Time-Spent-Following: $f_{pl} = 0.62$

EXHIBIT 15-26

$PTSF_{pl} = (89.8\ (1 + 3.4 + (0.62)(2) + ((1 + 0.62)/2)(3.6)))\ /\ 10 = 76.8\%$

EQUATION 15-20

4. Level of service is determined from the $PTSF_{pl}$ for Class II highways.

Level of Service: LOS = D (between 70 and 85)

EXHIBIT 15-3

HCS7

These screenshots show how Problem Two expands on Problem One adding the cross-weave information, followed by the results showing the reduced capacity as presented in the formatted report. All input data, intermediate computations, and final results are shown in logical order, leading to average travel speed, percent time-spent-following, and level of service.

Figure 3.79 *HCS7 TwoLane2 Input*

Passing Lane Analysis

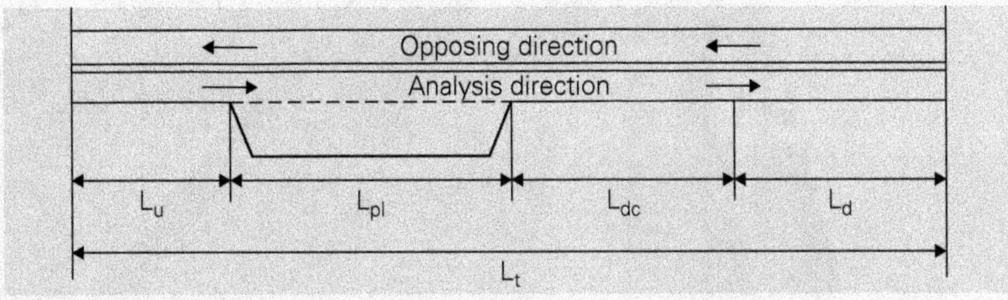

Passing Lane

Total length of analysis segment, L_t	10.0 mi
Length of two-lane highway upstream of passing lane, L_u	1.0 mi
Length of passing lane including tapers, L_{pl}	2.0 mi

Figure 3.80 *HCS7 TwoLane2 Output*

Average Travel Speed	
Length of the downstream highway segment within the effective length of passing lane for average travel speed, L_{de} (Exhibit 15-23)	*1.70*
Length of two-lane highway downstream of effective length of the passing lane for avg travel speed, L_d $L_d = L_t - (L_u + L_{pl} + L_{de})$	*5.30*
Adj. factor for the effect of passing lane on average speed, f_{pl} (Exhibit 15-28)	*1.11*
Average travel speed including passing lane[2], $ATS_{pl} = (ATS_d * L_t) / (L_u + L_d + (L_{pl}/f_{pl}) + (2L_{de}/(1+f_{pl,ATS})))$	*39.6*
Percent free flow speed including passing lane, $PFFS_{pl} = (ATS_{pl}/FFS)$	*69.4*
Percent Time-Spent-Following	
Length of the downstream highway segment within the effective length of passing lane for percent time-spent-following, L_{de} (Exhibit 15-23)	*3.60*
Length of two-lane highway downstream of effective length of the passing lane for percent-time-following, $L_d = L_t - (L_u + L_{pl} + L_{de})$	*3.40*
Adj. factor for the effect of passing lane on percent time-spent-following, $f_{pl,PTSF}$ (Exhibit 15-26)	*0.62*
Percent time-spent-following including passing lane[3], $PTSF_{pl}$(%) $PTSF_{pl} = PTSF_d[L_u + L_d + f_{pl,PTSF}L_{pl} + ((1+f_{pl,PTSF})/2)L_{de}]/L_t$	*76.8*
Level of Service and Other Performance Measures[4]	
Level of service including passing lane LOS_{pl} (Exhibit 15-3)	*D*

PART II
Interrupted Flow

This section deals with procedures for analyzing interrupted flow segments and facilities with each HCM chapter and HCS module described in detail. Each methodology is defined step-by-step, culminating in exercises to work problems by hand to demonstrate a full understanding and code them in software to learn how it will be done in practice.

Chapters and modules included in uninterrupted flow analysis include: Stop-Controlled Intersections; Roundabout Intersections and Segments; Signalized Intersections; Urban Street Segments; and Urban Street Facilities.

Stop-Controlled Intersections

The methodology for each HCM chapter is defined in detail to cover all intermediate calculations for a complete understanding of each procedure. For each HCM chapter, two example problems are presented showing every step in the computations, each followed by screen shots of the software data coding and results report.

Two-Way Stop Control

This procedure provides detailed explanations for computing conflicting flow for each movement and to discuss the critical and follow-up headway times from the HCM with the need for calibrating for local conditions. The calculation of capacity by movement is combined for shared lanes, then used to compute delay from the gap acceptance model equation. Delay is used to determine LOS, and the 95th percentile queue is also computed. Procedures are defined for determining pedestrian level of service.

All-Way Stop Control

This procedure involves an iterative process using the probabilities of each possible scenario occurring to converge on a value for departure headway. The number of scenarios is a function of approach lanes with 8 for single-lane approaches, 64 for two-lane approaches, and

512 for three-lane approaches. Delay is computed using departure headway for determining LOS with the 95th percentile queue also being calculated.

Roundabouts

This methodology covers all intermediate calculations for a complete understanding of each procedure. Two example problems will be presented showing every step in the computations, each followed by screen shots of the software data coding and results report.

Intersections

This procedure is a gap acceptance model following the newly updated capacity equations. Circulating flow is used for conflicting flow for most movements with exiting flow required for yielding bypass lanes. De facto lanes are considered in allocating flows as well as pedestrian impedance in capacity reduction. Delay is computed for all movements except non-yielding bypass lanes defined to have zero delay. LOS is determined from delay by lane, approach, and the overall intersection, and 95th percentile queue is computed by lane.

Segments

This procedure includes the definition of subsegments and the computation of circulating speed from the circle geometry for the computation of geometric delay. This value is added to the through movement delay and combined with the segment running time, then converted to speed using the segment length for the determination of LOS.

Signalized Intersections

This methodology is defined in detail to cover all intermediate calculations for a complete understanding of each procedure. Two example problems are presented showing every step in the computations, each followed by screen shots of the software data coding and results report.

Intersections

This most complex and most widely used procedure is the combination of the adjusted saturation flow rate, signal phasing, controller settings, and effective green times to compute capacity. The adjusted flow rate to capacity ratio provides for the computation of delay for determining LOS by lane group, approach, and the overall intersection. The calculation of back of queue to determine the queue storage ratio from the lane length provides a check for queue spillover and spillback. Procedures are defined for determining pedestrian and bicycle levels of service.

Traffic

The proper measurement of arrival rate is discussed as key to computing results for oversaturated signalized intersections that must be modeled using a multiple-period analysis. A common mistake that yields very underestimated delay and queue results is to use stop-bar counts and perform a single-period analysis, so this point is emphasized in detail. Other field data items that must be collected and are many times overlooked are listed and defined to ensure all required data are collected.

Phasing

The details of signal phasing consistent with the NEMA standard are described toward an appropriate definition of the signal operation. This must be correct for the phase duration model to accurately predict the average phase times for all actuated movements over the fifteen-minute analysis period. Care is taken to make sure the effective green times from this model are understood as contrasted with maximum green or phase split settings.

Urban Streets

This methodology covers all intermediate calculations for a complete understanding of each procedure. Two example problems are presented showing every step in the computations, each followed by screen shots of the software data coding and results report.

Segments

The base free-flow speed and running time are computed for use in quantifying the total time required to traverse the segment, defined as crossing the stop bar at the upstream signalized intersection to crossing the stop bar at the downstream signalized intersection. The restricted median and on-street parking percentages are used with traffic and access point densities to compute running time, which is converted to speed (knowing the segment length) to determine level of service. Procedures are defined for determining pedestrian, bicycle, and transit levels of service.

Facilities

This procedure includes the description of the flow profile model that tracks the platoon from one signal to the next computing dispersion and the proportion arriving on green at the downstream signal. This magnitude of this proportion dictates those with zero delay to quantify the effects of signal coordination on LOS. Access points (unsignalized side streets and driveways) impact the speed and integrity of the platoons (as well as the flow balancing) from signal to signal, which are uniquely modeled in this procedure. Procedures are defined for modeling travel time reliability with demand, weather, incidents, work zones, and special events varying for a distribution of results for hundreds of scenarios to produce travel-time reliability indices at three different percentiles. Procedures are defined for determining pedestrian, bicycle, and transit levels of service.

Interchanges and Alternative Intersections

This methodology covers all intermediate calculations for a complete understanding of each procedure. Example problems are presented showing every step in the computations, each followed by screen shots of the software data coding and results report.

Interchange Ramp Terminals

This procedure models the two signalized intersections that control the ramps at an interchange with a freeway. A special set of lane utilization adjustments predict pre-positioning; saturation flow adjustments for radius and traffic pressure are introduced; and two types (queue spillback and demand starvation) of additional lost time and employed to generate fourteen origin-destination results. Diverging Diamond Interchanges (DDI) are included to analyze this special design that includes major street split phasing and distance additional lost time.

Alternative Intersections

Three designs of distributed intersections are modeled, including Displaced Left Turn (DLT) intersections, Restricted Crossing U-Turn (RCUT) intersections, and Median U-Turn (MUT) intersections. Compared with conventional intersections, certain movements are distributed upstream of the main intersection for these strategic designs. Extra Distance Travel Time (EDTT) plays a major role in two (RCUT and MUT) of these, while modified signal phasing is required for two (DLT and RCUT) of these.

CHAPTER 4 Stop Control

Stop-Controlled Intersections can be one of two types: Two-Way Stop Control (TWSC) or All-Way Stop Control (AWSC). A TWSC intersection analysis models all minor-street and major-street left-turn movements, with potential delay to the major-street through and right-turn movements considered, including up to three through lanes on the major-street approaches and up to three total lanes on the minor-street approaches. An AWSC intersection analysis models all movements on all approaches, including up to three-lane approaches.

Two-Way Stop Control

The analysis of a TWSC intersection as defined in HCM Chapter 20 involves determining critical and follow-up headway values and computing total conflicting flow for all considered movements to compute capacity. The volume-to-capacity (*v/c*) ratios can be computed with the appropriate flow rates for use in calculating delay and the 95th percentile queue for each considered lane. Level of service is determined for all minor-street and major-street left-turn lanes as well as the minor-street approaches. Also, procedures are defined for determining pedestrian level of service.

Required Data

The required lane configuration includes the number of lanes on each approach with the movements allowed for each lane. Turning movement counts for all movements are needed with one peak hour factor applied to all movements and percent heavy vehicles for all movements for which delay will be computed, usually excluding the major-street through and right-turn movements. Major-street median type and minor-street flared right-turn storage values, as well as all approach grades, are required. If an upstream signal exists, the proportion time blocked (PTB) for each movement comes from an urban street analysis within the access point results.

Lane Configuration

The major- and minor-street directions are identified, including the number of lanes on each approach with exclusive and shared lanes. Up to three through lanes on the major street and three total lanes with no more than one lane per movement on the minor street are modeled. An exclusive right-turn lane on the major street must be identified as channelized if there is a triangular island with stop or yield control at the minor street. A flared right must be identified if the approach geometry allows right-turning vehicles to move around the queue. The median type is either divided or undivided.

Movement Ranks

Generally, major-street through traffic and right turns are Rank 1; major-street left turns and minor-street right turns are Rank 2; minor-street through traffic (for four-approach intersections) is Rank 3; and minor-street left turns are Rank 4 (Rank 3 for T-intersections).

Flow Rate

Turning movement data are converted to flow rates by dividing each by the intersection's peak hour factor using HCM Equation 20-1.

Figure 4.1 *HCM Equation 20-1*

$$v_i = \frac{V_i}{PHF}$$

where

v_i = demand flow rate for movement i (veh/h),

V_i = demand volume for movement i (veh/h), and

PHF = peak hour factor for the intersection

Conflicting Flows

For each Rank 2, 3, or 4 movement, the conflicting flows are defined by summing flows with higher rankings, subject to a few modifications to model actual driver behavior. The conflicting flows for each movement are illustrated in HCM Exhibits 20-7 through 20-11 with the computations for each defined in HCM Equations 20-2 through 20-29. While most are straightforward, a few require some explanation.

Figure 4.2 *HCM Exhibit 20-7*

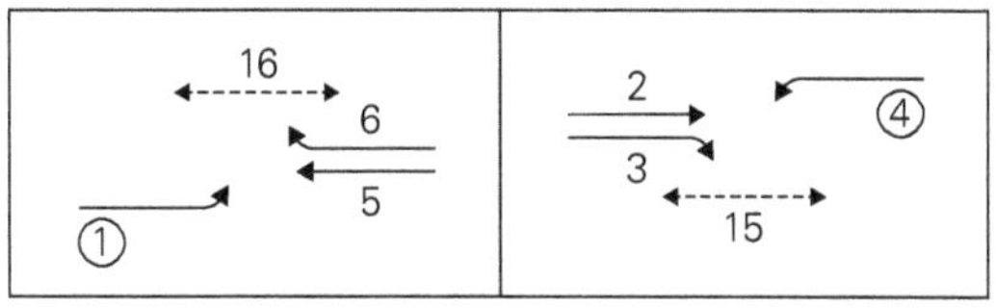

Figure 4.3 *HCM Equation 20-2*

$$v_{c,1} = v_5 + v_6 + v_{16}$$

Figure 4.4 *HCM Equation 20-3*

$$v_{c,4} = v_2 + v_3 + v_{15}$$

Figure 4.5 *HCM Exhibit 20-8*

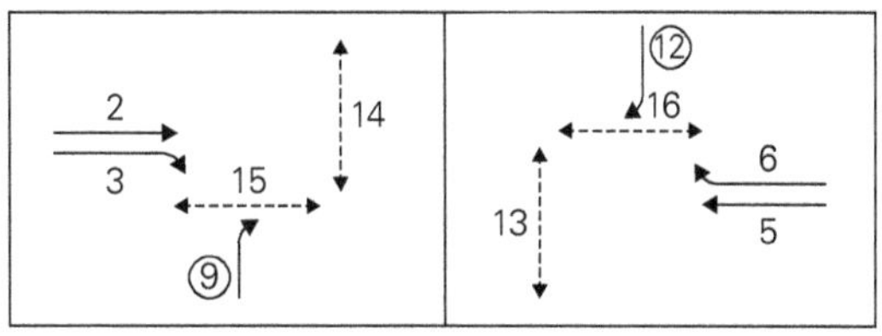

Figure 4.6 *HCM Equation 20-4*

Two-lane major streets:

$$v_{c,9} = v_2 + 0.5v_3 + v_{14} + v_{15}$$

Figure 4.7 *HCM Equation 20-5*

$$v_{c,12} = v_5 + 0.5v_6 + v_{13} + v_{16}$$

Figure 4.8 *HCM Equation 20-6*

Four-lane major streets:

$$v_{c,9} = 0.5v_2 + 0.5v_3 + v_{14} + v_{15}$$

Figure 4.9 *HCM Equation 20-7*

$$v_{c,12} = 0.5v_5 + 0.5v_6 + v_{13} + v_{16}$$

Figure 4.10 *HCM Equation 20-8*

Six-lane major streets:

$$v_{c,9} = 0.5v_2 + 0.5v_3 + v_{14} + v_{15}$$

Figure 4.11 *HCM Equation 20-9*

$$v_{c,12} = 0.5v_5 + 0.5v_6 + v_{13} + v_{16}$$

Figure 4.12 *HCM Exhibit 20-9*

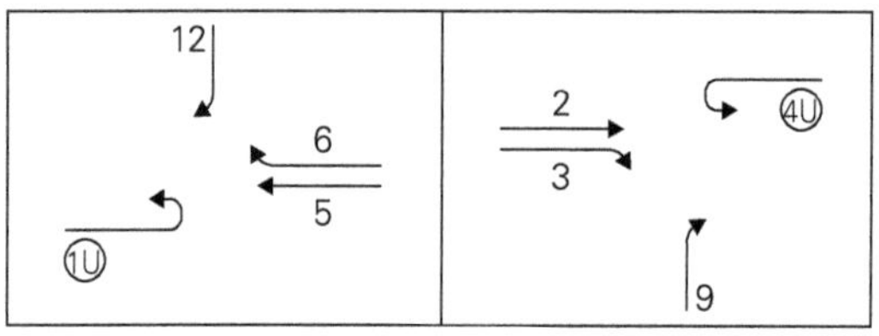

Figure 4.13 *HCM Equation 20-10*

Four-lane major streets:

$$v_{c,1U} = v_5 + v_6$$

Figure 4.14 *HCM Equation 20-11*

$$v_{c,4U} = v_2 + v_3$$

Figure 4.15 *HCM Equation 20-12*

Six-lane major streets:

$$v_{c,1U} = 0.73v_5 + 0.73v_6$$

Figure 4.16 *HCM Equation 20-13*

$$v_{c,4U} = 0.73v_2 + 0.73v_3$$

Figure 4.17 *HCM Exhibit 20-10*

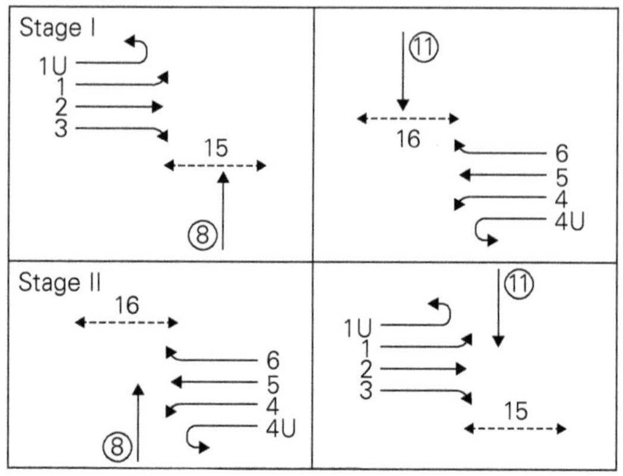

Figure 4.18 *HCM Equation 20-14*

$$v_{c,I,8} = 2(v_1 + v_{1U}) + v_2 + 0.5v_3 + v_{15}$$

Figure 4.19 *HCM Equation 20-15*

$$v_{c,I,11} = 2(v_4 + v_{4U}) + v_5 + 0.5v_6 + v_{16}$$

Figure 4.20 *HCM Equation 20-16*

$$v_{c,II,8} = 2(v_4 + v_{4U}) + v_5 + v_6 + v_{16}$$

Figure 4.21 *HCM Equation 20-17*

$$v_{c,II,11} = 2(v_1 + v_{1U}) + v_2 + v_3 + v_{15}$$

Figure 4.22 *HCM Exhibit 20-11*

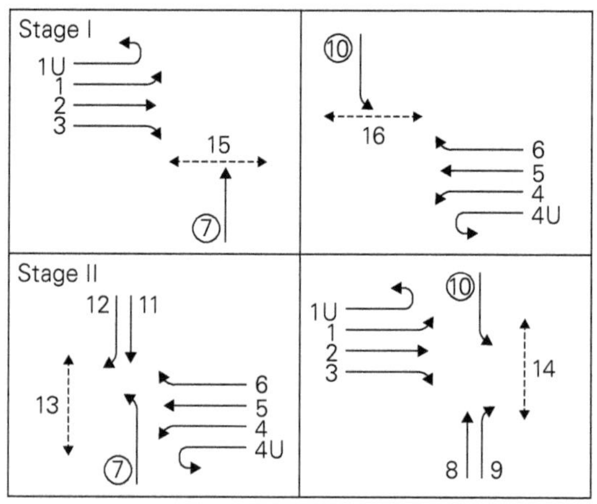

Figure 4.23 *HCM Equation 20-18*

Two-lane major streets:

$$v_{c,I,7} = 2v_1 + v_2 + 0.5v_3 + v_{15}$$

Figure 4.24 *HCM Equation 20-19*

$$v_{c,I,10} = 2v_4 + v_5 + 0.5v_6 + v_{16}$$

Figure 4.25 *HCM Equation 20-20*

Four-lane major streets:

$$v_{c,I,7} = 2(v_1 + v_{1U}) + v_2 + 0.5v_3 + v_{15}$$

Figure 4.26 *HCM Equation 20-21*

$$v_{c,I,10} = 2(v_4 + v_{4U}) + v_5 + 0.5v_6 + v_{16}$$

Figure 4.27 *HCM Equation 20-22*

Six-lane major streets:

$$v_{c,I,7} = 2(v_1 + v_{1U}) + v_2 + 0.5v_3 + v_{15}$$

Figure 4.28 *HCM Equation 20-23*

$$v_{c,I,10} = 2(v_4 + v_{4U}) + v_5 + 0.5v_6 + v_{16}$$

Figure 4.29 *HCM Equation 20-24*

Two-lane major streets:

$$v_{c,II,7} = 2v_4 + v_5 + 0.5v_6 + 0.5v_{12} + 0.5v_{11} + v_{13}$$

Figure 4.30 *HCM Equation 20-25*

$$v_{c,II,10} = 2v_1 + v_2 + 0.5v_3 + 0.5v_9 + 0.5v_8 + v_{14}$$

Figure 4.31 *HCM Equation 20-26*

Four-lane major streets:

$$v_{c,II,7} = 2(v_4 + v_{4U}) + 0.5v_5 + 0.5v_{11} + v_{13}$$

Figure 4.32 *HCM Equation 20-27*

$$v_{c,II,10} = 2(v_1 + v_{1U}) + 0.5v_2 + 0.5v_8 + v_{14}$$

Figure 4.33 *HCM Equation 20-28*

Six-lane major streets:

$$v_{c,II,7} = 2(v_4 + v_{4U}) + 0.4v_5 + 0.5v_{11} + v_{13}$$

Figure 4.34 *HCM Equation 20-29*

$$v_{c,II,10} = 2(v_1 + v_{1U}) + 0.4v_2 + 0.5v_8 + v_{14}$$

The minor-street right turn is not completely conflicted by the major-street right turn, but there is a partial conflict represented by the 0.5 multiplier. If the major-street right turns are in an exclusive lane, this value becomes zero.

Think about your own driving:
When you are turning right out of the side street and someone is turning right beside you from the main road, it really isn't a conflict, but you really don't trust them, even if the turn signal is on. You wait until the front bumper goes down (signaling they hit the brakes) and they start to make the turn (convinced they have committed). Then you go, not waiting for them to complete the turn—that's why it's only half of a conflict.

The minor-street right turn is not completely conflicted by the major-street through traffic if there is more than one lane in that direction, so the potential to turn alongside through traffic in the outer lane(s) is represented by the 0.5 multiplier for four- and six-lane major streets.

Think about your own driving:
Again, when you are turning right out of the side street onto a four- or six-lane main road, you must yield to through traffic in the rightmost lane, but you are willing to turn into the rightmost lane with traffic in the leftmost lane(s)—that's why it's half of a conflict.

The minor-street right turn is not conflicted by the major-street U-turn, even though the minor-street right turn has a stop sign, represented by the U-turn movement not included in the equation.

Think about your own driving:
Once more, when you are turning right out of the side street and someone is making a U-turn on the main road, you think they are turning left (which does not conflict with you) and you proceed. By the time you realize they are making a U-turn (if you ever do), you are 50 feet down the street well in front of them.

The minor-street left turn can make this movement in two stages if a median or two-way left-turn lane (TWTLT) exists on the major street, represented in the two-stage gap acceptance equations. Since two small headways are more available (for two stages) than one larger headway (for a single stage), two-stage gap acceptance will increase capacity. The minor-street through traffic cannot use two stages if there is a TWLTL and no median.

Think about your own driving:
When making a left turn from the side street, you have looked to your left to find a gap just to move to the median and stop. Then, you look to your right for a second gap to finish your turn. This works for left turns and through traffic if the median opening can store a vehicle going across the main road, but it only works for left turns if the refuge is a TWLTL, where often vehicles actually move, slowing in the TWLTL to merge downstream when a gap is found.

The major-street left turn and U-turn yield to the opposing major-street right turn, unless it is channelized where the right turn yields.

Headways

Critical headway is the minimum time between conflicting vehicles for the subject vehicle to proceed. This time differs by movement and major-street number of lanes, as well as number of stages for minor-street left turns and through movements. The default base values are provided in HCM Exhibit 20-12 and adjusted for heavy vehicles, grade, and T-intersections as defined in HCM Equation 20-30.

Figure 4.35 *HCM Exhibit 20-12*

Vehicle Movement	Base Critical Headway, $t_{c,base}$ (s)		
	Two Lanes	**Four Lanes**	**Six Lanes**
Left turn from major	4.1	4.1	5.3
U-turn from major	N/A	6.4 (wide*) 6.9 (narrow*)	5.6
Right turn from minor	6.2	6.9	7.1
Through traffic on minor	1-stage: 6.5 2-stage, Stage I: 5.5 2-stage, Stage II: 5.5	1-stage: 6.5 2-stage, Stage I: 5.5 2-stage, Stage II: 5.5	1-stage: 6.5** 2-stage, Stage I: 5.5** 2-stage, Stage II: 5.5**
Left turn from minor	1-stage: 7.1 2-stage, Stage I: 6.1 2-stage, Stage II: 6.1	1-stage: 7.5 2-stage, Stage I: 6.5 2-stage, Stage II: 6.5	1-stage: 6.4 2-stage, Stage I: 7.3 2-stage, Stage II: 6.7

Figure 4.36 *HCM Equation 20-30*

$$t_{c,x} = t_{c,\text{base}} + t_{c,HV}P_{HV} + t_{c,G}G - t_{3,LT}$$

where

$t_{c,x}$ = critical headway for movement *x* (s);

$t_{c,\text{base}}$ = base critical headway from Exhibit 20-12 (s);

$t_{c,HV}$ = adjustment factor for heavy vehicles (1.0 for major streets with one lane in each direction; 2.0 for major streets with two or three lanes in each direction) (s);

P_{HV} = proportion of heavy vehicles for movement (expressed as a decimal; e.g., P_{HV} = 0.02 for 2% heavy vehicles);

$t_{c,G}$ = adjustment factor for grade for given movement (0.1 for Movements 9 and 12; 0.2 for Movements 7, 8, 10, and 11) (s);

G = percent grade (expressed as an integer; e.g., $G = -2$ for a 2% downhill grade); and

$t_{3,LT}$ = adjustment factor for intersection geometry (0.7 for minor-street left-turn movement at three-leg intersections; 0.0 otherwise (s).

Follow-up headway is the time required for the second vehicle in queue to move up to become the first vehicle after the first vehicle ahead proceeds. The default base values are provided in HCM Exhibit 20-13 and adjusted for heavy vehicles as defined in HCM Equation 20-31.

Figure 4.37 *HCM Exhibit 20-13*

Vehicle Movement	Base Follow-Up Headway, $t_{f,\text{base}}$ (s) Two Lanes	Four Lanes	Six Lanes
Left turn from major	2.2	2.2	3.1
U-turn from major	N/5	2.5 (wide*) 3.1(narrow*)	2.3
Right turn from minor	3.3	3.3	3.9
Through traffic on minor	4.0	4.0	4.0
Left turn from minor	3.5	3.5	3.8

Figure 4.38 *HCM Equation 20-31*

$$t_{f,x} = t_{f,\text{base}} + t_{f,HV}P_{HV}$$

where

$t_{f,x}$ = follow-up headway for movement *x* (s),

$t_{f,\text{base}}$ = base follow-up headway from Exhibit 20-13 (s),

$t_{f,HV}$ = adjustment factor for heavy vehicles (0.9 for major streets with one lane in each direction, 1.0 for major streets with two or three lanes in each direction), and

P_{HV} = proportion of heavy vehicles for movement (expressed as a decimal; e.g., P_{HV} = 0.02 for 2% heavy vehicles).

The default values in these exhibits should be calibrated to local conditions, since gap acceptance can vary widely with prevailing conditions like population, wait time, and approach speed, which are not modeled within these procedures.

Think about your own driving:

Drivers in large cities are more aggressive, accepting shorter gaps than those in smaller towns. Your standard changes, and you accept shorter gaps after waiting even a minute. You demand longer gaps with higher approach speeds, since the risk is greater and the ability to perceive the gap is lessened.

The calibration process is fairly simple, involving counting the queue every twenty seconds or so while you are collecting other analysis data. A worksheet in the HCM allows you to convert these counts into control delay for comparison with the computed delay using the default headway values. Since the computed delay is typically higher than the measured delay, headway values are lowered in the software until the computed delay matches the measured delay to determine the actual headways being accepted.

Capacity

Potential capacity is computed for each movement using HCM Equation 20-32 with conflicting flow and headway values. This value is adjusted for the platooning effects of upstream signals and/or the possibility of a queue within a given conflicting flow if either situation exists.

Figure 4.39 *HCM Equation 20-32*

$$c_{p,x} = v_{c,x} \frac{e^{-v_{c,x}t_{c,x}/3,600}}{1 - e^{-v_{c,x}t_{f,x}/3,600}}$$

where

$c_{p,x}$ = potential capacity of movement x (veh/h),

$v_{c,x}$ = conflicting flow rate for movement x (veh/h),

$t_{c,x}$ = critical headway for minor movement x (s), and

$t_{f,x}$ = follow-up headway for minor movement x (s).

The urban streets procedure models the platooning between signals and can compute the proportion time blocked as any access point within the segment. This value is used in HCM Equations 20-33, 20-34, and 20-35 to adjust conflicting flow and capacity for the platoon effects that create more gaps for minor-street access.

Figure 4.40 *HCM Equation 20-33*

$$v_{c,u,x} = \begin{cases} \dfrac{v_{c,x} - 1.5v_{c,min}p_{b,x}}{1 - p_{b,x}} & \text{if } v_{c,x} > 1.5v_{c,min}p_{b,x} \\ 0 & \text{otherwise} \end{cases}$$

where

$v_{c,u,x}$ = conflicting flow for movement x during unblocked period (veh/h);

$v_{c,x}$ = total conflicting flow for movement x as determined from Step 3 (veh/h);

$v_{c,min}$ = minimum platooned flow rate (veh/h), assumed to be 1,000N, where N is the number of through lanes per direction on the major street; and

$p_{b,x}$ = proportion of time the subject movement x is blocked by the major-street platoon, which is determined from Exhibit 20-14.

Figure 4.41 *HCM Equation 20-34*

$$c_{p,x} = (1 - p_{b,x})c_{r,x}$$

where

$c_{p,x}$ = potential capacity of movement x (veh/h),

$p_{b,x}$ = proportion of time the movement x is blocked by a platoon, and

$c_{r,x}$ = capacity of movement x assuming random flow during the unblocked period.

Figure 4.42 *HCM Equation 20-35*

$$c_{r,x} = v_{c,u,x} \frac{e^{-v_{c,u,x}t_{c,x}/3,600}}{1 - e^{v_{c,u,x}t_{f,x}/3,600}}$$

Think about your own driving:

If you are waiting on a side street with an upstream signal nearby, you can see the platooned vehicles pass in front of you. While the grouping of traffic prevents you from moving, it also creates long periods of no traffic allowing you to move out more easily, increasing capacity overall.

Conflicting flows are used to compute capacity, but the effects on a given movement increase if the conflict has to wait in a queue and generates impedance. So, the probability of a queue forming is used to decrease capacity to account for this impedance, following HCM Equations 20-42, 20-46, and 20-47.

Think about your own driving:
Waiting the stop sign and yielding to a conflicting flow movement reduces your capacity and delays you, but if that movement sits in the median waiting for a gap in opposing flow, you are delayed even more.

Figure 4.43 *HCM Equation 20-42*

$$c_{SH} = \frac{\sum_y v_y}{\sum_y \frac{v_y}{c_{m,y}}}$$

where

c_{SH} = capacity of the shared lane (veh/h),

v_y = flow rate of the y movement in the subject shared lane (veh/h), and

$c_{m,y}$ = movement capacity of the y movement in the subject shared lane (veh/h).

Figure 4.44 *HCM Equation 20-46*

$$f_k = \prod_j p_{0,j}$$

where $\prod$ indicates the product of a series of terms and

$p_{0,j}$ = probability that conflicting Rank 2 movement j will operate in a queue-free state, and

k = Rank 3 movements.

Figure 4.45 *HCM Equation 20-47*

$$c_{m,k} = c_{p,k} \times f_k$$

Through this point in the procedure, all calculations are performed for each movement. From this point on, results are generated by lane. So, minor-street movement capacities are combined into lane capacities for any shared lanes, following HCM Equation 20-41.

Figure 4.46 *HCM Equation 20-41*

$$c_{SH} = \frac{\sum_y v_y}{\sum_y \frac{v_y}{c_{m,y}}}$$

where

c_{SH} = capacity of the shared lane (veh/h),

v_y = flow rate of the y movement in the subject shared lane (veh/h), and

$c_{m,y}$ = movement capacity of the y movement in the subject shared lane (veh/h).

Queue

The 95th percentile queue can be computed for each lane according to HCM Equation 20-68, not including major-street through and right-turn traffic.

Figure 4.47 *HCM Equation 20-68*

$$Q_{95} \approx 900T\left[\frac{v_x}{c_{m,x}} - 1 + \sqrt{\left(\frac{v_x}{c_{m,x}} - 1\right)^2 + \frac{\left(\frac{3{,}600}{c_{m,x}}\right)\left(\frac{v_x}{c_{m,x}}\right)}{150T}}\right]\left(\frac{c_{m,x}}{3{,}600}\right)$$

where

Q_{95} = 95th percentile queue (veh),

v_x = flow rate for movement x (veh/h),

$c_{m,x}$ = capacity of movement x (veh/h), and

T = analysis time period (0.25 h for a 15-min period) (h).

Delay

Control delay is computed for each modeled lane, not including major-street through and right-turn lanes, following HCM Equation 20-64. These values are combined for approach and intersection averages according to HCM Equations 20-66 and 20-67 with the analysis period equal to 0.25 hours (for 15 minutes).

Delay to Rank 1 vehicles is computed if there is major-street left-turn lane or if the median opening develops a queue beyond what can be stored; both situations can introduce delay to major-street through and right-turn traffic.

Figure 4.48 *HCM Equation 20-64*

$$d = \frac{3{,}600}{c_{m,x}} + 900T\left[\frac{v_x}{c_{m,x}} - 1 + \sqrt{\left(\frac{v_x}{c_{m,x}} - 1\right)^2 + \frac{\left(\frac{3{,}600}{c_{m,x}}\right)\left(\frac{v_x}{c_{m,x}}\right)}{450T}}\right] + 5$$

where

d = control delay (s/veh),
v_x = flow rate for movement x (veh/h),
$c_{m,x}$ = capacity of movement x (veh/h), and
T = analysis time period (equals 0.25 h for a 15-min period) (h).

Figure 4.49 *HCM Equation 20-66*

$$d_A = \frac{d_r v_r + d_t v_t + d_l v_l}{v_r + v_t + v_l}$$

where

d_A = control delay on the approach (s/veh);
d_r, d_t, d_l = computed control delay for the right-turn, through, and left-turn movements, respectively (s/veh); and
v_r, v_t, v_l = volume of flow rate of right-turn, through, and left-turn traffic on the approach, respectively (veh/h).

Figure 4.50 *HCM Equation 20-67*

$$d_I = \frac{d_{A,1}v_{A,1} + d_{A,2}v_{A,2} + d_{A,3}v_{A,3} + d_{A,4}v_{A,4}}{v_{A,1} + v_{A,2} + v_{A,3} + v_{A,4}}$$

where $d_{A,x}$ is the control delay on approach x (s/veh) and $v_{A,x}$ is the volume or flow rate on approach x (veh/h).

Figure 4.51 *HCM Equation 20-43*

$$p_{0,j}^{*} = 1-(1-p_{0,j})\left[\sqrt[(n_L+1)]{1+\frac{x_{i,1+2}^{(n_L+1)}}{1-x_{i,1+2}}}\right]$$

$$x_{i,1+2} = \frac{v_{i1}}{s_{i1}} + \frac{v_{i2}}{s_{i2}}$$

where

$p_{0,j}$ = probability of queue-free state for movement *j* assuming an exclusive left-turn lane on the major street (per Equation 20–42);

$p_{0,j}^{*}$ = probability of queue-free state for movement *j* assuming a shared left-turn lane on the major street;

j = 1 and 4 (major-street left-turning vehicular movements);

$i1$ = 2 and 5 (major-street through vehicular movements);

$i2$ = 3 and 6 (major-street right-turning vehicular movements);

$x_{i,1+2}$ = combined degree of saturation for the major-street through and right-turn movements;

s_{i1} = saturation flow rate for the major-street through movements (default assumed to be 1,800 veh/h; however, this parameter can be measured in the field);

s_{i2} = saturation flow rate for the major-street right-turn movements (default assumed to be 1,500 veh/h; however, this parameter can be measured in the field);

v_{i1} = major-street through-movement flow rate (veh/h);

v_{i2} = major-street right-turn flow rate (veh/h) (0 if an exclusive right-turn lane is provided); and

n_L = number of vehicles that can be stored in the left-turn pocket (see Exhibit 20-15.

Figure 4.52 *HCM Equation 20-65*

$$d_{Rank1} = \begin{cases} \dfrac{(1-p_{0,j}^{*})d_{M,LT}\left(\dfrac{v_{i,1}}{N}\right)}{v_{i,1}+v_{i,2}} & N>1 \\ (1-p_{0,j}^{*})d_{M,LT} & N=1 \end{cases}$$

where

d_{Rank1} = delay to Rank 1 vehicles (s/veh);

N = number of through lanes per direction on the major street;

$p_{0,j}^{*}$ = proportion of Rank 1 vehicles not blocked, from Equation 20-43;

$d_{M,LT}$ = delay to major left-turning vehicles, from Equation 20-64 (s/veh);

$v_{i,1}$ = major-street through vehicles in shared lane (veh/h); and

$v_{i,2}$ = major-street turning vehicles in shared lane (veh/h).

Figure 4.53 *HCM Exhibit 20-15*

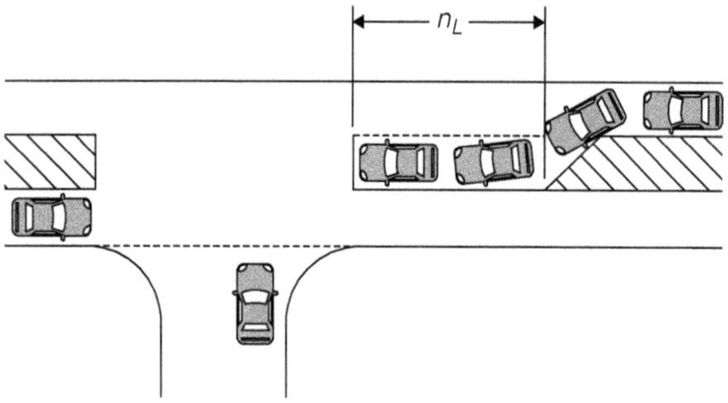

LOS

Level of service is determined in HCM Exhibit 20-2 for all minor-street lanes, but only the left-turn lanes on the major street. While delay values are computed for the major-street approach and the intersection, these are not used for LOS, since the majority of traffic has zero delay and the averages almost always result in LOS A—making these meaningless for LOS. Any lane with a volume-to-capacity ratio greater than one is assigned LOS F, regardless of its delay.

Figure 4.54 *HCM Exhibit 20-2*

Control Delay (s/veh)	LOS by Volume-to-Capacity Ratio	
	$v/c \leq 1.0$	$v/c > 1.0$
0–10	A	F
>10–15	B	F
>15–25	C	F
>25–35	D	F
>35–50	E	F
>50	F	F

Pedestrians

The procedure is defined to determine control delay for pedestrians at two-way stop controlled intersections and at mid-block crossings. Average control delay per pedestrian is the service measure for determining level of service. The method considers the effects of platooning pedestrians, the availability of a median refuge, and the probability that the motorist will yield.

REQUIRED DATA

Geometric information, including crosswalk length and width; major-street number of lanes; and whether a median is present, as well as demand data, including pedestrian flow rate; whether platooning exists; vehicular flow rate; pedestrian walking speed; pedestrian start-up and clearance times; and motorist yield rate are all required to perform the analysis.

PROCEDURE

The first step is to compute critical headway for a single pedestrian, following HCM Equation

Figure 4.55 *HCM Equation 20-77*

$$t_c = \frac{L}{S_p} + t_s$$

where

t_c = critical headway for a single pedestrian (s),
S_p = average pedestrian walking speed (ft/s),
L = crosswalk length (ft), and
t_s = pedestrian start-up time and end clearance time (s).

Figure 4.56 *HCM Equation 20-78*

$$N_p = \text{int}\left[\frac{8.0(N_c - 1)}{W_c}\right] + 1$$

where

N_p = spatial distribution of pedestrians (ped);

N_c = total number of pedestrians in the crossing platoon, from Equation 20-79 (ped);
W_c = crosswalk width (ft); and
8.0 = default clear effective width used by a single pedestrian to avoid interference when passing other pedestrians (ft).

Figure 4.57 *HCM Equation 20-79*

$$N_c = \frac{v_p e^{v_p t_c} + v e^{-v t_c}}{(v_p + v) e^{(v_p - v) t_c}}$$

where

N_c = total number of pedestrians in the crossing platoon (ped),
v_p = pedestrian flow rate (ped/s),
v = conflicting vehicular flow rate (veh/s) (combined flows for one-stage crossings; separate flows for two-stage crossings), and
t_c = single pedestrian critical headway (s).

Figure 4.58 *HCM Equation 20-80*

$$t_{c,G} = t_c + 2(N_p - 1)$$

where

$t_{c,G}$ = group critical headway (s),
t_c = critical headway for a single pedestrian (s), and
N_p = spatial distribution of pedestrian (ped).

20-77. If platooning is present, the distribution is calculated using HCM Equation 20-78 with this value becoming 1.0 if no groupings are observed. The groupings are accounted for in HCM Equation 20-79 to modeling this platooning using the group critical headway from HCM Equation 20-80.

Vehicular gaps are considered in the estimate of the probability of a delay crossing with no adequate gaps in traffic according to HCM Equations 20-81 and 20-82. Then, the average delay waiting for an adequate gap is computed using HCM Equation 20-83, including those not able to cross immediately in HCM Equation 20-84.

Figure 4.59 *HCM Equation 20-81*

$$P_b = 1 - e^{\frac{-t_{c,G} v}{N_L}}$$

Figure 4.60 *HCM Equation 20-82*

$$P_d = 1 - (1 - P_b)^{N_L}$$

where

P_b = probability of a blocked lane,
P_d = probability of a delayed crossing,
N_L = number of through lanes crossed,
$t_{c,G}$ = group critical headway (s), and
v = conflicting vehicular flow rate (veh/s) (combined flows for one-stage crossing; separate flows for two-stage crossing).

Figure 4.61 *HCM Equation 20-83*

$$d_g = \frac{1}{v}(e^{v t_{c,G}} - v t_{c,G} - 1)$$

where

d_g = average pedestrian gap delay (s),
$t_{c,G}$ = group critical headway (s), and
v = conflicting vehicular flow rate (veh/s) (combined flows for one-stage crossings; separate flows for two-stage crossings).

Figure 4.62 *HCM Equation 20-84*

$$d_{gd} = \frac{d_g}{P_d}$$

where

d_{gd} = average gap delay for pedestrians who incur nonzero delay,
d_g = average pedestrian gap delay (s), and
P_d = probability of a delayed crossing.

The availability of an adequate gap can be overcome somewhat as a function of the probability of motorists yielding by number of lanes crossed following HCM Equations 20-86 (one lane), 20-87, and 20-88 (two lanes); 20-89 (three lanes); and 20-90 (four lanes).

Figure 4.63 *HCM Equation 20-86*

$$P(Y_i) = P_d M_y (1 - M_y)^{i-1}$$

where

M_y = motorist yield rate (decimal), and

i = crossing event (i = 1 to n).

Figure 4.64 *HCM Equation 20-87*

$$P(Y_1) = 2P_b(1 - P_b)M_y + P_b^2 M_y^2$$

Figure 4.65 *HCM Equation 20-88*

$$P(Y_i) = \left[P_d - \sum_{j=0}^{i=1} P(Y_j)\right]\left[\frac{(2P_b[1 - P_b]M_y) + (P_b^2 M_y^2)}{P_d}\right]$$

Figure 4.66 *HCM Equation 20-89*

$$P(Y_i) = \left[P_d - \sum_{j=0}^{i-1} P(Y_j)\right]$$
$$\left[\frac{P_b^3 M_y^3 + 3P_b^2(1 - P_b)M_y^2 + 3P_b(1 - P_b)^2 M_y}{P_d}\right]$$

where $P(Y_0) = 0$.

Figure 4.67 *HCM Equation 20-90*

$$P(Y_i)$$
$$= \left[P_d - \sum_{j=0}^{i-1} P(Y_j)\right]$$
$$\times\left[\frac{P_b^4 M_y^4 + 4P_b^3(1 - P_b)M_y^3 + 6P_b^2(1 - P_b)^2 M_y^2 + 4P_b(1 - P_b)^3 M_y}{P_d}\right]$$

where $P(Y_0) = 0$.

Think about your own driving:

When you are navigating a stop-controlled intersection, you are assuming pedestrians will yield if you are on the major street. But when it gets more congested, the availability of gaps for pedestrians to cross goes to zero, However, when you are in a very slow (stop-and-go) platoon, you are very willing to yield to pedestrians and let them cross in front of you.

PEDESTRIAN LOS

This probability is included in the computation of average pedestrian delay in HCM Equation 20-85, which is used to determine level of service from HCM Exhibit 20-3.

Figure 4.68 *HCM Equation 20-85*

$$d_p = \sum_{i=1}^{n} h(i - 0.5)P(Y_i) + \left(P_d - \sum_{i=1}^{n} P(Y_i)\right) d_{gd}$$

where

d_p = average pedestrian delay (s),

i = crossing event (i = 1 to n),

h = average headway for each through lane = (N_L/v) (s),

$P(Y_i)$ = probability that motorists yield to pedestrian on crossing event i, and

n = int(d_{gd}/h), average number of crossing events before an adequate gap is avialable.

Figure 4.69 *HCM Exhibit 20-3*

LOS	Control Delay (s/ped)	Comments
A	0–5	Usually no conflicting traffic
B	5–10	Occasionally some delay due to conflicting traffic
C	10–20	Delay noticeable to pedestrians, but not inconveniencing
D	20–30	Delay noticeable and irritating, increased likelihood of risk taking
E	30–45	Delay approaches tolerance level, risk-taking behavior likely
F	>45	Delay exceeds tolerance level, high likelihood of pedestrian risk taking

PROBLEM ONE

Determine the northbound shared left-right lane LOS for a T-intersection with a major-street eastbound through-right turn shared lane and exclusive through and left-turn lanes westbound on an undivided street. The volumes (veh/h) are EBT = 500; EBR = 100; WBL = 50; WBT = 400; NBL = 70; and NBR = 100. The peak hour factor is 0.92 with 10% heavy vehicles in all movements and no pedestrians, grades, or upstream signals.

1. Compute the flow rates for all movements.

Flow Rate: $v_{EBT} = 500/0.92 = 543$ veh/h $v_{EBR} = 100/0.92 = 109$ veh/h $v_{WBL} = 50/0.92 = 54$ veh/h

EQUATION 20-1

$v_{WBT} = 400/0.92 = 435$ veh/h $v_{NBL} = 70/0.92 = 76$ veh/h $v_{NBR} = 100/0.92 = 109$ veh/h

2. Compute conflicting flow rates for the northbound left and right movements and westbound left movement.

Conflicting Flow: $v_{c,NBL} = 543 + 0.5(109) + 2(54) + 435 = 1140$ veh/h

EQUATIONS 20-18 & 20-24

$v_{c,NBR} = 543 + 0.5(109) = 598$ veh/h

EQUATION 20-4

$v_{c,WBL} = 543 + 109 = 652$ veh/h

EQUATION 20-3

3. Compute the headway times for the northbound left and right movements.

Critical Headway: $t_{c,NBL} = 7.1 + 1.0(0.10) - 0.7 = 6.50$ s

EXHIBIT 20-12

$t_{c,NBR} = 6.2 + 1.0(0.10) = 6.30$ s

EQUATION 20-30

$t_{c,WBL} = 4.1 + 1.0(0.10) = 4.20$ s

Follow-up Headway: $t_{f,NBL} = 3.5 + 0.9(0.10) = 3.59$ s

EXHIBIT 20-13

$t_{f,NBR} = 3.3 + 0.9(0.10) = 3.39$ s
EQUATION 20-31

$t_{f,WBL} = 2.2 + 0.9(0.10) = 2.29$ s

3. Compute the potential and movement capacities for the northbound lane.

Potential Capacity: $c_{p,NBL} = 1140(e^{-(1140)(6.50)/3600})/1 - (e^{-(1140)(3.59)/3600}) = 214$ veh/h
EQUATION 20-32

$c_{p,NBR} = 598(e^{-(598)(6.30)/3600})/1 - (e^{-(598)(3.39)/3600}) = 488$ veh/h
$c_{p,WBL} = 652(e^{-(652)(4.20)/3600})/1 - (e^{-(652)(2.29)/3600}) = 898$ veh/h
Impedance $p_{o,WBL} = 1 - 54/898 = 0.94$ $f_{NBL} = 0.94$
EQUATIONS 20-42 AND 20-46

Movement Capacity: $c_{m,NBL} = 214(0.94) = 201$
EQUATION 20-47

Shared-Lane Capacity: $c_{NB} = (76 + 109)/((76/201) + (109/488)) = 308$ veh/h
EQUATION 20-41

$v/c_{NB} = (76 + 109)/308 = 0.60$

3. Compute delay for the northbound lane.

Delay: $d_{NB} = (3600/308) + 900(0.25)((0.60) - 1 +((0.60 - 1)^2 + ((3600/308)(0.60))/((450)(0.25)))^{0.5})$
$+ 5 = 32.8$ s/veh
EQUATION 20-64

4. Determine LOS for the northbound lane.

LOS: LOS = D (between 25 and 35)
EXHIBIT 20-2

HCS7

These screenshots show how Problem One is coded into *HCS7* in a series of screens to break the data into small, intuitive sections, followed by the results as presented in the formatted report. All input data, intermediate computations, and final results are shown in logical order, leading to delay and level of service.

Figure 4.70 *HCS7 TWSC1 Input*

Lanes

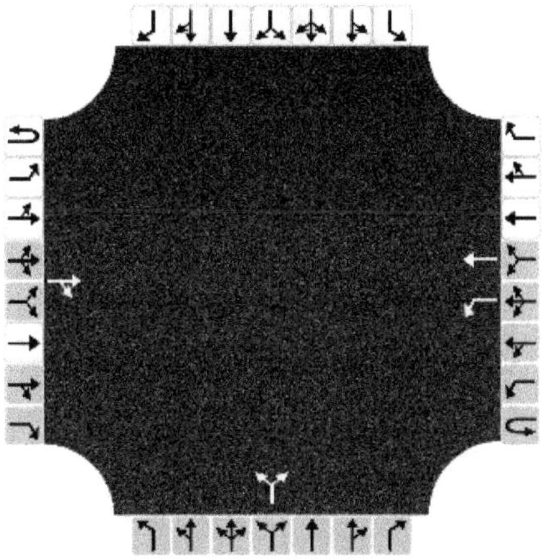

Major Street: East-West

	Eastbound	Westbound	Northbound	Southbound
Percent Grade (%)	0	0	0	0
Right Turn Channelized	Yes	Yes	Yes	Yes
Flared Minor-Street Approach	Yes	Yes	Yes	Yes
Flared Minor-Street Storage (veh)	1	1	1	1

Vehicle Volume and Adjustments

	Eastbound				Westbound				Northbound				Southbound			
	UTur	Left	Thru	Right	UTur	Left	Thru	Right	UTur	Left	Thru	Right	UTur	Left	Thru	Right
Priority	1U	1	2	3	4U	4	5	6		7	8	9		10	11	12
Volume (vph)	0	0	500	100	0	50	400	0	0	70	0	100	0	0	0	0
Percent Heavy Vehicles (%)	3	3	3	3	10	10	3	3	3	10	3	10	3	3	3	3
Short Left-Turn Pocket			Yes				Yes				Yes				Yes	
Left-Turn Storage			0				0				0				0	
Saturation Flow Rate (vph)			1800	1500			1800	1500			0	0			0	0
Percent Thrus Using Shared Lane (%)		50		50		50		50		50		50		50		50

Figure 4.71 *HCS7 TWSC1 Output*

Lanes

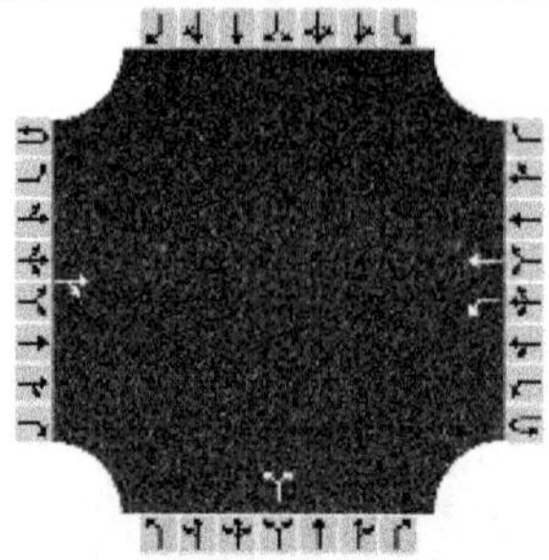

Vehicle Volumes and Adjustments

Approach	Eastbound				Westbound				Northbound				Southbound			
Movement	U	L	T	R	U	L	T	R	U	L	T	R	U	L	T	R
Priority	1U	1	2	3	4U	4	5	6		7	8	9		10	11	12
Number of Lanes	0	0	1	0	0	1	1	0		0	1	0		0	0	0
Configuration				TR		L	T				LR					
Volume, V (veh/h)			500	100		50	400			70		100				
Percent Heavy Vehicles (%)						10				10		10				
Proportion Time Blocked																
Percent Grade (%)									0							
Right Turn Channelized	No				No				No				No			
Median Type/Storage	Undivided															

Critical and Follow-up Headways

	U	L	T	R	U	L	T	R	U	L	T	R	U	L	T	R
Base Critical Headway (sec)						4.1				7.1		6.2				
Critical Headway (sec)						4.20				6.50		6.30				
Base Follow-Up Headway (sec)						2.2				3.5		3.3				
Follow-Up Headway (sec)						2.29				3.59		3.39				

Delay, Queue Length, and Level of Service

	U	L	T	R	U	L	T	R	U	L	T	R	U	L	T	R
Flow Rate, v (veh/h)						54					185					
Capacity, c (veh/h)						898					308					
v/c Ratio						0.06					0.60					
95% Queue Length, Q_{95} (veh)						0.2					3.6					
Control Delay (s/veh)						9.3					32.8					
Level of Service, LOS						A					D					
Approach Delay (s/veh)					1.0				32.8							
Approach LOS									D							

PROBLEM TWO

Building on Problem One, compute LOS with upstream signals that generate a Proportion Time Blocked value of 0.30 for the NBR movement.

1. Compute the modified conflicting flow rate for the northbound right-turn movement.

Conflicting Flow $v_{c,u,NBR}$ = (598 – 1.5(1000)0.30)/(1 – 0.30) = 211

EQUATION 20-33

2. Compute the revised capacity for the northbound right-turn movement and the lane.

Random Flow Capacity $c_{r,NBR} = 211(e^{-(211)(6.30)/3600})/1 - (e^{-(211)(3.39)/3600}) = 809$ veh/h
EQUATION 20-35

$c_{p,NBR} = (1 - 0.30)809 = 566$ veh/h
EQUATION 20-34

Shared-Lane Capacity $c_{NB} = (76 + 109)/((76/201) + (109/566)) = 324$ veh/h
EQUATION 20-59

$v/c_{NB} = (76 + 109)/324 = 0.57$

3. Compute revised delay for the northbound lane.

Delay $d_{NB} = (3600/324) + 900(0.25)((0.57) - 1 + ((0.57 - 1)^2 + ((3600/324)(0.57))/((450)(0.25)))^{0.5}) + 5 = 29.8$ s/veh
EQUATION 20-64

4. Determine LOS for the northbound lane.

LOS LOS = D (between 25 and 35)
EXHIBIT 20-2

HCS7

These screenshots show how Problem Two is coded into *HCS7* by adding data for Proportion Time Blocked within the Adjustment screen, followed by the updated results as presented in the formatted report. All input data, intermediate computations, and final results are shown in logical order, leading to delay and level of service.

All-Way Stop Control

The analysis of an AWSC intersection as defined in HCM Chapter 21 involves determining the probabilities of possible scenarios for the turning movement and lane configuration combinations, which are used in an iterative process

Figure 4.72 *HCS7 TWSC2 Input*

Upstream Signal Data				
Upstream Signal	☑ Yes			
	Left Thru Righ	Left Thru Righ	Left Thru Righ	Left Thru Righ
Proportion of Time Blocked	0.00 0.00 0.00	0.00 0.00 0.00	0.00 0.00 0.30	0.00 0.00 0.00

Figure 4.73 *HCS7 TWSC2 Output*

Vehicle Volumes and Adjustments																
Approach	Eastbound				Westbound				Northbound				Southbound			
Movement	U	L	T	R	U	L	T	R	U	L	T	R	U	L	T	R
Priority	1U	1	2	3	4U	4	5	6		7	8	9		10	11	12
Number of Lanes	0	0	1	0	0	1	1	0		0	1	0		0	0	0
Configuration				TR		L	T				LR					
Volume, V (veh/h)			500	100		50	400			70		100				
Percent Heavy Vehicles (%)						3				3		3				
Proportion Time Blocked						0.000				0.000		0.300				
Percent Grade (%)									0							
Right Turn Channelized	No				No				No				No			
Median Type/Storage	Undivided															
Critical and Follow-up Headways																
Base Critical Headway (sec)						4.1				7.1		6.2				
Critical Headway (sec)						4.13				6.43		6.23				
Base Follow-Up Headway (sec)						2.2				3.5		3.3				
Follow-Up Headway (sec)						2.23				3.53		3.33				
Delay, Queue Length, and Level of Service																
Flow Rate, v (veh/h)						54					185					
Capacity, c (veh/h)						929					334					
v/c Ratio						0.06					0.55					
95% Queue Length, Q_{95} (veh)						0.2					3.2					
Control Delay (s/veh)						9.1					28.3					
Level of Service, LOS						A					D					
Approach Delay (s/veh)					1.0				28.3							
Approach LOS									D							

to determine departure headway. With the flow rate for each movement, this can be converted into degree of utilization for computing delay and the 95th percentile queue for each considered lane. Level of service is determined for all lanes, then expanded to approaches and the intersection using flow-rate weighted averages.

Required Data

The required lane configuration includes the number of lanes on each approach with the movements allowed for each lane. Turning movement counts for all movements are needed with one peak hour factor applied to all movements and percent heavy vehicles for all movements.

Lane Configuration

The number of lanes (up to three) on each approach are identified with exclusive and shared lanes.

Flow Rate

Turning movement data are converted to flow rates by dividing each by the intersection's peak hour factor using HCM Equation 21-12.

Geometry Group

Geometry groups are identified for each approach as functions of the intersection configuration with number of lanes on the subject and opposing and conflicting approaches, following HCM Exhibit 21-11.

Figure 4.74 *HCM Equation 21-12*

$$v_i = \frac{V_i}{PHF}$$

where

v_i = demand flow rate for movement i (veh/h),

V_i = demand volume for movement i (veh/h), and

PHF = intersection peak-hour factor.

Saturation Headway

Saturation headway adjustments are computed for turns and heavy vehicles using HCM Equation 21-13 with values from HCM Exhibit 21-12 applied to the appropriate proportions.

Probability States

For each degree-of-conflict case, the probability is determined using HCM Equations 21-16 through 21-20, following HCM Exhibit 21-14. These are converted into adjustments using HCM Equations 21-21 through 21-26. The number of possible scenarios increases with the number of lanes on each approach—from 8 with single-lane approaches, to 64 with two-lane approaches, and to 512 with three-lane approaches.

Saturation Headway

The saturation headway is the time between success vehicles departing for a given case with a queue in place. Saturation headway is a sum of the base values from HCM Exhibit 21-15 applying the adjustment using HCM Equation 21-27.

Figure 4.75 *HCM Exhibit 21-11*

Intersection Configuration	Number of Lanes: Subject Approach	Opposing Approach	Conflicting Approaches[a]	Geometry Group
Four leg or T	1	0 or 1	1	1
Four leg or T	1	0 or 1	2	2
Four leg or T	1	2	1	3a/4a
T	1	2	2	3b
Four leg	1	2	2	4b
Four leg or T	1	0 or 1	3	5
	1	3	1	
	2	0, 1, or 2	1 or 2	
	3	0 or 1	1	
	3	0 or 1	2 or 3	
	3	2 or 3	1	
Four leg or T	1	3	2	6
	1	2	3	
	1	3	3	
	2	3	1, 2, or 3	
	2	0, 1, 2 or 3	3	
	3	2 or 3	2 or 3	

Figure 4.76 *HCM Equation 21-13*

$$h_{adj} = h_{LT,adj}P_{LT} + h_{RT,adj}P_{RT} + h_{HV,adj}P_{HV}$$

where

h_{adj} = headway adjustment (s),

$h_{LT,adj}$ = headway adjustment for left turns (see Exhibit 21-12) (s),

$h_{RT,adj}$ = headway adjustment for right turns (see Exhibit 21-12) (s),

$h_{HV,adj}$ = headway adjustment for heavy vehicles (see Exhibit 21-12) (s),

P_{LT} = proportion of left-turning vehicles in the lene,

P_{RT} = proportion of right-turning vehicles in the lane, and

P_{HV} = proportion of heavy vehicles in the lane.

Figure 4.77 *HCM Exhibit 21-12*

Factor	Saturation Headway Adjustment (s): Group 1	Group 2	Group 3a	Group 3b	Group 4a	Group 4b	Group 5	Group 6
LT	0.2	0.2	0.2	0.2	0.2	0.2	0.5	0.5
RT	−0.6	−0.6	−0.6	−0.6	−0.6	−0.6	−0.7	−0.7
HV	1.7	1.7	1.7	1.7	1.7	1.7	1.7	1.7

Figure 4.78 *HCM Equation 21-16*

$$P(C_1) = P(1)$$

Figure 4.79 *HCM Equation 21-17*

$$P(C_2) = \sum_{i=2}^{4} P(i)$$

Figure 4.80 *HCM Equation 21-18*

$$P(C_3) = \sum_{i=5}^{10} P(i)$$

Figure 4.81 *HCM Equation 21-19*

$$P(C_4) = \sum_{i=11}^{37} P(i)$$

Figure 4.82 *HCM Equation 21-20*

$$P(C_5) = \sum_{i=38}^{64} P(i)$$

Figure 4.83 *HCM Exhibit 21-14*

i	Doc Case	Number of Vehicles	Opposing Approach L1	Opposing Approach L2	Conflicting Left Approach L1	Conflicting Left Approach L2	Conflicting Right Approach L1	Conflicting Right Approach L2
1	1	0	0	0	0	0	0	0
2	2	1	1	0	0	0	0	0
3			0	1	0	0	0	0
4		2	1	1	0	0	0	0
5	3	1	0	0	1	0	0	0
6			0	0	0	1	0	0
7			0	0	0	0	1	0
8			0	0	0	0	0	1
9		2	0	0	1	1	0	0
10			0	0	0	0	1	1
11	4	2	0	0	0	1	0	1
12			0	0	1	0	0	1
13			0	0	1	0	1	0
14			0	0	0	1	1	0
15			0	1	0	1	0	0
16			1	0	1	0	0	0
17			0	1	0	0	1	0
18			1	0	0	1	0	0
19			0	1	1	0	0	0
20			0	1	0	0	0	1
21			1	0	0	0	1	0
22			1	0	0	0	0	1
23		3	0	0	0	1	1	1
24			0	0	1	1	0	1
25			0	0	1	1	1	0
26			1	0	1	1	0	0
27			1	1	1	0	0	0
28			1	1	0	0	1	0
29			1	1	0	0	0	1
30			0	1	1	1	0	0
31			1	0	0	0	1	1
32			0	0	1	0	1	1
33			1	1	0	1	0	0
34			0	1	0	0	1	1
35		4	1	1	0	0	1	1
36			0	0	1	1	1	1
37			1	1	1	1	0	0
38	5	3	0	1	0	1	0	1
39			1	0	0	1	1	0
40			0	1	1	0	1	0
41			0	1	0	1	1	0
42			0	1	1	0	0	1
43			1	0	1	0	0	1
44			1	0	0	1	0	1
45			1	0	1	0	1	0
46		4	1	0	0	1	1	1
47			0	1	1	1	1	0
48			0	1	1	1	0	1
49			1	0	1	0	1	1
50			1	0	1	1	1	0
51			0	1	0	1	1	1
52			1	1	1	0	0	1
53			1	0	1	1	0	1
54			0	1	1	0	1	1
55			1	1	0	1	1	0
56			1	1	0	1	0	1
57			1	1	1	0	1	0
58		5	1	0	1	1	1	1
59			1	1	0	1	1	1
60			1	1	1	0	1	1
61			0	1	1	1	1	1
62			1	1	1	1	1	1
63			1	1	1	1	0	0
64		6	1	1	1	1	1	1

Figure 4.84 *HCM Equation 21-21*

$$AdjP(1) = \alpha[P(C_2) + 2P(C_3) + 3P(C_4) + 4P(C_5)$$

Figure 4.85 *HCM Equation 21-22*

$$AdjP(2) \text{ through } AdjP(4) = \alpha[P(C_3) + 2P(C_4) +$$

Figure 4.86 *HCM Equation 21-23*

$$AdjP(5) \text{ through } AdjP(10) = \alpha[P(C_4) + 2P(C_5) -$$

Figure 4.87 *HCM Equation 21-24*

$$AdjP(11) \text{ through } Adj\,P(37) = \alpha[P(C_5) - 6P(C_4)]/27$$

Figure 4.88 *HCM Equation 21-25*

$$AdjP(38) \text{through } AdjP(64) = -\alpha[10P(C_5)]/27$$

Figure 4.89 *HCM Equation 21-26*

$$P'(i) = P(i) + AdjP(i)$$

Departure Headway

The departure headway is the average time between successive vehicles on a given approach.

Starting with an initial departure headway value of 3.2 s, the final departure headway is computed through iterative process using the probabilities of each possible scenario for a given set of lanes existing, following HCM Equation

Figure 4.91 *HCM Equation 21-27*

$$h_{si} = h_{base} + h_{adj}$$

21-28. The calculated headway replaces the initial headway until they converge to within 0.1 s.

Service Time

Service time is computed following HCM Equation 21-29 using the final departure headway value with the move-up time (2.0 s for geometry groups 1 through 4 and 2.3s for geometry groups 5 and 6 from HCM Exhibit 21-11).

Figure 4.92 *HCM Equation 21-28*

$$h_d = \sum_{i=1}^{64} P'(i)h_{si}$$

Figure 4.90 *HCM Exhibit 21-15*

		Base Saturation Headway, h_{base} (s)							
Case	No. of Veh.	Group 1	Group 2	Group 3a	Group 3b	Group 4a	Group 4b	Group 5	Group 6
1	0	3.9	3.9	4.0	4.3	4.0	4.5	4.5	4.5
2	1	4.7	4.7	4.8	5.1	4.8	5.3	5.0	6.0
	2							6.2	6.8
	≥3								7.4
3	1	5.8	5.8	5.9	6.2	5.9	6.4	6.4	6.6
	2							7.2	7.3
	≥3								7.8
4	2	7.0	7.0	7.1	7.4	7.1	7.6	7.6	8.1
	3							7.8	8.7
	4							9.0	9.6
	≥5								12.3
5	3	9.6	9.6	9.7	10.0	9.7	10.2	9.7	10.0
	4							9.7	11.1
	5							10.0	11.4
	≥6							11.5	13.3

Degree of Utilization

The degree of utilization is computed from HCM Equation 21-14 for use in calculating queue and delay.

Figure 4.94 *HCM Equation 21-14*

$$x = \frac{vh_d}{3{,}600}$$

Queue

Figure 4.93 *HCM Equation 21-29*

$$t_s = h_d - m$$

where t_s is the service time, h_d is the departure headway, and m is the move-up time (2.0 s for Geometry Groups 1 through 4; 2.3 s for Geometry Groups 5 and 6.

The 95th percentile queue can be computed for each lane according to HCM Equation 21-33.

Delay

Delay for each lane is computed following HCM Equation 21-30. Approach and intersection delays are computed using HCM Equations 21-31 and 21-32, respectively.

LOS

Figure 4.95 *HCM Equation 21-33*

$$Q_{95} \approx \frac{900T}{h_d}\left[(x-1)+\sqrt{(x-1)^2+\frac{h_d x}{150T}}\right]$$

where

Q_{95} = 95th percentile queue (veh),

$x = vh_d/3{,}600$ = degree of utilization (unitless),

h_d = depature headway (s), and

T = length of analysis period (h).

Level of service is determined in HCM Exhibit 21-8 for all lanes, approaches, and the intersection. Any lane with a volume-to-capacity ratio greater than one is assigned LOS F, regardless of its delay.

Figure 4.93a *HCM Exhibit 21-11*

Intersection Configuration	Number of Lanes Subject Approach	Opposing Approach	Conflicting Appoaches[a]	Geometry Group
Four leg or T	1	0 or 1	1	1
Four leg or T	1	0 or 1	2	2
Four leg or T	1	2	1	3a/4a
T	1	2	2	3b
Four leg	1	2	2	4b
Four leg or T	1	0 or 1	3	5
	1	3	1	
	2	0, 1, or 2	1 or 2	
	3	0 or 1	1	
	3	0 or 1	2 or 3	
	3	2 or 3	1	
Four leg or T	1	3	2	6
	1	2	3	
	1	3	3	
	2	3	1, 2, or 3	
	2	0, 1, 2 or 3	3	
	3	2 or 3	2 or 3	

Figure 4.96 *HCM Equation 21-30*

$$d = t_s + 900T\left[(x-1) + \sqrt{(x-1)^2 + \frac{h_d x}{450T}}\right] + 5$$

where

d = average control delay (s/veh),

$x = vh_d/3{,}600$ = degree of utilization (unitless),

t_s = service time (s),

h_d = departure headway (s), and

T = length of analysis period (h).

Figure 4.97 *HCM Equation 21-31*

$$d_a = \frac{\sum d_i v_i}{\sum v_i}$$

where

d_a = control delay for the approach (s/veh),

d_i = control delay for lane i (s/veh), and

v_i = flow rate for lane i (veh/h).

Figure 4.98 *HCM Equation 21-32*

$$d_{\text{intersection}} = \frac{\sum d_a v_a}{\sum v_a}$$

where

$d_{\text{intersection}}$ = control delay for the entire intersection (s/veh),

d_a = control delay for approach a (s/veh), and

v_a = flow rate for approach a (veh/h).

PROBLEM ONE

Since the procedures for this analysis do not lend themselves to performing manually, this problem is presented for an *HCS7* solution only. Determine the LOS for an intersection with all single-lane approaches with the following traffic volume (veh/h) data: EBL = 10; EBT = 100; EBR = 20; WBL = 30; WBT = 300; WBR = 40; NBL = 20; NBT = 200; NBR = 30; SBL = 40; SBT = 400; and SBR = 50. The peak hour factor is 0.92 with 2% heavy vehicles in all movements.

Figure 4.99 *HCM Exhibit 21-8*

Control Delay (s/veh)	Los by Volume-to-Capacity Ratio*	
	$v/c \le 1.0$	$v/c > 1.0$
0–10	A	F
>10–15	B	F
>15–25	C	F
>25–35	D	F
>35–50	E	F
>50	F	F

HCS7

These screenshots show how Problem One is coded into *HCS7* in two screens to break the data into small, intuitive sections, followed by the results as presented in the formatted report. All input data, intermediate computations, and final results are shown in logical order, leading to delay and level of service. Notice the delay and LOS results with single-lane approaches.

PROBLEM TWO

HCS7

These screenshots show how Problem Two is coded into *HCS7* to expand the number of lanes on each approach in two screens to break the data into small, intuitive sections, followed by the results as presented in the formatted report. All input data, intermediate computations, and final results are shown in logical order, leading to delay and level of service. Notice the delay and LOS results with three-lane approaches.

Figure 4.100 *HCS7 AWSC1 Input*

Quick Lanes

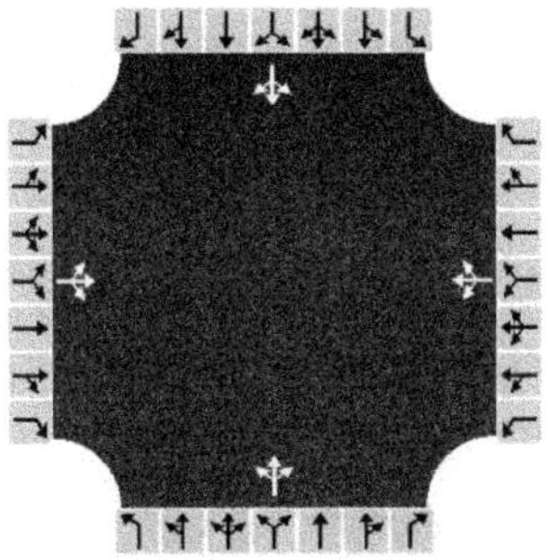

Vehicle Volume and Adjustments

	Eastbound			Westbound			Northbound			Southbound		
	Left	Thru	Right	Left	Thru	Right	Left	Thru	Right	Left	Thru	Right
Volume (vph)	10	100	20	30	300	40	20	200	30	40	400	50
Percent Thrus Using Shared Lane (%)												
	Lane 1	Lane 2	Lane 3	Lane 1	Lane 2	Lane 3	Lane 1	Lane 2	Lane 3	Lane 1	Lane 2	Lane 3
Percent Heavy Vehicles (%)	2			2			2			2		

Figure 4.101 *HCS7 AWSC1 Output*

Vehicle Volume and Adjustments												
Approach	Eastbound			Westbound			Northbound			Southbound		
Movement	L	T	R	L	T	R	L	T	R	L	T	R
Volume	10	100	20	30	300	40	20	200	30	40	400	50
% Thrus in Shared Lane												
Lane	L1	L2	L3	L1	L2	L3	L1	L2	L3	L1	L2	L3
Configuration	LTR			LTR			LTR			LTR		
Flow Rate, v (veh/h)	141			402			272			533		
Percent Heavy Vehicles	2			2			2			2		
Departure Headway and Service Time												
Initial Departure Headway, hd (s)	3.20			3.20			3.20			3.20		
Initial Degree of Utilization, x	0.126			0.357			0.242			0.473		
Final Departure Headway, hd (s)	8.11			7.16			7.41			6.74		
Final Degree of Utilization, x	0.318			0.800			0.559			0.997		
Move-Up Time, m (s)	2.0			2.0			2.0			2.0		
Service Time, ts (s)	6.11			5.16			5.41			4.74		
Capacity, Delay and Level of Service												
Flow Rate, v (veh/h)	141			402			272			533		
Capacity	444			503			486			534		
95% Queue Length, Q_{95} (veh)	1.4			7.5			3.4			14.0		
Control Delay (s/veh)	14.8			33.0			19.4			64.0		
Level of Service, LOS	B			D			C			F		
Approach Delay (s/veh)	14.8			33.0			19.4			64.0		
Approach LOS	B			D			C			F		
Intersection Delay, s/veh | LOS	40.6						E					

Figure 4.102 *HCS7 AWSC2 Input*

Vehicle Volume and Adjustments

	Eastbound			Westbound			Northbound			Southbound		
	Left	Thru	Right	Left	Thru	Right	Left	Thru	Right	Left	Thru	Right
Volume (vph)	10	100	20	30	300	40	20	200	30	40	400	50
Percent Thrus Using Shared Lane (%)												
	Lane 1	Lane 2	Lane 3	Lane 1	Lane 2	Lane 3	Lane 1	Lane 2	Lane 3	Lane 1	Lane 2	Lane 3
Percent Heavy Vehicles (%)	2	2	2	2	2	2	2	2	2	2	2	2

Figure 4.103 *HCS7 AWSC2 Output*

Vehicle Volume and Adjustments												
Approach	Eastbound			Westbound			Northbound			Southbound		
Movement	L	T	R	L	T	R	L	T	R	L	T	R
Volume	10	100	20	30	300	40	20	200	30	40	400	50
% Thrus in Shared Lane												
Lane	L1	L2	L3	L1	L2	L3	L1	L2	L3	L1	L2	L3
Configuration	L	T	R	L	T	R	L	T	R	L	T	R
Flow Rate, v (veh/h)	11	109	22	33	326	43	22	217	33	43	435	54
Percent Heavy Vehicles	2	2	2	2	2	2	2	2	2	2	2	2
Departure Headway and Service Time												
Initial Departure Headway, hd (s)	3.20	3.20	3.20	3.20	3.20	3.20	3.20	3.20	3.20	3.20	3.20	3.20
Initial Degree of Utilization, x	0.010	0.097	0.019	0.029	0.290	0.039	0.019	0.193	0.029	0.039	0.386	0.048
Final Departure Headway, hd (s)	9.00	8.50	7.80	8.28	7.78	7.08	8.54	8.04	7.34	7.92	7.42	6.72
Final Degree of Utilization, x	0.027	0.257	0.047	0.075	0.704	0.085	0.052	0.485	0.066	0.096	0.896	0.101
Move-Up Time, m (s)	2.3	2.3	2.3	2.3	2.3	2.3	2.3	2.3	2.3	2.3	2.3	2.3
Service Time, ts (s)	6.70	6.20	5.50	5.98	5.48	4.78	6.24	5.74	5.04	5.62	5.12	4.42
Capacity, Delay and Level of Service												
Flow Rate, v (veh/h)	11	109	22	33	326	43	22	217	33	43	435	54
Capacity	400	423	461	435	463	509	422	448	491	454	485	536
95% Queue Length, Q_{95} (veh)	0.1	1.0	0.1	0.2	5.4	0.3	0.2	2.6	0.2	0.3	10.0	0.3
Control Delay (s/veh)	12.0	14.1	10.9	11.6	27.0	10.4	11.7	18.1	10.6	11.5	46.3	10.2
Level of Service, LOS	B	B	B	B	D	B	B	C	B	B	E	B
Approach Delay (s/veh)	13.5			23.9			16.7			39.8		
Approach LOS	B			C			C			E		
Intersection Delay, s/veh | LOS	27.6						D					

CHAPTER 5 Roundabouts

Roundabouts can be modeled as isolated intersections or part of an urban street that includes the segments between them. As an intersection, the procedure models all movements on one- or two-lane approaches to one or two circulating lanes through the roundabout, including yielding or non-yielding bypass lanes. Along an urban street, the procedure is expanded to include the running time along the segments and geometric delay through the roundabout intersections.

Intersections

The analysis of a roundabout intersection as defined in HCM Chapter 22 involves determining entering and conflicting flow rates for all considered movements to compute capacity, with exiting flow rates required for yielding bypass lanes. The volume-to-capacity ratios can be computed with the appropriate flow rates for use in calculating delay and the 95th percentile queue for each considered lane. Level of service is determined for all lanes, approaches, and the intersection.

Required Data

The required lane configuration includes the number of lanes on each approach with the movements allowed for each lane and any bypass lanes defined as yielding or non-yielding. Turning movement counts for all movements with percent heavy vehicles are needed with one peak hour factor applied to all movements. Turning movement data are converted to flow rates by dividing each by the intersection's peak hour factor using HCM Equation 22-8, then applying the heavy vehicle proportion with the passenger-car equivalent using HCM Equations 22-9 and 22-10 with HCM Exhibit 22-11.

Figure 5.1 *HCM Equation 22-8*

$$v_i = \frac{V_i}{PHF}$$

where

v_i = demand flow rate for movement *i* (veh/h),

V_i = demand volume for movement *i* (veh/h), and

PHF = peak hour factor.

Figure 5.2 *HCM Equation 22-9*

$$v_{i,pce} = \frac{v_i}{f_{HV}}$$

Figure 5.3 *HCM Equation 22-10*

$$f_{HV} = \frac{1}{1 + P_T(E_T - 1)}$$

where

$v_{i,pce}$ = demand flow rate for movement *i* (pc/h),

v_i = demand flow rate for movement *i* (veh/h),

f_{HV} = heavy-vehicle adjustment factor,

P_T = proportion of demand volume that consists of heavy vehicles, and

E_T = passenger car equivalent for heavy vehicles.

Figure 5.4 *HCM Exhibit 22-11*

Vehicle Type	Passenger Car Equivalent, E_T
Passenger car	1.0
Heavy Vehicle	2.0

Conflicting Flows

For most movements, the circulating flow is the conflicting flow as illustrated in HCM Exhibit 22-12 and computed in HCM Equation 22-11. For yielding bypass lanes, the exiting flow is the conflicting flow as illustrated in HCM Exhibit 22-13 and computed in HCM Equation 22-12. For non-yielding bypass lanes, there is no conflicting flow, capacity is not computed, and delay is assumed to be zero.

Figure 5.5 *HCM Exhibit 22-12*

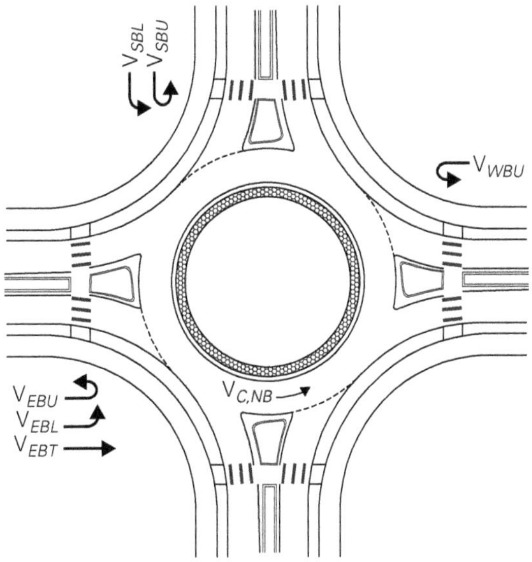

Figure 5.6 *HCM Equation 22-11*

$$v_{c,NB,pce} = v_{WBU,pce} + v_{SBL,pce} + v_{SBL,pce} + v_{EBT,pce} + v_{EBL,pce} + v_{EBU,pce}$$

Figure 5.7 *HCM Exhibit 22-13*

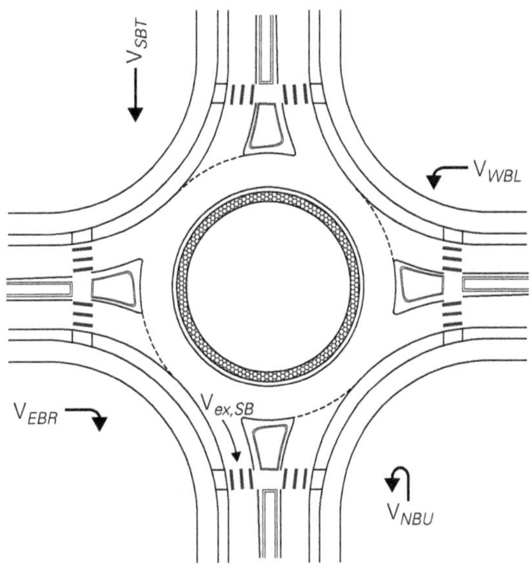

Figure 5.8 *HCM Equation 22-12*

$$v_{ex,pce} = v_{NBU,pce} + v_{WBL,pce} + v_{SBT,pce} + v_{EBR,pce} - v_{EBR,pce,bypass}$$

Lane Assignments

The relative proportions of movement flow rates can impact how a lane is used, so the procedure models this for two-lane approaches. Even though movements are allowed in a particular lane, drivers may choose to avoid that lane, creating de facto lane configurations that function differently from the striping—maybe by time of day when proportions change. HCM Exhibit 22-14 provides rules for implementing these modifications. Also, field data may indicate specific percentages for particular movement flow rates by lane, which can be used according to HCM Exhibit 22-15.

Figure 5.9 *HCM Exhibit 22-14*

Designated Lane Assignment	Assumed Lane Assignment
LT, TR	If $v_U + v_L > v_T + v_{R,e}$: L, TR (de facto left-turn lane) If $v_{R,e} > v_U + v_L + v_T$: LT, R (de facto right-turn lane) Else LT, TR
L, LTR	If $v_T + v_{R,e} > v_U + v_L$: L, TR (de facto through-right lane) Else L, LTR
LTR, R	If $v_U + v_L + v_T > v_{R,e}$: LT, R (de facto left-through lane) Else LTR, R

Figure 5.10 *HCM Exhibit 22-15*

Case	Assumed Lane Assignment	Left Lane	Right Lane
1	L, TR	$v_U + v_L$	$V_T + v_{R,e}$
2	LT, R	$v_U + v_L + v_T$	$v_{R,e}$
3	LT, TR	$(\%LL)v_e$	$(\%RL)v_e$
4	L, LTR	$(\%LL)v_e$	$(\%RL)v_e$
5	LTR, R	$(\%LL)v_e$	$(\%RL)v_e$

Capacity

Capacity is computed using the appropriate equation assigned to each of seven entry and conflicting lane combinations. HCM Exhibits 22-2 through 22-5 and 22-7 illustrate the various lane combinations with HCM Equations 22-1 through 22-7 applied to each.

Figure 5.11 *HCM Exhibit 22-2*

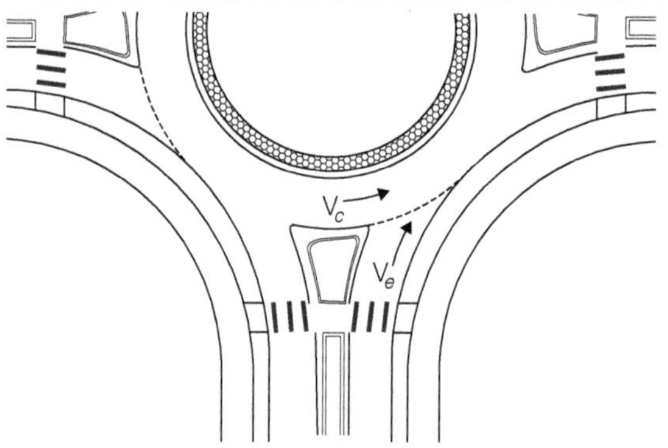

Figure 5.12 *HCM Equation 22-1*

$$c_{e,pce} = 1{,}380e^{(-1.02\times10^{-3})v_{c,pce}}$$

where

$c_{e,pce}$ = lane capacity, adjusted for heavy vehicles (pc/h), and

$v_{c,pce}$ = conflicting flow rate (pc/h).

Figure 5.13 *HCM Exhibit 22-3*

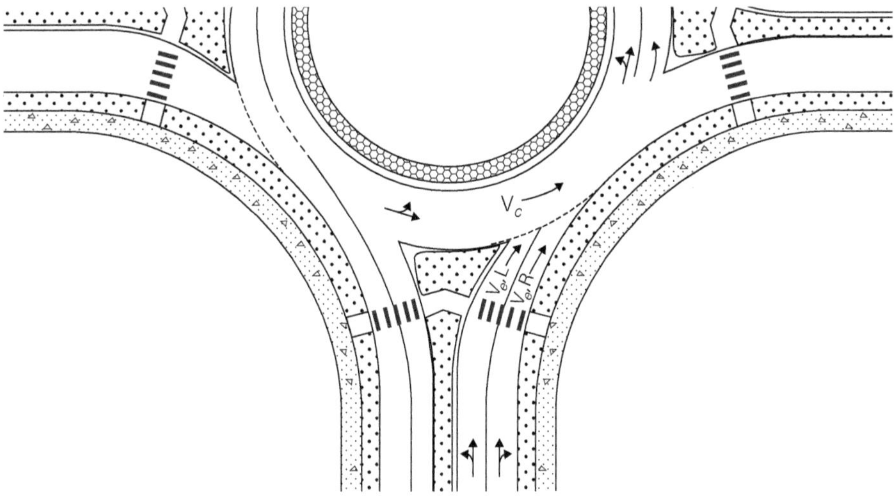

Figure 5.14 *HCM Equation 22-2*

$$c_{e,pce} = 1{,}420e^{(-0.91\times10^{-3})v_{c,pce}}$$

Figure 5.15 *HCM Exhibit 22-4*

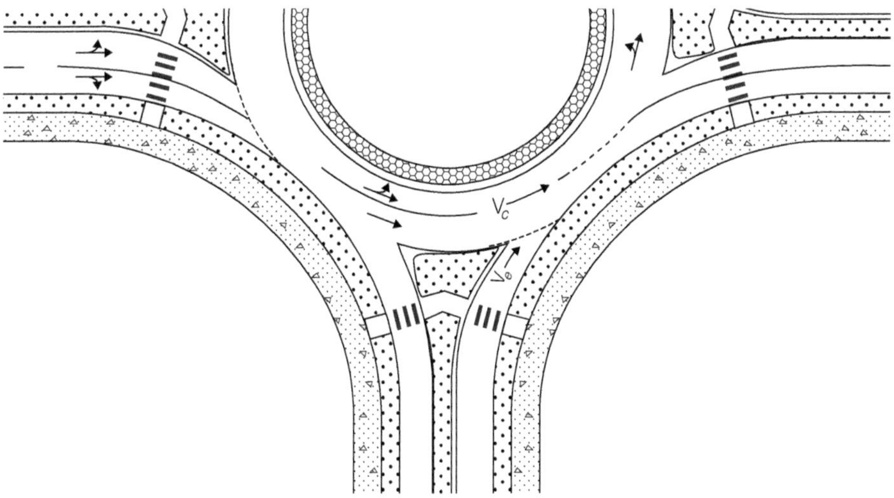

Figure 5.16 *HCM Equation 22-3*

$$c_{e,pce} = 1{,}420e^{(-0.85\times10^{-3})v_{c,pce}}$$

Figure 5.17 *HCM Exhibit 22-5*

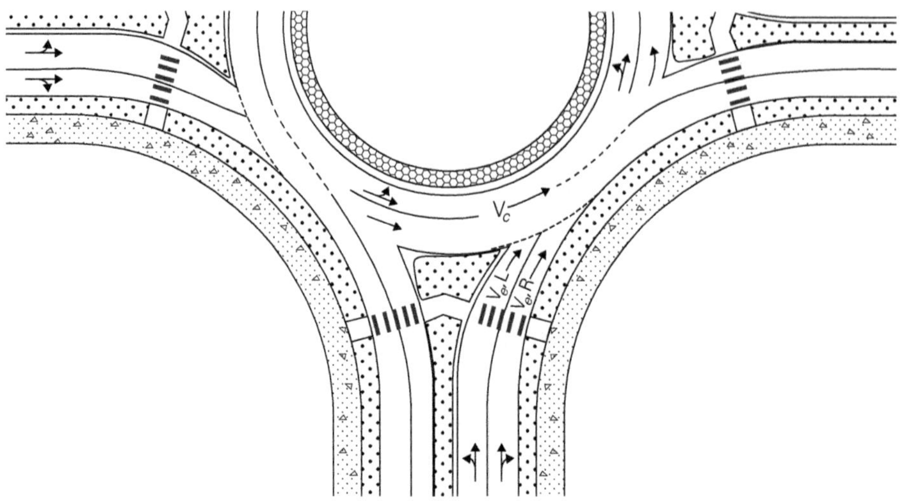

Figure 5.18 *HCM Equation 22-4*

$$c_{e,R,pce} = 1{,}420e^{(-0.85\times10^{-3})v_{c,pce}}$$

Figure 5.19 *HCM Equation 22-5*

$$c_{e,L,pce} = 1{,}350e^{(-0.92\times10^{-3})v_{c,pce}}$$

where

$c_{e,R,pce}$ = capacity of the right entry lane, adjusted for heavy vehicles (pc/h),
$c_{e,L,pce}$ = capacity of the left entry lane, adjusted for heavy vehicles (pc/h), and
$v_{c,pce}$ = conflicting flow rate (total of both lanes) (pc/h).

Figure 5.20 *HCM Exhibit 22-7*

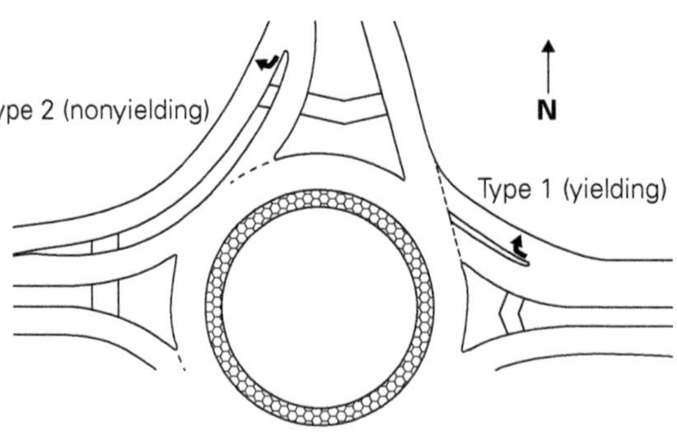

Figure 5.21 *HCM Equation 22-6*

$$c_{bypass,pce} = 1{,}380e^{(-1.02\times10^{-3})v_{ex,pce}}$$

Figure 5.22 *HCM Equation 22-7*

$$c_{\text{bypass},pce} = 1{,}420e^{(-0.85\times10^{-3})v_{ex,pce}}$$

where

$c_{\text{bypass},pce}$ = capacity of the bypass lane, adjusted for heavy vehicles (pc/h); and

$v_{ex,pce}$ = conflicting exiting flow rate (pc/h).

Pedestrians

Pedestrians can impact a vehicle's ability to enter the roundabout, but only if the conflicting flow is low enough so as not to require a stop, allowing the pedestrian to walk behind the spotted vehicle, causing no impact. Otherwise, this impact results in a reduction in capacity quantified as a function of the number of pedestrians

Figure 5.23 *HCM Exhibit 22-18*

Case	One-Lane Entry Capacity Adjustment Factor for Pedestrians
If $v_{c,pce} > 881$	$f_{ped} = 1$
Else of $n_{ped} \le 101$	$f_{ped} = 1 - 0.000137n_{ped}$
Else	$f_{ped} = \dfrac{1{,}119.5 - 0.715v_{c,pce} - 0.644n_{ped} + 0.00073v_{c,pce}n_{ped}}{1{,}068.6 - 0.654v_{c,pce}}$

where

f_{ped} = entry capacity adjustment factor for pedestrians,

n_{ped} = number of conflicting pedestrians per hour (p/h), and

$v_{c,pce}$ = conflicting vehicular flow rate in the circulatory roadway (pc/h).

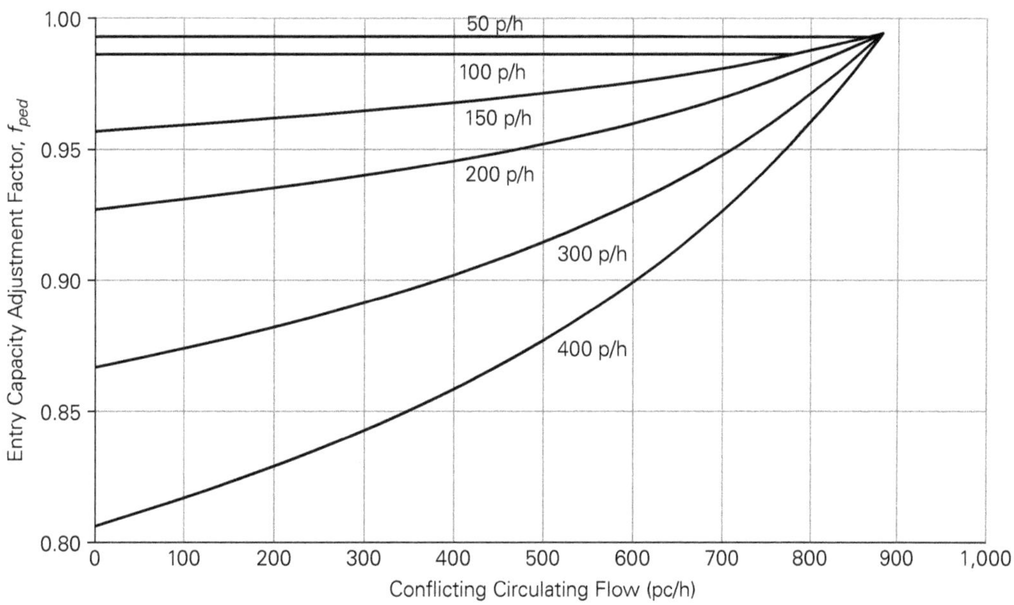

Figure 5.24 *HCM Exhibit 22-19*

Figure 5.25 *HCM Exhibit 22-20*

Case	Two-Lane Entry Capacity Adjustment Factor for Pedestrians
If $n_{ped} < 100$	$f_{ped} = \min\left[1 - \frac{n_{ped}}{100}\left(1 - \frac{1{,}260.6 - 0.329v_{c,pce} - 0.381 \times 100}{1{,}380 - 0.5v_{c,pce}}\right), 1\right]$
Else	$f_{ped} = \min\left[\frac{1{,}260.6 - 0.329v_{c,pce} - 0.381n_{ped}}{1{,}380 - 0.5v_{c,pce}}, 1\right]$

where

f_{ped} = entry capacity adjustment factor for pedestrians,

n_{ped} = number of conflicting pedestrians (p/h), and

$v_{c,pce}$ = conflicting vehicular flow rate in the circulatory roadway (pc/h).

Figure 5.26 *HCM Exhibit 22-21*

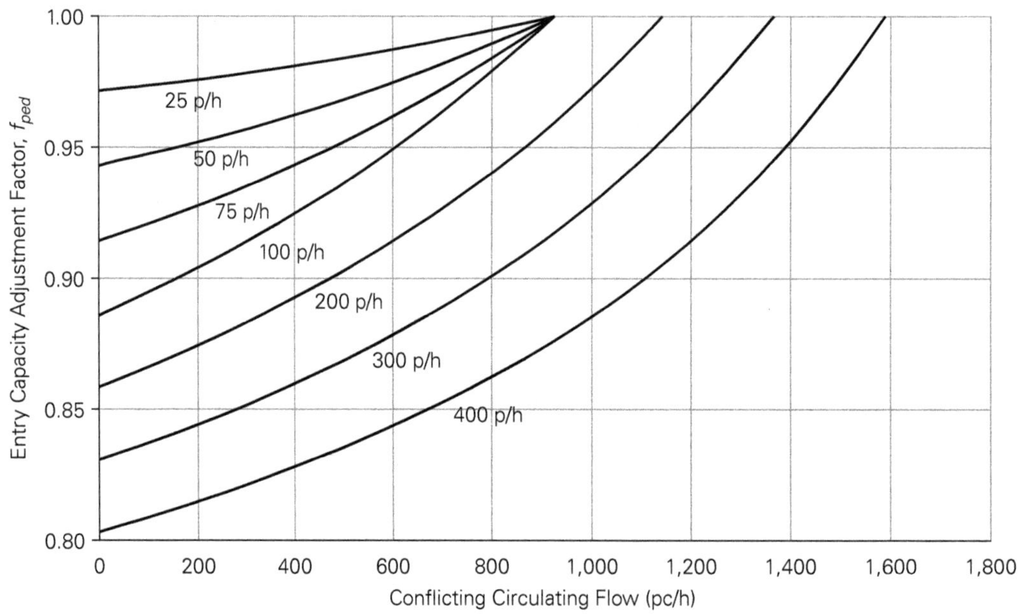

and the number of entry lanes as defined in HCM Exhibits 22-18 and 22-19 for one lane and HCM Exhibits 22-20 and 22-21 for two lanes.

Queue

The 95th percentile queue can be computed for each lane according to HCM Equation 22-20.

Figure 5.27 *HCM Equation 22-20*

$$Q_{95} = 900T\left[x - 1 + \sqrt{(1-x)^2 + \frac{\left(\frac{3{,}600}{c}\right)x}{150T}}\right]\left(\frac{c}{3{,}600}\right)$$

where

Q_{95} = 95th percentile queue (veh),

x = volume-to-capacity ratio of the subject lane,

c = capacity of the subject lane (veh/h), and

T = time period (h) (T = 1 for a 1-h analysis, T = 0.25 for a 15-min analysis).

Delay

Control delay is computed for each lane following HCM Equation 22-17 using volume-to-ratios from HCM Equations 22-13 through 22-16 that accommodate conversion from passenger cars to vehicles. These values are combined for approach and intersection averages according to HCM Equations 22-18 and 22-19 with the analysis period equal to 0.25 hours (for 15 minutes).

Figure 5.28 *HCM Equation 22-13*

$$v_i = v_{i,PCE} f_{HV,e}$$

where

v_i = flow rate for lane i (veh/h),

$v_{i,PCE}$ = flow rate for lane i (pc/h), and

$v_{HV,e}$ = heavy-vehicle adjustment factor for the lane (see below).

Figure 5.29 *HCM Equation 22-14*

$$c_i = c_{i,PCE} f_{HV,e} f_{ped}$$

where

c_i = capacity for lane i (veh/h),

$c_{i,PCE}$ = capacity for lane i (pc/h),

$f_{HV,e}$ = heavy-vehicle adjustment factor for the lane (see below), and

f_{ped} = pedestrian impedance factor.

Figure 5.30 *HCM Equation 22-15*

$$f_{HV,e} = \frac{f_{HV,U}v_{U,PCE} + f_{HV,L}v_{L,PCE} + f_{HV,T}v_{T,PCE} + f_{HV,R,e}v_{R,e,PCE}}{v_{U,PCE} + v_{L,PCE} + v_{T,PCE} + v_{R,e,PCE}}$$

where

$f_{HV,e}$ = heavy-vehicle adjustment factor for the entry lane,

$f_{HV,i}$ = heavy-vehicle adjustment factor for movement i, and

$v_{i,PCE}$ = demand flow rate for movement i (pc/h).

Figure 5.31 *HCM Equation 22-16*

$$x_i = \frac{v_i}{c_i}$$

where

x_i = volume-to-capacity ratio of the subject lane i,

v_i = demand flow rate of the subject lane i (veh/h), and

c_i = capacity of the subject lane i (veh/h).

Figure 5.32 *HCM Equation 22-17*

$$d = \frac{3{,}600}{c} + 900T\left[x - 1 + \sqrt{(x-1)^2 + \frac{\left(\frac{3{,}600}{c}\right)x}{450T}}\right] + 5 \times \min[x, 1]$$

where

d = average control delay (s/veh),

x = volume-to-capacity ratio of the subject lane,

c = capacity of the subject lane (veh/h), and

T = time period (h) (T = 0.25 h for a 15-min analysis).

Figure 5.33 *HCM Equation 22-18*

$$d_{\text{approach}} = \frac{d_{LL}v_{LL} + d_{RL}v_{RL} + d_{\text{bypass}}v_{\text{bypass}}}{v_{LL} + v_{RL} + v_{\text{bypass}}}$$

Figure 5.34 *HCM Equation 22-19*

$$d_{\text{intersection}} = \frac{\sum d_i v_i}{\sum v_i}$$

where

$d_{\text{intersection}}$ = control delay for the entire intersection (s/veh),

d_i = control delay for approach i (s/veh), and

v_i = flow rate for approach i (veh/h).

LOS

Level of service is determined in HCM Exhibit 22-8 for all lanes, approaches, and the intersection. Any lane with a volume-to-capacity ratio greater than one is assigned LOF, regardless of its delay.

Figure 5.35 *HCM Exhibit 22-8*

Control Delay (s/veh)	LOS by Volume-to-Capacity Ratio[a] v/c ≤ 1.0	v/c > 1.0
0–10	A	F
>10–15	B	F
>15–25	C	F
>25–35	D	F
>35–50	E	F
>50	F	F

PROBLEM ONE

Determine the northbound lane LOS for a single-lane roundabout with single-lane approaches; volumes (veh/h) are EBL = 50; EBT = 300; EBR = 20; WBL = 60; WBT = 400; WBR= 40; NBL = 30; NBT = 500; NBR = 250; SBL = 10; SBT = 300; and SBR = 70. The peak hour factor is 0.90 with 2% heavy vehicles in all movements and no U-turns or pedestrians.

1. Compute the flow rates for all movements.

Flow Rate

EQUATION 22-8

$v_{EBL} = 50/0.90 = 55$ veh/h $v_{EBT} = 300/0.90 = 333$ veh/h $v_{EBR} = 20/0.90 = 22$ veh/h
$v_{WBL} = 60/0.90 = 67$ veh/h $v_{WBT} = 400/0.90 = 444$ veh/h $v_{WBR} = 40/0.90 = 44$ veh/h
$v_{NBL} = 30/0.90 = 33$ veh/h $v_{NBT} = 500/0.90 = 556$ veh/h $v_{NBR} = 250/0.90 = 278$ veh/h
$v_{SBL} = 10/0.90 = 11$ veh/h $v_{SBT} = 300/0.90 = 333$ veh/h $v_{SBR} = 70/0.90 = 78$ veh/h
$f_{HV} = 1/(1 + 0.02(2.0 - 1)) = 0.98$

EQUATIONS 22-10 & 22-9

$v_{EBL} = 55/0.98 = 57$ pc/h $v_{EBT} = 333/0.98 = 340$ pc/h $v_{EBR} = 22/0.98 = 23$ pc/h
$v_{WBL} = 67/0.98 = 68$ pc/h $v_{WBT} = 444/0.98 = 453$ pc/h $v_{WBR} = 44/0.98 = 45$ pc/h
$v_{NBL} = 33/0.98 = 34$ pc/h $v_{NBT} = 556/0.98 = 567$ pc/h $v_{NBR} = 278/0.98 = 283$ pc/h
$v_{SBL} = 11/0.98 = 11$ pc/h $v_{SBT} = 333/0.98 = 340$ pc/h $v_{SBR} = 78/0.98 = 79$ pc/h

2. Compute conflicting flow rate for the northbound lane.

Conflicting Flow: $v_{c,NB} = 0 + 11 + 0 + 340 + 57 + 0 = 408$ pc/h

EQUATION 22-11

3. Compute the capacity for the northbound lane.

Capacity: $c_{pce,NB} = 1380(e^{(-1.02)(0.001)(408)}) = 910$ pc/h

EQUATION 22-1

$c_{NB} = 910(0.98) = 892$ veh/h

EQUATION 22-14

$v_{NB} = 33 + 556 + 278 = 867$ veh/h
$v/c = 867/892 = 0.97$

EQUATION 22-16

4. Compute delay for the northbound lane.

Delay: $d_{NB} = (3600/892) + ((900)(0.25)((0.97) - 1 + ((0.97 - 1)^2 + ((3600/892)(0.97))/((450)(0.25)))^{0.5} + (5)(0.97) = 44.9$ s/veh

EQUATION 22-17

5. Determine LOS for the northbound lane.

LOS: LOS = E (between 35 and 50)

EXHIBIT 22-8

HCS7

These screenshots show how Problem One is coded into *HCS7* in a series of screens to break the data into small, intuitive sections, followed

Figure 5.36 *HCS7 Roundabouts1 Input*

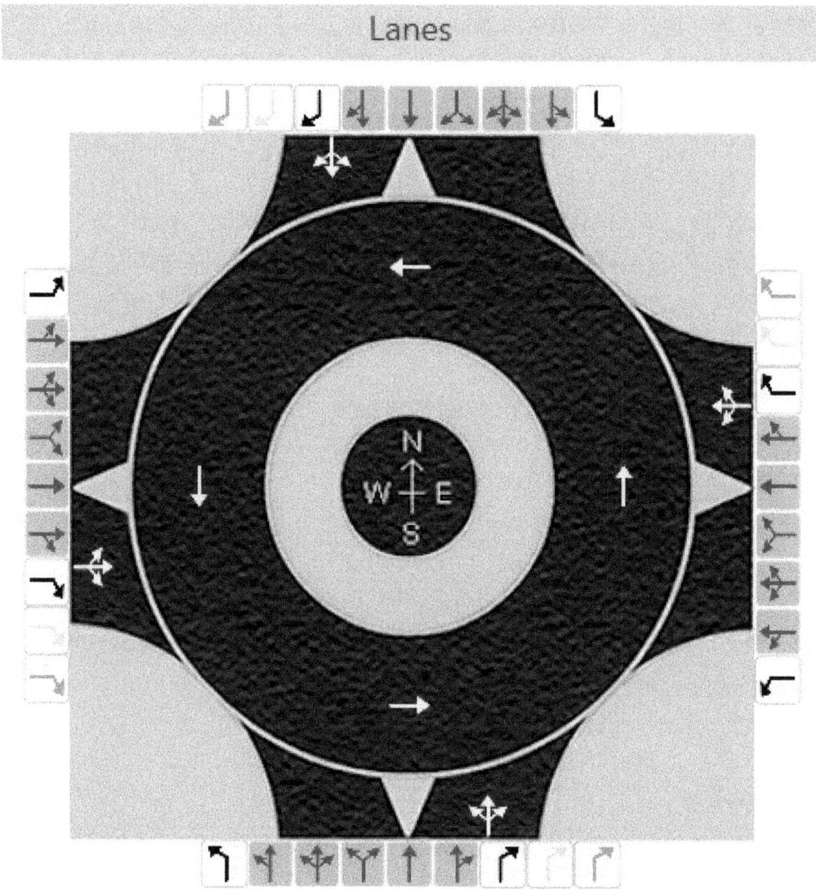

Vehicle Volumes and Adjustments

	EastBound				WestBound				NorthBound				SouthBound			
	Uturn	Left	Thru	Right	Uturn	Left	Thru	Right	Uturn	Left	Thru	Right	Uturn	Left	Thru	Right
Volume (veh/h)	0	50	300	20	0	60	400	40	0	30	500	250	0	10	300	70
Percent Heavy Vehicles (%)	2	2	2	2	2	2	2	2	2	2	2	2	2	2	2	2
Passenger Car Equivalent	2.0	2.0	2.0	2.0	2.0	2.0	2.0	2.0	2.0	2.0	2.0	2.0	2.0	2.0	2.0	2.0
Per. of Entry Vehicles Using Left Lane (%)																
Pedestrians Crossing Entry (p/h)	0				0				0				0			

Figure 5.37 *HCS7 Roundabouts1 Output*

Volume Adjustments and Site Characteristics

Approach	EB				WB				NB				SB			
Movement	U	L	T	R	U	L	T	R	U	L	T	R	U	L	T	R
Number of Lanes (N)	0	0	1	0	0	0	1	0	0	0	1	0	0	0	1	0
Lane Assignment			LTR				LTR				LTR					LTR
Volume (V), veh/h	0	50	300	20	0	60	400	40	0	30	500	250	0	10	300	70
Percent Heavy Vehicles, %	2	2	2	2	2	2	2	2	2	2	2	2	2	2	2	2
Flow Rate (v_{PCE}), pc/h	0	57	340	23	0	68	453	45	0	34	567	283	0	11	340	79
Right-Turn Bypass	None				None				None				None			
Conflicting Lanes	1				1				1				1			
Pedestrians Crossing, p/h	0				0				0				0			

Critical and Follow-Up Headway Adjustment

Approach	EB			WB			NB			SB		
Lane	Left	Right	Bypass	Left	Right	Bypass	Left	Right	Bypass	Left	Right	Bypass
Critical Headway (s)		4.9763			4.9763			4.9763			4.9763	
Follow-Up Headway (s)		2.6087			2.6087			2.6087			2.6087	

Flow Computations, Capacity and v/c Ratios

Approach	EB			WB			NB			SB		
Lane	Left	Right	Bypass	Left	Right	Bypass	Left	Right	Bypass	Left	Right	Bypass
Entry Flow (v_e), pc/h		420			566			884			430	
Entry Volume veh/h		412			555			867			422	
Circulating Flow (v_c), pc/h		419			658			408			555	
Exiting Flow (v_{ex}), pc/h		634			566			669			431	
Capacity (c_{pce}), pc/h		900			705			910			783	
Capacity (c), veh/h		882			692			892			768	
v/c Ratio (x)		0.47			0.80			0.97			0.55	

Delay and Level of Service

Approach	EB			WB			NB			SB		
Lane	Left	Right	Bypass	Left	Right	Bypass	Left	Right	Bypass	Left	Right	Bypass
Lane Control Delay (d), s/veh		9.9			26.9			44.9			13.0	
Lane LOS		A			D			E			B	
95% Queue, veh		2.5			8.2			16.5			3.4	
Approach Delay, s/veh		9.9			26.9			44.9			13.0	
Approach LOS		A			D			E			B	
Intersection Delay, s/veh | LOS	28.1						D					

by the results as presented in the formatted report. All input data, intermediate computations, and final results are shown in logical order, leading to delay and level of service.

PROBLEM TWO

Building on Problem One, compute LOS with an NBR yielding bypass lane added.

1. Compute the flow rates for all movements.

Flow Rate

EQUATION 22-8

$v_{EBL} = 50/0.90 = 55$ veh/h $v_{EBT} = 300/0.90 = 333$ veh/h $v_{EBR} = 20/0.90 = 22$ veh/h
$v_{WBL} = 60/0.90 = 67$ veh/h $v_{WBT} = 400/0.90 = 444$ veh/h $v_{WBR} = 40/0.90 = 44$ veh/h
$v_{NBL} = 30/0.90 = 33$ veh/h $v_{NBT} = 500/0.90 = 556$ veh/h $v_{NBR} = 250/0.90 = 278$ veh/h
$v_{SBL} = 10/0.90 = 11$ veh/h $v_{sBT} = 300/0.90 = 333$ veh/h $v_{ESBR} = 70/0.90 = 78$ veh/h
$f_{HV} = 1/(1 + 0.02(2.0 - 1)) = 0.98$

EQUATIONS 22-9 & 22-10

$v_{EBL} = 55/0.98 = 57$ pc/h $v_{EBT} = 333/0.98 = 340$ pc/h $v_{EBR} = 22/0.98 = 23$ pc/h
$v_{WBL} = 67/0.98 = 68$ pc/h $v_{WBT} = 444/0.98 = 453$ pc/h $v_{WBR} = 44/0.98 = 45$ pc/h
$v_{NBL} = 33/0.98 = 34$ pc/h $v_{NBT} = 556/0.98 = 567$ pc/h $v_{NBR} = 278/0.98 = 283$ pc/h
$v_{SBL} = 11/0.98 = 11$ pc/h $v_{SBT} = 333/0.98 = 340$ pc/h $v_{SBR} = 78/0.98 = 79$ pc/h

2. Compute conflicting flow rate for the northbound lane.

Conflicting Flow: $v_{c,NB} = 0 + 11 + 0 + 340 + 57 + 0 = 408$ pc/h

EQUATION 22-11

Exiting Flow: $v_{ex,EB} = 340 + 11 + 0 + 0 = 351$ pc/h

EQUATION 22-12

3. Compute the capacity for the northbound lane.

Capacity: $c_{pce,NB} = 1380(e^{(-1.02)(0.001)(408)}) = 910$ pc/h

EQUATION 22-1

$c_{NB} = 910(0.98) = 892$ veh/h

EQUATION 22-14

$v_{NB} = 33 + 556 = 589$ veh/h $v/c = 589/892 = 0.66$

EQUATION 22-16

$c_{pce,bypass,NB} = 1380(e^{(-1.02)(0.001)(351)}) = 965$ pc/h

EQUATION 22-6

$c_{bypass,NB} = 965(0.98) = 946$ veh/h

EQUATION 22-14

$v_{NBR} = 277$ veh/h $v/c = 277/946 = 0.29$

EQUATION 22-16

3. Compute delay for the northbound lanes.

Delay:

$$d_{NB} = (3600/892) + ((900)(0.25)((0.66) - 1 + ((0.66 - 1)^2 + ((3600/892)(0.66))/((450)(0.25)))^{0.5} + (5)(0.66) = 14.8 \text{ s/veh}$$

Figure 5.38 *HCS7 Roundabouts2 Input*

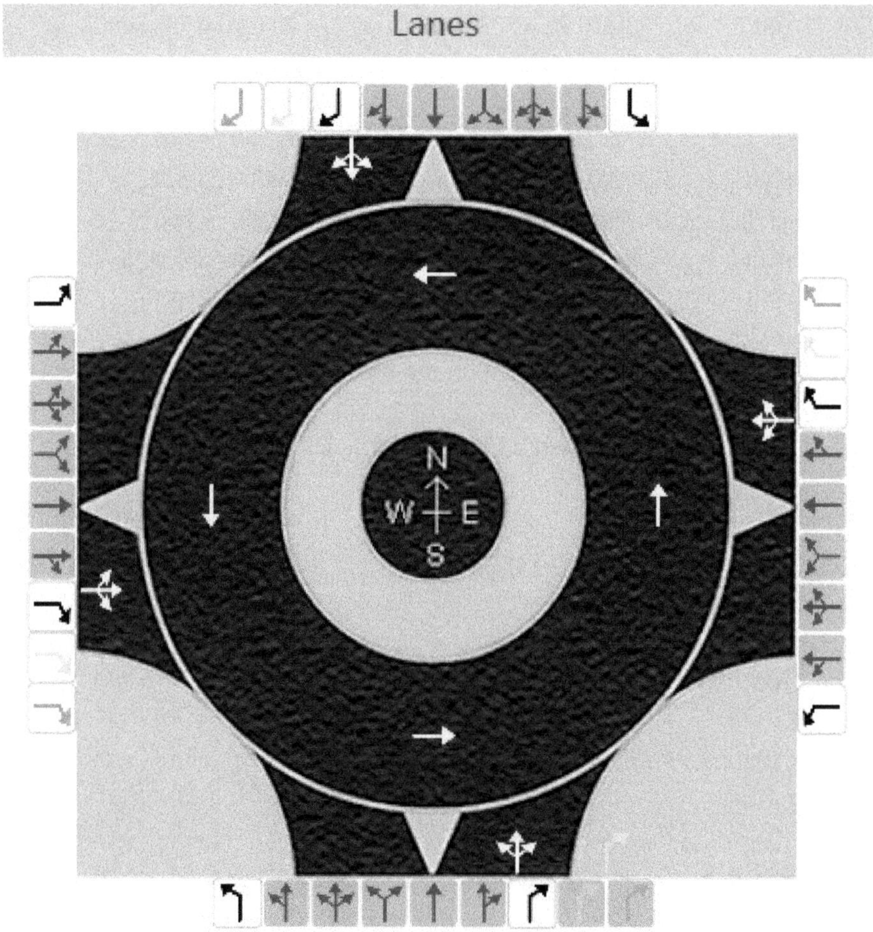

Vehicle Volumes and Adjustments

	EastBound				WestBound				NorthBound				SouthBound			
	Uturn	Left	Thru	Right	Uturn	Left	Thru	Right	Uturn	Left	Thru	Right	Uturn	Left	Thru	Right
Volume (veh/h)	0	50	300	20	0	60	400	40	0	30	500	250	0	10	300	70
Percent Heavy Vehicles (%)	2	2	2	2	2	2	2	2	2	2	2	2	2	2	2	2
Passenger Car Equivalent	2.0	2.0	2.0	2.0	2.0	2.0	2.0	2.0	2.0	2.0	2.0	2.0	2.0	2.0	2.0	2.0
Per. of Entry Vehicles Using Left Lane (%)																
Pedestrians Crossing Entry (p/h)	0				0				0				0			

Figure 5.39 *HCS7 Roundabouts2 Output*

Volume Adjustments and Site Characteristics

Approach	EB				WB				NB				SB			
Movement	U	L	T	R	U	L	T	R	U	L	T	R	U	L	T	R
Number of Lanes (N)	0	0	1	0	0	0	1	0	0	0	1	0	0	0	1	0
Lane Assignment			LTR				LTR				LTR				LTR	
Volume (V), veh/h	0	50	300	20	0	60	400	40	0	30	500	250	0	10	300	70
Percent Heavy Vehicles, %	2	2	2	2	2	2	2	2	2	2	2	2	2	2	2	2
Flow Rate (v_{PCE}), pc/h	0	57	340	23	0	68	453	45	0	34	567	283	0	11	340	79
Right-Turn Bypass	None				None				Yielding				None			
Conflicting Lanes	1				1				1				1			
Pedestrians Crossing, p/h	0				0				0				0			

Critical and Follow-Up Headway Adjustment

Approach	EB			WB			NB			SB		
Lane	Left	Right	Bypass	Left	Right	Bypass	Left	Right	Bypass	Left	Right	Bypass
Critical Headway (s)		4.9763			4.9763			4.9763	4.9763		4.9763	
Follow-Up Headway (s)		2.6087			2.6087			2.6087	2.6087		2.6087	

Flow Computations, Capacity and v/c Ratios

Approach	EB			WB			NB			SB		
Lane	Left	Right	Bypass	Left	Right	Bypass	Left	Right	Bypass	Left	Right	Bypass
Entry Flow (v_e), pc/h		420			566			601	283		430	
Entry Volume veh/h		412			555			589	277		422	
Circulating Flow (v_c), pc/h	419			658			408			555		
Exiting Flow (v_{ex}), pc/h	351			566			669			431		
Capacity (c_{pce}), pc/h		900			705			910	965		783	
Capacity (c), veh/h		882			692			892	946		768	
v/c Ratio (x)		0.47			0.80			0.66	0.29		0.55	

Delay and Level of Service

Approach	EB			WB			NB			SB		
Lane	Left	Right	Bypass	Left	Right	Bypass	Left	Right	Bypass	Left	Right	Bypass
Lane Control Delay (d), s/veh		9.9			26.9			14.8	6.8		13.0	
Lane LOS		A			D			B	A		B	
95% Queue, veh		2.5			8.2			5.1	1.2		3.4	
Approach Delay, s/veh	9.9			26.9			12.3			13.0		
Approach LOS	A			D			B			B		
Intersection Delay, s/veh | LOS	15.6						C					

$$d_{bypass,NB} = (3600/946) + ((900)(0.25)((0.29) - 1 + \quad ((0.29-1)^2 + ((3600/946)(0.29))/((450)(0.25)))^{0.5} +$$

(5)(0.29) = 6.8 s/veh

EQUATION 22-17

4. Determine LOS for the northbound approach.

Delay: d_{NBR} = ((14.8)(589) + (6.8)(277)) / (589 + 277) = 12.3 s/veh

EQUATION 22-18

LOS: LOS = B (between 10 and 15)

EXHIBIT 22-8

HCS7

These screenshots show how Problem Two is coded into *HCS7* by adding the yielding bypass lane, followed by the updated results as presented in the formatted report. All input data, intermediate computations, and final results are shown in logical order, leading to delay and level of service.

Figure 5.40 *HCM Exhibit 30-42*

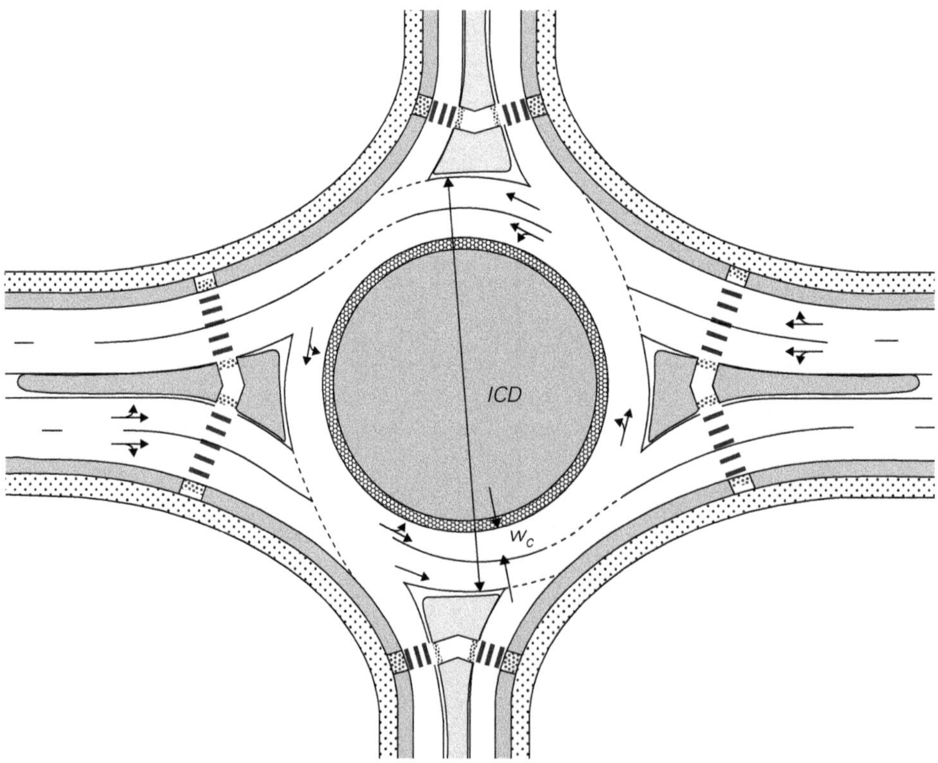

Figure 5.41 *HCM Exhibit 30-44]*

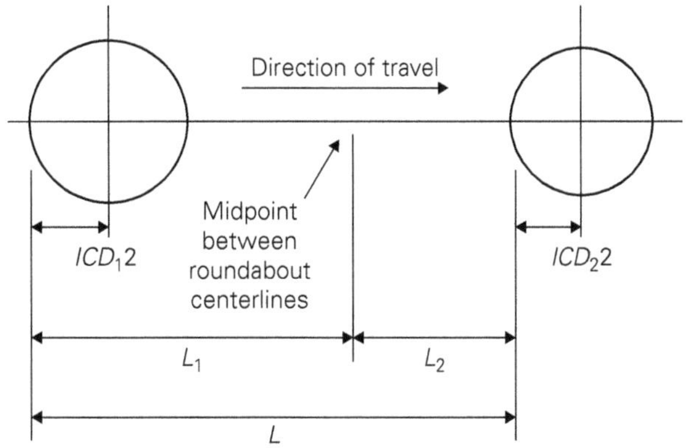

Figure 5.42 *HCM Equation 30-76*

$$L_1 = \frac{1}{2}\left(L - \frac{ICD_1}{2} + \frac{ICD_2}{2}\right) + \frac{ICD_1}{2}$$

Figure 5.43 *HCM Equation 30-77*

$$L_2 = L - L_1$$

where

L_1 = length of subsegment 1 (ft),
L_2 = length of subsegment 2 (ft),
L = length of segment (ft),
ICD_1 = inscribed circle diameter of roundabout 1 (ft), and
ICD_2 = inscribed circle diameter of roundabout 2 (ft).

Figure 5.44 *HCM Equation 30-75*

$$S_c = 3.4614 r_{c,th}^{\ 0.3673}$$

where

S_c = circulating speed (mi/h), and
$r_{c,th}$ = average radius of circulating path of through movement (ft).

Figure 5.45 *HCM Equation 30-78*

$$S_{f,1,\text{initial}} = 14.6 + 0.0039 L_1 + 0.48 S_{PL} + 0.02 CID_1$$

where

$S_{f,1,\text{initial}}$ = initial free-flow speed for Subsegment 1 (mi/h),
L_1 = length of Subsegment 1 (ft),
S_{pL} = posted speed limit (mi/h), and
CID_1 = central island diameter for roundabout at upstream end of Subsegment 1 (ft).

Figure 5.46 *HCM Equation 30-79*

$$RIA_1 = -149.8 + 31.4S_{f,1,\text{initial}} - 22.5S_{c,1}$$

where

RIA_1 = roundabout influence area for Subsegment 1 (ft),

$S_{f,1,\text{initial}}$ = initial free-flow speed for Subsegment 1 (mi/h), and

$S_{c,1}$ = through movement circulating speed for roundabout at upstream end of segment (mi/h).

Figure 5.47 *HCM Equation 30-80*

$$S_{f,1} = S_{f,1,\text{initial}} - 4.43 \text{ if } RIA_1 \geq L_1, \text{else}$$
$$S_{f,1} = S_{f,1,\text{initial}}$$

where $S_{f,1}$ is the free-flow speed for Subsegment 1 (mi/h).

Figure 5.48 *HCM Equation 30-81*

$$S_{f,2,\text{initial}} = 15.1 + 0.0037L_2 + 0.43S_{PL} + 0.05CID_2$$

where

$S_{f2,\text{ initial}}$ = initial free-flow speed for Subsegment 2 (mi/h),

L = length of subsegment (ft),

S_{PL} = posted speed limit (mi/h), and

CID_2 = central island diameter for roundabout at downstream end of subsegment (ft).

Figure 5.49 *HCM Equation 30-82*

$$RIA_2 = 165.9 + 13.8S_{f,2,\text{initial}} - 21.1S_{c,2}$$

where

RIA_2 = roundabout influence area for Subsegment 2 (ft),

$S_{f,2,\text{initial}}$ = initial free-flow speed for Subsegment 2 (mi/h), and

$S_{c,2}$ = through movement circulating speed for roundabout at downstream end of subsegment (mi/h).

Figure 5.50 *HCM Equation 30-83*

$$S_{f,2} = S_{f,2,\text{initial}} - 4.73 \text{ if } RIA_2 \geq L_2, \text{else}$$
$$S_{f,2} = S_{f,2,\text{initial}}$$

Figure 5.51 *HCM Equation 30-72*

$$S_{fo} = S_{calib} + S_0 + f_{cs} + f_A + f_{pk}$$

where

S_{fo} = base free-flow speed (mi/h),

S_{calib} = base free-flow speed calibration factor (mi/h),

S_0 = speed constant (mi/h),

f_{cs} = adjustment for cross section (mi/h),

f_A = adjustment for access points (mi/h), and

f_{pk} = adjustment of on-street parking (mi/h).

Figure 5.52 *HCM Exhibit 30-43*

Speed Limit (mi/h)	Speed Constant S_0 (mi/h)[a]	Median Type	Percent with Restrictive Median (%)	Adjustment for Cross Section f_{cs} (mi/h)[b] No Curb	Curb
25	37.4	Restrictive	20	0.3	−0.9
30	39.7		40	0.6	−1.4
35	42.1		60	0.9	−1.8
40	44.4		80	1.2	−2.2
45	46.8		100	1.5	−2.7
50	49.1	Nonrestrictive	Not applicable	0.0	−0.5
		No median	Not applicable	0.0	−0.5

Access Density D_a (points/mi)	Adjustment for Access Points f_A by Lanes N_{th} (mi/h)[c] 1 Lane	2 Lanes	3 Lanes	Percent with On-Street Parking (%)	Adjustment for Parking (mi/h)[d]
0	0.0	0.0	0.0	0	0.0
2	−0.2	−0.1	−0.1	20	−0.6
4	−0.3	−0.2	−0.1	40	−1.2
10	−0.8	−0.4	−0.3	60	−1.8
20	−1.6	−0.8	−0.5	80	−2.4
40	−3.1	−1.6	−1.0	100	−3.0
60	−4.7	−2.3	−1.6		

Figure 5.53 *HCM Equation 30-84*

$$f_L = 1.02 - 4.7\frac{S_{fo} - 19.5}{\max(L_s, 400)} \leq 1.0$$

where

f_L = boundary intersection spacing adjustment factor,

S_{fo} = base free-flow speed (mi/h), and

L_s = distance between adacent boundary intersections that (*a*) bracket the subject segment and (*b*) each have a type of control that can impose on the subject through movement a legal requirement to stop or yield, such as a roundabout (ft).

Figure 5.54 *HCM Equation 30-85*

$$S_{f,non-rbt} = S_{fo} f_L \geq S_{pl}$$

Segments

With multiple roundabouts, segments between the intersections are modeled to generate running time and geometric delay. Running time can be combined with geometric and control delays for the through movement to be converted to travel speed for determining level of service.

Required Data

Beyond the requirements to compute control delay, more geometric information is needed, including segment length; mid-segment lane number and flow rate; average circulating lane width; and inscribed diameter. For free-flow speed along the segment, the speed limit; percent restrictive median with or without curb; access point density; and percent on-street parking is also used.

Figure 5.55 *HCM Equation 30-86*

$$S_f = \min[S_{f,1}, S_{f,2}, S_{f,non-rbt}]$$

Figure 5.56 *HCM Equation 30-87*

$$t_R = \frac{3.5}{0.0025L}\min\left[\frac{v_{th}}{c_{th}}, 1.00\right] + \frac{3{,}600\,L}{5{,}280\,S_f} f_v + \sum_{i=1}^{N_{ap}} d_{ap,i} + d_{other}$$

where

t_R = segment running time (s);
L = segment length (ft);
v_{th} = through-demand flow rate (veh/h);
c_{th} = through-movement capacity (veh/h);
f_v = proximity adjustment factor;
$d_{ap,i}$ = delay due to left and right turns from the street into access point intersection i (s/veh);
N_{ap} = number of influential access point approaches along the segment = $N_{ap,s}$ + $P_{ap,lt} N_{ap,o}$ (points);
$N_{ap,s}$ = number of access point approaches on the right side in the subject direction of travel (points);
$N_{ap,o}$ = number of access point approaches on the right side in the opposing direction of travel (points);
$p_{ap,lt}$ = proportion of $N_{ap,o}$ that can be accessed by a left turn from the subject direction of travel;
d_{other} = delay due to other sources along the segment (e.g., curb parking or pedestrians) (s/veh); and

Figure 5.57 *HCM Equation 18-6*

$$f_v = \frac{2}{1 + \left(1 - \frac{v_m}{52.8\,N_{th} S_f}\right)^{0.21}}$$

where

f_v = proximity adjustment factor,
v_m = midsegment demand flow rate (veh/h),
N_{th} = number of through lanes on the segment in the subject direction of travel (ln), and
S_f = free-flow speed (mi/h).

Geometry

The inscribed diameter and average circulating lane width is used to compute the central island diameter and radius, following HCM Exhibit 30-42. The radius values for both boundary intersections are used with the segment length to compute the two subsegment lengths between them using HCM Exhibit 30-44 and HCM Equations 30-76 and 30-77.

Figure 5.58 *HCM Equation 30-89*

$$d_{geom,1} = \max\left[-2.63 + 0.09S_f + 0.625ICD_1\left(\frac{1}{S_{c,1}} - \frac{1}{S_f}\right), 0\right]$$

Figure 5.59 *HCM Equation 30-90*

$$d_{geom,2} = \max[1.57 + 0.11S_f - 0.21S_{c,2}, 0]$$

Figure 5.60 *HCM Equation 30-91*

$$d_t = d_{control,t} + d_{geom,1} + d_{geom,2}$$

Speeds

The circulating speed is calculated from the radius using HCM Equation 30-75. Free-flow speed is computed from HCM Equations 30-78, 30-79, and 30-80 for Subsegment 1 and HCM Equations 30-81, 30-82, and 30-83 for Subsegment 2.

Figure 5.61 *HCM Equation 18-15*

$$S_{T,seg} = \frac{3{,}600\ L}{5{,}280(t_R + d_t)}$$

where

$S_{T,seg}$ = travel speed of through vehicles for the segment (mi/h),

L = segment length (ft),

t_R = segment running time (s), and

d_t = through delay (s/veh).

Figure 5.62 *HCM Exhibit 18-1*

	Travel Speed Threshold by Base Free-Flow Speed (mi/h)							Volume-to-Capacity Ratio*
LOS	55	50	45	40	35	30	25	
A	>44	>40	>36	>32	>28	>24	>20	≤ 1.0
B	>37	>34	>30	>27	>23	>20	>17	
C	>28	>25	>23	>20	>18	>15	>13	
D	>22	>20	>18	>16	>14	>12	>10	
E	>17	>15	>14	>12	>11	>9	>8	
F	≤17	≤15	≤14	≤12	≤11	≤9	≤8	
F				any				> 1.0

Base free-flow speed without roundabout influence is calculated using HCM Equation 30-72 populated by factors from HCM Exhibit 30-43, then adjusted in HCM Equation 30-85 using HCM Equation 30-84 to account for spacing—not allowing free-flow speed to be lower than the posted speed limit.

Then, HCM Equation 30-86 is used to determine the free-flow speed as the minimum among the three just computed. Finally, running time is computed from HCM Equation 30-87 with the proximity adjustment factor coming from the HCM Urban Streets procedure using HCM Equation 18-6.

Delay

Using the segment free-flow speed and the circulating speed for each subsegment, the geometric delay is computed for Subsegment 1 and Subsegment 2, following HCM Equations 30-89 and 30-90, respectively. The final through delay is calculated as the sum of the control delay for the through movement and the geometric delays for the two subsegments using HCM Equation 30-91.

Think about your own driving: Approaching a signalized intersection at midnight with the light green, you move through the intersection without slowing down and with zero delay. Approaching a roundabout at midnight, even though you are all alone, you must slow to 15 mi/h and navigate the circle, creating geometric delay to go through the intersection.

LOS

Level of service is determined from the HCM Urban Streets procedure using HCM Exhibit 18-1 with travel speed by base free-flow speed thresholds computed, following HCM Equation 18-15. Any lane with a volume-to-capacity ratio greater than one is assigned LOS F.

PROBLEM ONE

Determine the northbound LOS for the segment created by using the isolated intersection in Intersection Problem One as the downstream boundary intersection for this 1000-foot single-lane segment with a speed limit of 40 mi/h, 20% restricted median with curb, 10 access points per mile, and 60% on-street parking. Additional data for each intersection include an inscribed diameter of 100 feet with 12-foot circulating lane width.

1. Compute base free-flow speed and circulating speed.

$$S_{fo} = 44.4 - 0.9 - 0.8 - 1.8 = 40.9 \text{ mi/h}$$

EQUATION **30-72** AND EXHIBIT **30-43**

Figure 5.63 *HCS7 RoundaboutSegments1 Input*

Segment Properties			
Name			
Segment Length (ft)	1000	Median Type	Undivided
	Forward		Reverse
BFFS Calibration Factor	0.0		0.0
Restrictive Median (ft)	20		20
Right-Hand Curb (%)	20		20
Right-Hand Access Points (pts/mi)	1		1
On-Street Parking (%)	60		60
Mid-Segment Demand Flow Rate	590		590

$$CID = 100 - 2(1)(12) = 76 \text{ ft}$$

Figure 5.64 *HCS7 RoundaboutSegments1 Output*

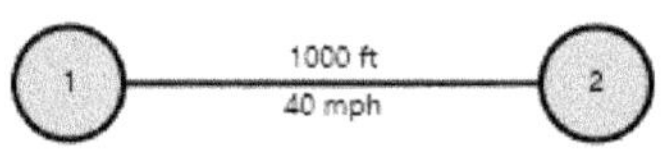

Basic Segment Information

Segment	Speed Limit		Through Lanes		Segment Length		ICD		CID		Length RM		Percent Curb	
	NB	SB	NB	SB	NB	SB	NB	SB	NB	SB	NB	SB	NB	SB
1	40	40	1	1	1000	1000	130	100	90.00	76.00	20	20	20	20

Segment Ouput Data

Segment		NB	SB
1	Base Free-Flow Speed, mi/h	41.58	41.58
1	Delay due to Turns, s	0.26	0.26
1	Running Time, s	22.63	22.35
1	Running Speed, mi/h	30.13	30.50
1	Through Delay, s/veh	49.69	18.30
1	Travel Time, s	72.3	40.7
1	Through Stop Rate, stops/veh	0.74	0.55
1	Travel Speed, mi/h	9.4	16.8
1	Spatial Stop Rate, stops/mi	3.93	2.90
1	Through v/c Ratio	0.74	0.55
1	Percent of Base FFS	22.68	40.34
1	Level of Service	F	D

EQUATION **30-73**

$r_{c,th} = (76/2) + ((1)(12))/2 = 44$ ft

EQUATION 30-74

$S_{c1} = S_{c2} = 3.4614(44)^{0.3673} = 13.9$ mi/h

EQUATION 30-75

2. Compute subsegment lengths.

$L_1 = ½(1000 – 100/2 + 100/2) + 100/2 = 550$ ft $L_2 = 1000 – 550 = 450$ ft

EQUATIONS 30-76 AND 30-77

3. Compute free-flow speed.

$S_{f,1,initial} = 14.6 + 0.0039(550) + 0.48(40) + 0.02(76) = 37.47$ mi/h

EQUATION 30-78

$RIA_1 = –149.8 + 31.4(37.47) – 22.5(13.9) = 714 \text{ ft} > 550$ ft

EQUATION 30-79

$S_{f,1} = 37.47 – 4.43 = 33.0$ mi/h

EQUATION 30-80

$S_{f,2,initial} = 15.1 + 0.0037(450) + 0.43(40) + 0.05(76) = 37.77$ mi/h

EQUATION 30-81

$RIA_2 = 165.9 + 13.8(37.77) – 21.1(13.9) = 395 \text{ ft} < 450$ ft

EQUATION 30-82

$S_{f,2} = 37.77 = 37.8$ mi/h

EQUATION 30-83

$f_L = 1.02 – 4.7((40.9 – 19.5)/1000) = 0.919$

EQUATION 30-84

$S_{f,non\text{-}rbt} = 40.9(0.919) = 37.6 \text{ mi/h} < 40$ mi/h

EQUATION 30-85

$S_f = \min(33.0, 37.8, 40.0) = 33.0$ mi/h

EQUATION 30-86

4. Compute running time.

$f_v = 2/(((1 + ((1 - 867/(52.8)(1)(33.0))^{0.21}) = 1.07$

EQUATION 18-6

$t_R = (3.5/((0.0025)(1000)))(0.97) + ((3600)(1000))/((5280)(33.0))(1.07) = 23.5 \text{ s}$

EQUATION 30-87

5. Compute geometric delay.

$d_{geom,1} = -2.63 + (0.09)(33.0) + (0.625)(100)((1/13.9) - (1/33.0)) = 2.9 \text{ s/veh}$

EQUATION 30-89

$d_{geom,2} = 1.57 + (0.11)(33.0) - (0.21)(13.9) = 2.3 \text{ s/veh}$

EQUATION 30-90

6. Compute travel speed.

$d_t = 44.9 + 2.9 + 2.3 = 50.1 \text{ s/veh}$

PROBLEM 1 AND EQUATION 30-91

$S_T = ((3600)(1000))/((5280)(23.5 + 50.1)) = 9.3 \text{ mi/h}$

EQUATION 18-15

7. Determine LOS.

LOS = F ($S_{fo} = 40.9$ and $S_T = 9.3$)

EXHIBIT 18-1

HCS7

These screenshots show how Problem One is coded as northbound into *HCS7* in a few screens to break the data into small, intuitive sections, followed by the results as presented in the formatted report. All input data, intermediate computations, and final results are shown in logical order, leading to delay and level of service.

Figure 5.65 *HCS7 RoundaboutSegments2 Input*

Segment Properties			
Name			
Segment Length (ft)	2000	Median Type	Undivided
	Forward	Reverse	
BFFS Calibration Factor	0.0	0.0	
Restrictive Median (ft)	20	20	
Right-Hand Curb (%)	20	20	
Right-Hand Access Points (pts/mi)	4	4	
On-Street Parking (%)	60	60	
Mid-Segment Demand Flow Rate	590	590	

Figure 5.66 *HCS7 RoundaboutSegments1 Output*

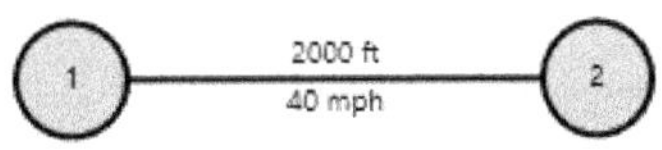

Basic Segment Information

Segment	Speed Limit		Through Lanes		Segment Length		ICD		CID		Length RM		Percent Curb	
	NB	SB	NB	SB	NB	SB	NB	SB	NB	SB	NB	SB	NB	SB
1	40	40	1	1	2000	2000	130	100	90.00	76.00	20	20	20	20

Segment Ouput Data

Segment		NB	SB
1	Base Free-Flow Speed, mi/h	40.75	40.75
1	Delay due to Turns, s	0.26	0.26
1	Running Time, s	36.43	36.29
1	Running Speed, mi/h	37.43	37.58
1	Through Delay, s/veh	51.32	19.85
1	Travel Time, s	87.7	56.1
1	Through Stop Rate, stops/veh	0.74	0.55
1	Travel Speed, mi/h	15.5	24.3
1	Spatial Stop Rate, stops/mi	1.96	1.45
1	Through v/c Ratio	0.74	0.55
1	Percent of Base FFS	38.13	59.61
1	Level of Service	E	C

Chapter 6 Signals

Signalized Intersections can be modeled as isolated intersections or part of an urban street that includes the segments between them. As an intersection, the procedure models all movements combined into lane groups by computing saturation flow rate controlled by signal phases for appropriate effective green times for each to generate capacity for computing delay that determines level of service. Procedures are defined for determining pedestrian and bicycle levels of service as well. Along an urban street, the procedure is expanded to include the running time along the segments with through movement delay to compute travel speed for level of service, which is covered in Chapter Seven.

Traffic

From the traffic perspective, the analysis of a signalized intersection as defined in HCM Chapter 19 will typically be an operational, multiple-period analysis. This allows for peaking characteristics to be modeled without the peak hour factor and for unmet demand to be considered as the subsequent period initial queue for oversaturated conditions.

Required Data

The lane configuration (including turn-lane storage lengths) and turning movement demands (including right turns on red) are required in fifteen-minute increments. Since saturation flow rate is a key component to this analysis, data required for some adjustments are needed that include uneven lane distribution; heavy vehicles; stopping buses; on-street parking maneuvers; and pedestrians and bicycles. If upstream signals present, the volume-to-capacity ratios for the three movements contributing to the segment flow are needed for the upstream filtering factor, as well as the proportion arriving on green to model coordinate effects on approach delay. More details on these data are discussed in the saturation flow adjustment section to provide additional information on how they are used.

Lane Groups

The lane configuration is defined by the number of lanes on each approach with the movements allowed in each lane. From this, lane groups can be determined using HCM Exhibit 19-19. Capacity, delay, and level of service are computed by lane group before averaging for the approaches and intersection.

Arrival Demand

When collecting demand data, the arrival rate must be measured, not just the departure flow (stop-bar counts), for all time periods that are congested, where queues are forming and vehicles are waiting more than one cycle to move through the intersection.

This requires quantifying the queue as the beginning of red during the last cycle of the fifteen-minute cycle for each lane group. These values are added to the fifteen-minute stop-bar counts (subtracting the last period queue) to represent the arrival demand. Counting at the stop-bar only will essentially yield capacity, but for oversaturated conditions, the demand exceeds capacity.

In the procedure, the unmet demand from one period becomes the initial queue for the subsequent period that is used to compute the third term of the delay equation and account for the additional delay created by the initial queue. This will make a (many times significant) difference in total control delay.

FLOW RATE

Flow rate for each lane group is computed by dividing the hourly volumes by the intersection peak hour factor for a single-period analysis. For a multiple-period analysis, the fifteen-minute flow rate is used directly with no peak

Figure 6.1 *HCM Exhibit 19-19*

Number of Lanes	Movements by Lanes	Movement Groups (MG)	Lane Groups (LG)
1	Left, thru., & right:	MG 1:	LG 1:
2	Exclusive left: Thru. & right:	MG 1: MG 2:	LG 1: LG 2:
2	Left & thru.: Thru. & right:	MG 1:	LG 1: LG 2:
3	Exclusive left: Exclusive left: Through: Through: Thru. & right:	MG 1: MG 2:	LG 1: LG 2: LG 3:

hour factor needed. The fifteen-minute analysis period (*T*) is expressed as 0.25 hours.

SHARED LANES

For approaches where movement has a choice between an exclusive lane and a shared lane as shown in HCM Exhibit 31-6, the shared lane model predicts the distribution of traffic by adjusting the flow rates between the lane groups. More lane changing occurs when there is motivation (to gain an advantage) and opportunity (not too crowded) as illustrated in HCM Exhibit 31-5.

Saturation Flow

The computation of the saturation flow rate is (along with effective green time, discussed later) key to this methodology. A base saturation flow rate is first defined with default values of 1900 pc/h/ln for areas with a population of

Figure 6.1a *HCM Exhibit 31-6*

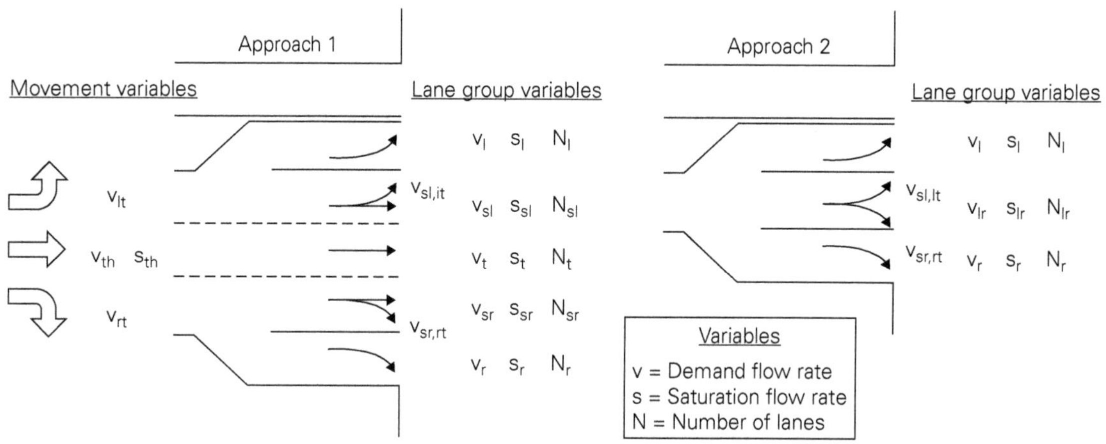

Figure 6.1b *HCM Exhibit 31-5*

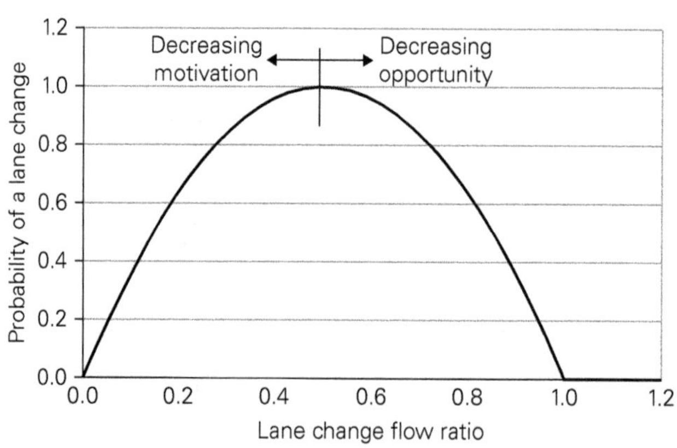

250,000 or more and 1750 pc/h/ln. However, since many areas and time periods operate differently, this value should be calibrated to local conditions. Some big cities have rates well over 2000, and many smaller towns have rates below 1500, so this value can be critical to this very important parameter.

This base value is adjusted to model several elements affecting the headways (and therefore flow rate) navigating a signalized intersection. The following section will go through each with the rules and some practical considerations. While each adjustment equation or table is somewhat simple, the cumulative value is quite defendable. The saturation flow rate equation is shown in HCM Equation 19-8 with most adjustment factors generated by separate equations or tables.

LANE WIDTH

Saturation flow can be reduced if the average width of lanes in a lane group is below 10 feet, and it can actually increase if the lane width is over 12.9 feet, according to HCM Exhibit 19-20.

Figure 6.3 *HCM Exhibit 19-20*

Average Lane Width (ft)	Adjustment Factor (f_w)
<10.0[a]	0.96
≥10.0–12.9	1.00
>12.9	1.04

LANE UTILIZATION

Unequal lane distribution reduces saturation flow rate because all lanes are not being fully used. File data for the flow rate in the heaviest lane of a multiple-lane group can be used in HCM Equation 19-7 to quantify this loss. If no data are provided, the adjustment default values are used according to HCM Exhibit 19-15.

Figure 6.2 *HCM Equation 19-8*

$$s = s_o f_w f_{HVg} f_p f_{bb} f_a f_{LU} f_{LT} f_{RT} f_{Lpb} f_{Rpb} f_{wz} f_{ms} f_{sp}$$

where

s = adjusted saturation flow rate (veh/h/ln),
s_o = base saturation flow rate (pc/h/ln),
f_w = adjustment factor for lane width,
f_{HVg} = adjustment factor for heavy vehicles and grade,
f_p = adjustment factor for existence of a parking lane and parking activity adjacent to lane group,
f_{bb} = adjustment factor for blocking effect of local buses that stop within intersection area,
f_a = adjustment factor for area type,
f_{LU} = adjustment factor for lane utilization,
f_{LT} = adjustment factor for left-turn vehicle presence in a lane group,
f_{RT} = adjustment factor for right-turn vehicle presence in a lane group,
f_{Lpb} = pedestrian adjustment factor for left-turn groups,
f_{Rpb} = pedestrian-bicycle adjustment factor for right-turn groups,
f_{wz} = adjustment factor for work zone presence at the intersection,
f_{ms} = adjustment factor for downstream lane blockage, and
f_{sp} = adjustment factor for sustained spillback.

Figure 6.3a *HCM Exhibit 19-15*

Movement Group	Number of Lanes in Movement Group (ln)	Traffic in Most Heavily Traveled Lane (%)	Lane Utilization Adjustment Factor f_{LU}
Exclusive through	1	100.0	1.000
	2	52.5	0.952
	3[a]	36.7	0.908
Exclusive left turn	1	100.0	1.000
	2[a]	51.5	0.971
Exclusive right turn	1	100.0	1.000
	2[a]	56.5	0.885

Think about your own driving:
There are situations where you are at a signal, knowing what you want to do at the next signal, so you preposition yourself in the appropriate lane upstream. If your movements are typical of most drivers for a particular pair of signals (or even with a downstream lane drop), you will be joined by most others—putting most of the traffic in one lane as compared with the other adjacent lane because of the situation.

Figure 6.4 *HCM Equation 19-7*

$$f_{LU} = \frac{v_g}{N_e\, v_{g1}}$$

where

f_{LU} = adjustment factor for lane utilization,
v_g = demand flow rate for movement group (veh/h),
V_{g1} = demand flow rate in the single exclusive lane with the highest flow rate of all exclusive lanes in movement group (veh/h/ln), and
N_e = number of exclusive lanes in movement group (ln).

HEAVY VEHICLES AND GRADE

Headway values increase with the presence of heavy vehicles in the traffic stream, which can be affected by grade as well. The percent heavy vehicles and the approach grade are used in HCM Equation 19-9 (for negative grades) or HCM Equation 19-10 (for non-negative grades).

Figure 6.5 *HCM Equation 19-9*

$$f_{HVg} = \frac{100 - 0.79\, P_{HV} - 2.07\, P_g}{100}$$

Figure 6.6 *HCM Equation 19-10*

$$f_{HVg} = \frac{100 - 0.78\, P_{HV} - 0.31\, P_g^2}{100}$$

where

P_{HV} = percent heavy vehicles in the corresponding movement group (%), and
P_g = approach grade for the corresponding movement group (%).

PARKING

For on-street parking within 250 feet of the stop bar, saturation flow is reduced by 10% for the presence and further reduced as a function of

Figure 6.7 *HCM Equation 19-11*

$$f_p = \frac{N - 0.1 - \frac{18\, N_m}{3{,}600}}{N} \geq 0.050$$

where

N_{m} = parking maneuver rate adjacent to lane group (maneuvers/h), and
N = number of lanes in lane group (ln).

parking maneuvers per hour, following HCM Equation 19-11. The adjustment assumes 18 seconds per maneuver, but should be modified for lower values for angle parking, etc.

> ***Think about your own driving:***
> Approaching a signalized intersection in the rightmost lane, a row of parked cars to your right changes how you drive (as compared with just a curb and sidewalk to your right). You are wary of a car door opening or a car pulling out, so you provide a little cushion by increasing your headway time.

BUS BLOCKAGE

Buses stopping in the travel lane within 250 feet of the intersection, near side or far side, affect saturation flow rate as a function of the number of stopping buses as defined by HCM Equation 19-12. This adjustment assumes a 14.4-second stopped time, but should be adjusted using field data for actual average times.

Figure 6.8 *HCM Equation 19-12*

$$f_{bb} = \frac{N - \frac{14.4\,N_b}{3{,}600}}{N} \geq 0.050$$

where N is the number of lanes in lane group (ln) and N_b is the bus stopping rate on the subject approach (buses/h).

> ***Think about your own driving:***
> When driving in the right lane on a multiple-lane arterial, you move to the left lane if there is a bus in front of you, knowing that it will stop at some point and hold you up. But not everyone thinks to or can move over, so some get stuck behind the bus approaching a signalized intersection—just not you!

AREA TYPE

If the intersection is within a central business district (CBD) with additional distractions like double-parking, jay walking, delivery vehicles, etc., an adjustment of 0.9 is applied to account for the longer headways generated by this friction. Not all CBD areas have this distracting environment, and the environment that warrants this adjustment is not limited to CBD areas, so judgment is required in applying this adjustment.

RIGHT TURNS

Right-turning vehicles must slow down to navigate the radius of the turn, which increases headways and lowers saturation flow rate. The assumed radius value is 32 feet for the 1.18 equivalent value used in HCM Equation 19-13 but should be adjusted for larger or skewed intersections.

Figure 6.9 *HCM Equation 19-13*

$$f_{RT} = \frac{1}{E_R}$$

where E_R is the equivalent number of through cars for a protected right-turning vehicle (= 1.18).

LEFT TURNS

Left-turning vehicles must slow down to navigate the radius of the turn, which increases headways and lowers saturation flow rate. The assumed radius value is 112 feet for the 1.05 equivalent value used in HCM Equation 19-14 but should be adjusted for larger or skewed intersections.

Permitted left turns from an exclusive lane are modeled as gap-acceptance movements, following HCM Equation 31-100 to compute

Figure 6.10 *HCM Equation 19-14*

$$f_{LT} = \frac{1}{E_L}$$

where E_L is the equivalent number of through cars for a protected left-turning vehicle (= 1.05).

Figure 6.11 *HCM Equation 31-100*

$$s_p = \frac{v_o e^{-v_o t_{cg}/3{,}600}}{1 - e^{-v_o t_{fh}/3{,}600}}$$

where

- s_p = saturation flow rate of a permitted left-turn movement (veh/h/ln);
- v_o = opposing demand flow rate (veh/h);
- t_{cg} = critical headway = 4.5 (s); and
- t_{fh} = follow-up headway = 2.5 (s).

Figure 6.12 *HCM Equation 31-110*

$$s_l = s_p f_w f_{HVg} f_p f_{bb} f_a f_{LU} f_{Lpb} f_{wz} f_{ms} f_{sp}$$

the saturation flow rate that is further adjusted using HCM Equation 31-110.

PEDESTRIANS AND BICYCLES

A separate procedure for the saturation flow rate adjustments for pedestrian and bicycle influence on turning vehicles is presented in Section 2 of HCM Chapter 31.

WORK ZONES

The presence of work zones within 250 feet of the stop bar generates caution, which results in a lower saturation flow rate as illustrated in HCM Exhibit 31-9. The adjustment is defined in HCM Equations 31-89, 31-90, and 31-91 with HCM Exhibit 31-10. The reduction is significant with no reduction in approach width or number of lanes, but is reduced further if either of those situations exists.

Figure 6.13 *HCM Exhibit 31-9*

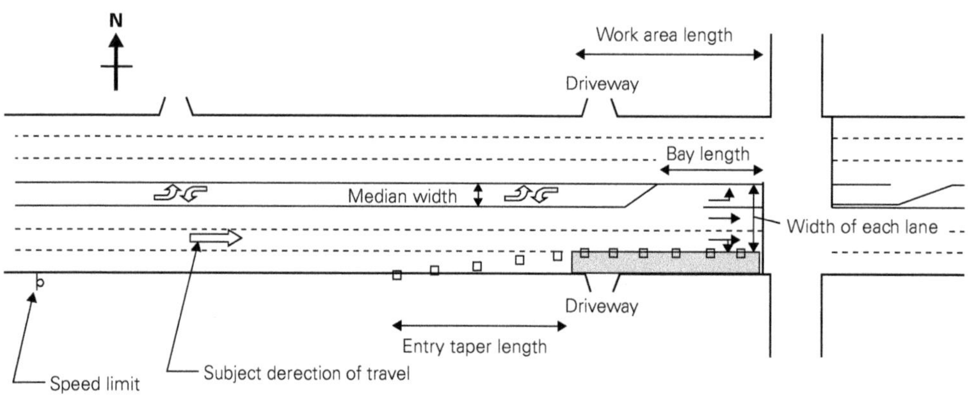

Figure 6.14 *HCM Equation 31-89*

$$f_{wz} = 0.858 \times f_{wid} \times f_{reduce} \quad \leq 1.0$$

Figure 6.15 *HCM Equation 31-90*

$$f_{wid} = \frac{1}{1 - 0.0057(a_w - 12)}$$

Figure 6.16 *HCM Equation 31-91*

$$f_{reduce} = \frac{1}{1 + 0.0402(n_o - n_{wz})}$$

where

f_{wz} = adjustment factor for work zone presence at the intersection;

f_{wid} = adjustment factor for approach width;

f_{reduce} = adjustment factor for reducing lanes during work zone presence;

a_w = approach lane width during work zone (= total width of all open left-turn, through, and right-turn lanes), ft;

n_o = number of left-turn and through lanes open during normal operation, ln; and

n_{wz} = number of left-turn and through lanes open during work zone presence, ln.

Figure 6.17 *HCM Exhibit 31-10*

Input Data Element and Units	Basis
Number of lanes open on the approach in the work zone (ln)	Approach
Approach lane width during work zone (ft)	Approach

Think about your own driving:
Whenever you encounter orange cones or barrels while driving, you generally slow down and exercise more caution. This is exactly the purpose of the advance warning signs and the barriers themselves, since the normally expected environment has been changed.

Timing

From the control perspective, the analysis of a signalized intersection as defined in HCM Chapter 19 will typically include actuated phases using the phase duration model. Phases are defined following the National Electrical Manufacturers Association (NEMA) standard that can include permitted left turns and right-turn overlaps that extend beyond NEMA definitions.

Required Data

The operation type must be defined among pretimed, fully actuated, semi-actuated, or coordinated-actuated for the analysis to properly apply the controller and timing parameters. All phases are specified, following the NEMA standard with Phase 2 and 4 directions provided. Data provided for each phase include: maximum and minimum green times; yellow change and red clearance times; passage times; leading or lagging sequences; side-street split phasing sequences; recall mode as minimum, maximum, pedestrian, or off; dual entry and simultaneous gap settings; and Dallas phasing if present. The approach speed limit is needed to compute maximum allowable headway time used in the phase duration model. For coordinated-actuated operation, the cycle length is specified along with force mode as fixed or floating, and phase split replaces maximum green time data.

Operation

The type of signal operation dictates how the intersection is modeled as a function of which phases are pretimed and which are actuated.

This combination is defined in four types as pretimed, fully actuated, semi-actuated, or coordinated-actuated.

PRETIMED

Pretimed operation provides set times for all phases and ignores the demand, so no actuated data like passage time or recall mode are needed. Instead, the provided phase times are fixed and used as the average phase times for the purposes of computing effective green time.

FULLY ACTUATED

Fully actuated operation allows all phase times to vary as functions of demand, honoring the minimum and maximum green times. The phase duration model is applied to all phases to compute the average times for multiple cycles within the fifteen-minute analysis period.

SEMI-ACTUATED

Semi-actuated operation allows all phase times (except the major-street through phases whose time are fixed as non-actuated) to vary as functions of demand, honoring the minimum and maximum green times. The phase duration model is applied to all actuated phases to compute the average times for multiple cycles within the fifteen-minute analysis period.

COORDINATED-ACTUATED

Coordinated-actuated operation allows all actuated phase times to vary as functions of demand, honoring the minimum and maximum green times. The major-street through phases are non-actuated, but vary to honor the cycle length and not as a function of through-movement demand. The phase duration model is applied to all actuated phases to compute the average times for multiple cycles within the fifteen-minute analysis period; then, the sum is subtracted from the cycle length to compute the non-actuated phase times.

Phasing

Phasing is dictated by NEMA as illustrated by the example shown in HCM Exhibit 19-2. Phases are numbered 1 through 8, with Phases 2 and 6 for major-street through movements;

Figure 6.18 *HCM Exhibit 19-2*

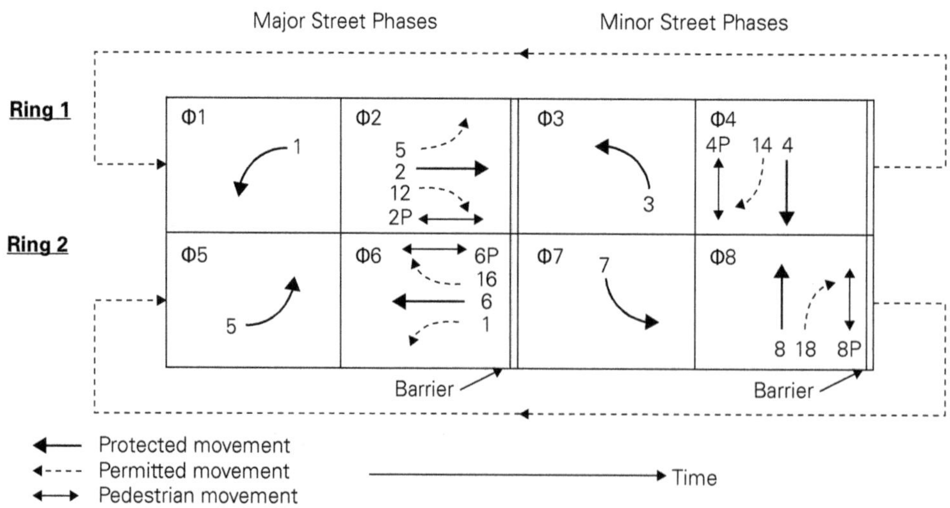

Phases 4 and 8 for minor-street through movements; Phases 1 and 5 for major-street left-turn movements; and Phases 3 and 7 for minor-street left-turn phases. Phases 1, 2, 3, and 4 are in Ring 1, and Phases 5, 6, 7, and 8 are in Ring 2 with only one phase from each ring allowed to run simultaneously.

LEFT TURNS

Left-turn phases can be permitted, protected, or protected-plus-permitted for either the major street or the side street. The left-turn phases can be leading (ahead of the adjacent through phases), lagging (after the adjacent through phases), or in a lead-lag sequence (with one direction leading and the opposite direction lagging).

In a lead-lag sequence, the left turns are normally protected only to prevent the left-turn (yellow) trap. Or Dallas phasing can be implemented with the leading left-turn phase extending the permitted portion of the phase into the lagging left-turn sequence to provide for lead-lag and protected-permitted phasing without the trap.

Phase Duration

An iterative process is implemented to compute the average phase duration for all actuated phases using the lane configuration, movement demands, available phases, and controller settings. Actuated phase times will typically vary cycle to cycle, but the average time over the multiple cycles in the fifteen-minute analysis period must be determined in order to compute the effective green time for actuated phases.

EFFECTIVE GREEN TIME

The effective green time is computed from the components of the phase duration process as shown in HCM Exhibit 19-7. Start-up Lost Time (l_1) represents the additional time it takes the first few cars to get up to the saturation flow rate from being stopped.

Clearance Lost Time (l_2) provides time for the vehicles using the ending phase to clear the intersection for the subsequent phase to begin. Queue Service Time (g_s) is the time required to process the vehicles in queue at the beginning of the phase. Green Extension Time

Figure 6.19 *HCM Exhibit 19-7]*

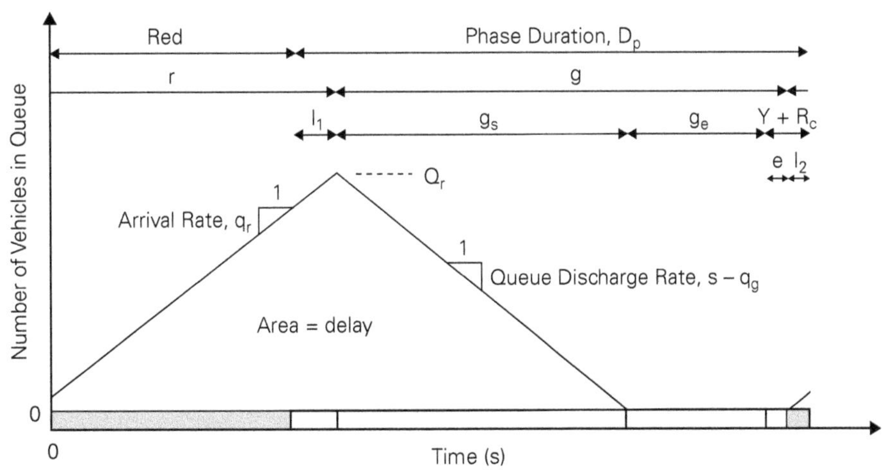

(g_e) provides additional green time for other vehicles arriving before the passage time ends. Extension of Effective Green (e) represents the amount of yellow time used as green.

Think about your own driving:

You sometimes have to go through the intersection on a yellow light because it came up when you were too close to stop. But sometimes you go through on the yellow because the intersection is congested and you just don't want to sit through another cycle. As a signalized intersection approaches oversaturation (v/c ratio approaches 1.0), drivers "push" the yellow time this way, and the extension of effective green increases.

The effective green time is the time vehicles are moving, which includes queue service time, green extension time, and extension of effective green time—but not start-up lost time and clearance lost time. Start-up lost time is assumed to be two seconds, and clearance lost time is yellow plus red minus extension of effective green time. From the phase duration components in HCM Equation 19-2, effective green can be computed using HCM Equation 19-3 by either adding the time moving or subtracting lost times from the phase duration.

Figure 6.20 *HCM Equation 19-2*

$$D_p = l_1 + g_s + g_e + Y + R_c$$

where

D_p = phase duration (s),
l_l = start-up lost time = 2.0 (s),
g_s = queue service time (s),
g_e = green extension time (s),
Y = yellow change interval (s), and
R_c = red clearance interval (s).

Figure 6.21 *HCM Equation 19-3*

$$\begin{aligned} g &= D_p - l_1 - l_2 \\ &= g_s + g_e + e \end{aligned}$$

where

l_2 = clearance lost time = $Y + R_c - e$ (s),
e = extension of effective green = 2.0 (s),

Capacity

After computing the adjusted saturation flow rate and effective green time for each lane group, capacity is computed using HCM Equation 19-16. The flow rate and capacity for each lane group are used to compute the volume-to-capacity (v/c) ratio, following HCM Equation 19-17, needed to compute delay.

Figure 6.22 *HCM Equation 19-16*

$$c = N s \frac{g}{C}$$

Figure 6.23 *HCM Equation 19-17*

$$X = \frac{v}{c}$$

where

X = volume-to-capacity ratio,
v = demand flow rate (veh/h), and
c = capacity (veh/h).

Delay

Delay is computed in three terms (control, incremental, and initial queue delay), then combined to generate control delay to determine level of service: first by lane group, then averaged for approaches and the intersection, following HCM Equation 19-18.

Figure 6.24 *HCM Equation 19-18*

$$d = d_1 + d_2 + d_3$$

where

d = control delay (s/veh),

d_1 = uniform delay (s/veh),

d_2 = incremental delay (s/veh), and

d_3 = initial queue delay (s/veh).

Uniform

Uniform delay (d_1) can be computed using HCM Equation 19-19, which assumes arrivals and departures are constant. The progression factor is needed for this process and is determined as a function of the proportion of vehicles arriving on green by using HCM Equations 19-5, 19-20, and 19-21, which increases with better coordination from the upstream signal. The higher the proportion arriving on green, the smaller the queue will be, with more vehicles arriving during the green time and fewer arriving during the red time that generates the queue.

Figure 6.25 *HCM Equation 19-19*

$$d_1 = PF \frac{0.5\,C(1-g/C)^2}{1-[\min(1,X)\,g/C]}$$

Figure 6.26 *HCM Equation 19-5*

$$R_p = \frac{P}{(g/C)}$$

where

R_p = platoon ratio,

P = proportion of vehicles arriving during the green indication (decimal),

g = effective green time (s), and

C = cycle length (s).

Figure 6.27 *HCM Equation 19-20*

$$PF = \frac{1-P}{1-g/C} \times \frac{1-y}{1-\min(1,X)P} \times \left[1 + y\frac{1-P\,C/g}{1-g/C}\right]$$

Figure 6.28 *HCM Equation 19-21*

$$y = \min(1,X)\,g/C$$

where

PF = progression adjustment factor,

y = flow ratio,

P = proportion of vehicles arriving during the green indication (decimal),

g = effective green time (s),

If arrivals and/or departures are not constant, the incremental queue accumulation method is used as defined by HCM Exhibit 19-24 and HCM Equations 19-34 and 19-35. This process accounts for varying arrival and/or departure rates for a more accurate modeling of multiple platoon arrivals and protected-permitted phasing departures.

Figure 6.29 *HCM Exhibit 19-24*

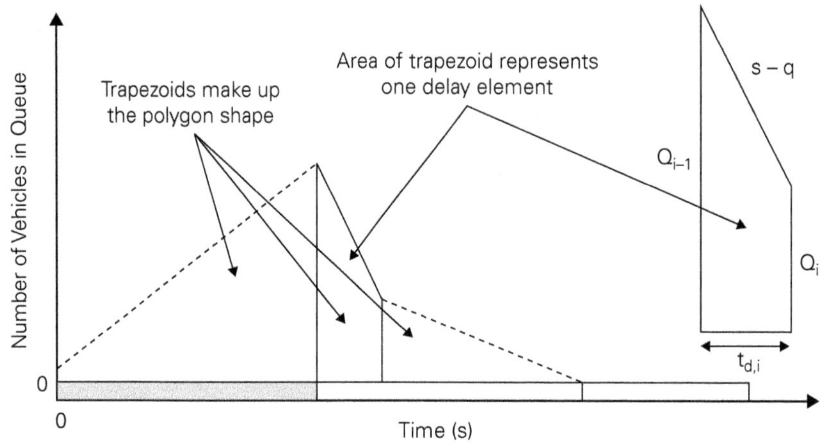

Figure 6.30 *HCM Equation 19-34*

$$Q_i = Q_{i-1} - \left(\frac{s}{3{,}600} - \frac{q}{N}\right)t_{d,i} \qquad \geq 0.0$$

where

Q_i = queue size at the end of interval i (veh),

N = the number of lanes in lane group (ln),

s = adjusted saturation flow rate (veh/h/ln),

$t_{d,i}$ = duration of time interval i during which the arrival flow rate and saturation flow rate are constant (s), and

Figure 6.31 *HCM Equation 19-35*

$$d_1 = \frac{0.5\Sigma(Q_{i-1} + Q_i)\, t_{t,i}}{q\,C}$$

$$t_{t,i} = \min(t_{d,i}, Q_{i-1}/w_q)$$

where

d_1 = uniform delay (s/veh),

$t_{t,i}$ = duration of trapezoid or triangle in interval i (s),

w_q = queue change rate (i.e., slope of the upper boundary of the trapezoid or triangle) (veh/s), and

Incremental

Incremental delay (d_2) is computed following HCM Equation 19-26 and requires the upstream filtering adjustment factor (I) from HCM Equation 19-6 and the incremental delay factor (k) using HCM Equations 19-22 through 19-25. The I-factor is assumed to be 1.0 for adjacent signals more than 0.6 miles away, and the k-value is assumed to be 0.5 for non-actuated movements.

Figure 6.32 *HCM Equation 19-26*

$$d_2 = 900\,T\left[(X_A - 1) + \sqrt{(X_A - 1)^2 + \frac{8\,k\,I\,X_A}{c_A\,T}}\right]$$

Figure 6.33 *HCM Equation 19-6*

$$I = 1.0 - 0.91\,X_u^{2.68} \qquad \geq 0.090$$

where

I = upstream filtering adjustment factor, and

X_u = weighted volume-to-capacity ratio for all upstream movements contributing to the volume in the subject movement group.

Figure 6.34 *HCM Equation 19-22*

$$k = (1 - 2\,k_{min})(v/c_a - 0.5) + k_{min} \qquad \leq 0.50$$

Figure 6.35 *HCM Equation 19-23*

$$k_{min} = -0.375 + 0.354\,PT - 0.0910\,PT^2 + 0.00889\,PT^3$$
$$\geq 0.04$$

Figure 6.36 *HCM Equation 19-24*

$$c_a = 3{,}600\frac{g_a\,s\,N}{C}$$

Figure 6.37 *HCM Equation 19-25*

$$g_a = G_{max} + Y + R_c - l_1 - l_2$$

where

k = incremental delay factor,

c_a = available capacity for a lane group served by an actuated phase (veh/h),

k_{min} = minimum incremental delay factor,

PT = passage time setting (s),

G_{max} = maximum green setting (s),

g_a = available effective green time (s)

This term is modeling the probability of cycle failure causing vehicles to wait through a subsequent cycle, which increases delay exponentially. As the v/c ratio approaches 1.0, this value rises quickly and can be quite significant for oversaturated conditions.

Initial Queue

Initial queue delay (d_3) is calculated from HCM Equation 19-44 for lane groups with unmet demand from the previous period generating an initial queue for the analysis period. This term can be very high as oversaturated conditions build and dissipate during a peak period analysis.

Figure 6.38 *HCM Equation 19-44*

$$d_3 = \frac{3{,}600}{vT}\left(t_A\frac{Q_b + Q_e - Q_{eo}}{2} + \frac{Q_e^2 - Q_{eo}^2}{2\,c_A} - \frac{Q_b^2}{2c_A}\right)$$

where

t_A = adjusted duration of unmet demand in the analysis period (h),

Q_e = queue at the end of the analysis period (veh),

Q_{eo} = queue at the end of the analysis period when $v \geq C_A$ and Q_b = 0.0 (veh)

The modeling of the congestion effects of one period spilling over to the subsequent period is extremely important to replicate field conditions. A multiple-period analysis is the only way to model a congested signalized intersection.

Averages

After computing the control delay for all lane groups, weighted averages facilitate the consolidated approach and intersection delays, following HCM Equations 19-28 and 19-29. All delays are computed in seconds per vehicle, which does not show differences relative to flow rate variance, so computing vehicle-hours of delay can be useful in comparing analyses for project priority purposes.

Figure 6.38a *HCM Equation 19-28*

$$d_{A,j} = \frac{\sum_{i=1}^{m_j} d_i v_i}{\sum_{i=1}^{m_j} v_i}$$

where

$d_{A,j}$ = approach control delay approach j (s/veh),
d_i = control delay for lane group i (s/veh),
m_j = number of lane groups on approach j, and

Figure 6.38b *HCM Equation 19-29*

$$d_I = \frac{\sum d_i v_i}{\sum v_i}$$

where d_I is the intersection control delay (s/veh)

Unsignalized movements can be considered in these averages if the delay for those movements can be computed or estimated outside of the HCM procedures, normally from field data or simulation tools.

Level of Service

Level of service for signalized intersections is based on the eighty-second scale shown in HCM Exhibit 19-8 and is applied to lane groups, approaches, and the intersection averages. If any lane group has a v/c ratio greater than 1.0, LOS F is applied, regardless of the delay value.

Figure 6.38c *HCM Exhibit 19-8*

Control Delay (s/veh)	LOS by Volume-to-Capacity Ratio[a] ≤1.0	>1.0
≤10	A	F
>10–20	B	F
>20–35	C	F
>35–55	D	F
>55–80	E	F
>80	F	F

Think about your own driving:
Of course, if you arrive at a signal on green and don't even slow down, that would be LOS A. And if you are in congested traffic and wait for more than one cycle to get through, that would be LOS F. In between, after about forty-five or fifty seconds, you get a little impatient, looking at the light, maybe moving up a little just to do something—that's when LOS D is turning into LOS E. For most of us, the scale is fairly reflective of our comfort and convenience levels.

Back of Queue

The cycle-to-cycle back of queue is computed for the average or a given percentile over the fifteen-minute analysis period according to HCM Equations 31-130 from the procedure detailed in HCM Chapter 31, Section 4. The queue storage ratio is computed, following HCM Equations 31-154 and 31-155 with percentiles using HCM Equation 31-156.

If the queue storage ratio is greater than 1.0 for a left- or right-turn lane group, the queue is longer than the available storage, causing turning vehicles to spill into the adjacent through lane. This interaction that reduces the through lane capacity is not modeled in the HCM procedures and must be estimated using an alternative tool such as simulation.

Figure 6.39 *HCM Equation 31-130*

$$Q = Q_1 + Q_2 + Q_3$$

where

Q = back-of-queue size (veh/ln),
Q_1 = first-term back-of-queue size (veh/ln),
Q_2 = second-term back-of-queue size (veh/ln), and
Q_3 = third-term back-of-queue size (veh/ln).

Figure 6.40 *HCM Equation 31-154*

$$R_Q = \frac{L_h Q}{L_a}$$

Figure 6.41 *HCM Equation 31-155*

$$L_h = L_{pc}(1 - 0.01\,P_{HV}) + 0.01\,L_{HV} P_{HV}$$

where

R_Q = queue storage ratio,
L_a = available queue storage distance (ft/ln),
L_h = average vehicle spacing in stationary queue (ft/veh),
L_{pc} = stored passenger car lane length = 25 (ft),
L_{HV} = stored heavy-vehicle lane length = 45 (ft), and
P_{HV} = percent heavy vehicles in the corresponding movement group (%).

Figure 6.42 *HCM Equation 31-156*

$$R_{Q\%} = \frac{L_h Q_{\%}}{L_a}$$

Pedestrians

The procedure uses circulation area to compute pedestrian space as an indicator of the ability to move to rate that against a descriptive scale, but is not used in the level of service process that is based on an index score compared to the LOS scale. This index involves pedestrian delay as a function of signal timing and geometry, including intersection lanes, traffic, and approach speed.

REQUIRED DATA

For the level of service process, signal timing parameters are needed to compute pedestrian delay, including effective walk time; pedestrian clearance time; phase duration; cycle length; and yellow change and red clearance intervals. For other index terms, intersection geometry and demand information is needed, including number of traffic lanes crossed; channelized islands, if any; vehicular major-street demand; and the 85th percentile mid-segment approach speed.

PROCEDURE

Pedestrian delay is computed in average seconds per pedestrian using HCM Equation 19-70 from the signal cycle length and effective walk time that comes from HCM Equations 19-54, 19-55, and 19-56.

Figure 6.43 *HCM Equation 19-70*

$$d_p = \frac{(C - g_{Walk,mi})^2}{2C}$$

Figure 6.44 *HCM Equation 19-54*

The following guidance is provided to estimate the effective walk time on the basis of the aforementioned research findings. If the phase providing service to the pedestrians is either (*a*) actuated with a pedestrian signal head and rest-in-walk not enabled or (*b*) pretimed with a pedestrian signal head, then

$$g_{walk,mi} = Walk_{mi} + 4.0$$

Figure 6.45 *HCM Equation 19-55*

If the phase providing service to the pedestrians is actuated with a pedestrian signal head and rest-in-walk enabled, then

$$g_{Walk,mi} = D_{p,mi} - Y_{mi} - R_{c,mi} - PC_{mi} + 4.0$$

Figure 6.46 *HCM Equation 19-56*

Otherwise (i.e., no pedestrian signal head)

$$g_{Walk,mi} = D_{p,mi} - Y_{mi} - R_{c,mi}$$

where

$g_{Walk,mi}$ = effective walk time for the phase serving the minor-street through movement (s),
$Walk_{mi}$ = pedestrian walk setting for the phase serving the minor-street through movement (s),
PC_{mi} = pedestrian clear setting for the phase serving the minor-street through movement (s),
$D_{p,mi}$ = duration of the phase serving the minor-street through movement (s),
Y_{mi} = yellow change interval of the phase serving the minor-street through movement (s), and
$R_{c,mi}$ = red clearance interval of the phase serving the minor-street through movement (s).

The cross-section adjustment factor is computed using HCM Equation 19-72. The motorized vehicle volume adjustment factor is computed using HCM Equation 19-73. The motorized vehicle speed adjustment factor is computed using HCM Equations 19-74 and 19-76. The pedestrian delay adjustment factor is computed by employing HCM Equation 19-75.

Figure 6.47 *HCM Equation 19-72*

$$F_w = 0.681\,(N_d)^{0.514}$$

Figure 6.48 *HCM Equation 19-73*

$$F_v = 0.00569\left(\frac{v_{rtor} + v_{lt,perm}}{4}\right) - N_{rtci,d}(0.0027\, n_{15,mj} - 0.1946)$$

Figure 6.49 *HCM Equation 19-74*

$$F_S = 0.00013\, n_{15,mj}\, S_{85,mj}$$

Figure 6.50 *HCM Equation 19-76*

$$n_{15,mj} = \frac{0.25}{N_d} \sum_{i \in m_d} v_i$$

Figure 6.51 *HCM Equation 19-75*

$$F_{delay} = 0.0401\, ln(d_{p,d})$$

PEDESTRIAN LOS

These are combined in HCM Equation 19-71 to generate the index score used in HCM Exhibit 19-9 for each crosswalk.

Figure 6.52 *HCM Equation 19-71*

$$I_{p,int} = 0.5997 + F_w + F_v + F_S + F_{delay}$$

where

$I_{p,int}$ = pedestrian LOS score for intersection,
F_w = cross-section adjustment factor,
F_v = motorized vehicle volume adjustment factor,
F_s = motorized vehicle speed adjustment factor,
F_{delay} = pedestrian delay adjustment factor,
$\ln_{(x)}$ = natural logarithm of x,
N_d = number of traffic lanes crossed when traversing Crosswalk D (ln),
$N_{rtci,d}$ = number of right-turn channelizing islands along Crosswalk D,
$n_{15,mj}$ = count of vehicles traveling on the major street during a 15-min period (veh/ln),
v_i = demand flow rate for movement i (veh/h),
m_d = set of all motorized vehicle movements that cross Crosswalk D (see figure in margin);
$S_{85,mj}$ = 85th percentile speed at a midsegment location on the major street

(min/h), and
$d_{p,d}$ = pedestrian delay when traversing Crosswalk D (s/p).

Figure 6.53 *HCM Exhibit 19-9*

LOS	LOS Scrore
A	≤2.00
B	>2.00–2.75
C	>2.75–3.50
D	>3.50–4.25
E	>4.25–5.00
F	>5.00

Bicycles

The procedure uses uniform delay similar to that computed for the motorized vehicle, but it is not used to determine LOS. This index is calculated as a function of geometry, including intersection lanes, traffic, and parking.

REQUIRED DATA

The effective green time of the adjacent vehicle lane is needed only to compute bicycle delay. For other index terms used in computing the index, intersection geometry and demand information are needed, including curb-to-curb width; width of outside through lane, bicycle lane, paved shoulder, and striped parking lane; number of through traffic lanes; vehicular left, through, and right; and the proportion of on-street parking occupied.

PROCEDURE

Bicycle delay is computed in average seconds per bicycle, HCM Equation 19-78, which used the bicycle capacity from HCM Equation 19-77.

Figure 6.54 *HCM Equation 19-78*

$$d_b = \frac{0.5\,C\,(1 - g_b/C)^2}{1 - \min\left[\frac{v_{bic}}{c_b}, 1.0\right]\frac{g_b}{C}}$$

where d_b is bicycle delay (s/bicycle), v_{bic} is bicycle flow rate (bicycles/h).

Figure 6.55 *HCM Equation 19-77*

$$c_b = s_b \frac{g_b}{C}$$

where

c_b = capacity of the bicycle lane (bicycles/h),
s_b = saturation flow rate of the bicycle lane = 2,000 (bicycles/h),
g_b = effective green time for the bicycle lane (s), and
C = cycle length (s).

The cross-section adjustment factor is computed using HCM Equation 19-80 using the width parameters combined in HCM Equation 19-82. The motorized vehicle volume adjustment factor is computed using HCM Equation 19-81.

Figure 6.56 *HCM Equation 19-80*

$$F_w = 0.0153\,W_{cd} - 0.2144\,W_t$$

Figure 6.57 *HCM Equation 19-82*

$$W_t = W_{ol} + W_{bl} + W_{os}^{*} + I_{pk}W_{pk}$$

Figure 6.58 *HCM Equation 19-81*

$$F_v = 0.0066\frac{v_{lt} + v_{th} + v_{rt}}{4\,N_{th}}$$

Figure 6.59 *HCM Equation 19-79*

$$I_{b,int} = 4.1324 + F_w + F_v$$

where

$I_{b,nt}$ = bicycle LOS score for intersection;
F_w = cross-section adjustment factor,
F_v = motorized vehicle volume adjustment factor,
W_{cd} = curb-to-curb width of the cross street (ft);
W_t = total width of the outside through lane, bicycle lane, and paved shoulder (ft);
v_{lt} = left-turn demand flow rate (veh/h);
v_{th} = through demand flow rate (veh/h);
v_{rt} = right-turn demand flow rate (veh/h);
N_{th} = number of through lanes (shared or exclusive) (ln);
W_{ol} = width of the outside through lane (ft);
W_{bl} = width of the bicycle lane = 0.0 if bicycle lane not provided (ft);
W_{pk} = width of striped parking lane (ft);
I_{pk} = indicator variable for on-street parking occupancy = 0 if $p_{pk} > 0.0$, 1 otherwise;
p_{pk} = proportion of on-street parking occupied (decimal);
W_{os} = width of paved outside shoulder (ft); and
W^{*}_{os} = adjusted width of paved outside shoulder; if curb is present $W^{*}_{os} = W_{os} - 1.5 \geq 0.0$, otherwise $W^{*}_{os} = W_{os}$ (ft).

BICYCLE LOS

These are combined in HCM Equation 19-79 to generate the index score used in HCM Exhibit 19-9 for each crosswalk.

Figure 6.60 *HCM Exhibit 19-9*

LOS	LOS Scrore
A	≤2.00
B	>2.00–2.75
C	>2.75–3.50
D	>3.50–4.25
E	>4.25–5.00
F	>5.00

PROBLEM ONE

Determine the eastbound right-turn (EBR) lane group LOS for an isolated and undersaturated pretimed signalized intersection in the CBD of an area with 150,000 population. The EBR 13-foot lane on a 3% grade has a demand of 400 veh/h (including 20 RTOR) with a peak hour factor of 0.90. There are 10% heavy vehicles; 5 buses/h (stopping an average of 30 s); 20 parking maneuvers/h; and no pedestrians or bicycles. The signal operates on a 100 s cycle length with 55 s green, 4 s yellow, and 1 s red east-west and 35 s green, 4 s yellow, and 1 s red north-south, including 2 s start-up lost times and 2 s extension of effective green times.

1. Compute the EBR flow rate.

Flow Rate: $n_{60} = 400 - 20(\text{RTOR}) = 380$ veh/h $v_{EBR} = (380)/0.90 = 422$ veh/h
EQUATION 19-4

2. Compute the EBR saturation flow rate.

Base Saturation Flow Rate: Metro Population < 250,000 $s_o = 1750$ pc/h/ln
EXHIBIT 19-11

Lane Width 13-foot lane: $f_{lw} = 1.04$
EXHIBIT 19-20

Heavy Vehicles and Grade 10% heavy vehicles 3% grade:
$f_{HVg} = (100 - 0.78(10) - 0.31(3)^2)/100 = 0.89$
EQUATION 19-10

Parking 20 maneuvers: $f_p = (1 - 0.1 - 18(20)/3600)/1 = 0.80$
EQUATION 19-11

Bus Blockage 5 buses stop for 30 s: $f_{bb} = (1 - 30(5)/3600)/1 = 0.96$
EQUATION 19-12

Area Type CBD: $f_a = 0.90$
HCM PAGE 19-47

Lane Utilization One EBR lane: $f_{LU} = 1.0$
EQUATION 19-7

Right Turns: $E_R = 1.18$ $f_{RT} = 1/1.18 = 0.85$
EQUATION 19-13

Left Turns No LT in EBR lane: $f_{LT} = 1.00$
EQUATION 19-14

Adjusted Saturation Flow Rate: s = (1750)(1.04)(0.89)(0.80)(0.96)(0.90)(0.85) = 952 pc/h/ln
EQUATION 19-8

3. Compute the EBR effective green time.

Phase Duration: Phase duration = 55 + 4 + 1 = 60 s
EQUATION 19-2

Effective Green Time: $l_1 = 2$ s $l_2 = 4 + 1 - 2 = 3$ s $g = 60 - 2 - 3 = 55$ s
EQUATION 19-3

4. Compute the EBR capacity.

Capacity: $g/C = 55/100 = 0.55$ $c = 952(0.55) = 523$ veh/h
EQUATION 19-16

5. Compute the EBR volume-to-capacity ratio.

Volume-to-Capacity Ratio: $v/c = 422/523 = 0.81$
EQUATION 19-16

6. Determine the EBR progression factor.

Platoon Ratio Isolated = Random Arrivals: $R_p = 1.0$
EXHIBIT 19-13

Proportion Arriving on Green: $P = 1.0(55/100) = 0.55$
EQUATION 19-15

Progression Factor: $PF = 1.0$
EQUATIONS 19-20 & 19-21

7. Compute EBR delay.

Uniform Delay: $d_1 = 1.0((0.5)(100)(1–0.55)^2)/(1–(0.81)(0.55)) = 18.2$ s/veh
EQUATION 19-19

Incremental Delay:

$d_2 = 900(0.25)((0.81 - 1) + ((0.81-1)^2 + 8(0.5)(1.0)(0.81)/523(0.25))^{0.5} = 12.5$ s/veh

EQUATION 19-26

Initial Queue Delay: $d_3 = 0$ s No initial queue

EQUATION 19-44

Control Delay: $d = 18.2 + 12.5 = 30.7$ s/veh

EQUATION 19-18

8. Determine the EBR level of service.

LOS: LOS = C (between 20 and 35)

EXHIBIT 19-8

HCS7

These screenshots show how Problem One is coded into *HCS7*, followed by the results as presented in the formatted report. All input data, intermediate computations, and final results are shown in logical order, leading to delay and level of service.

Figure 6.61 *HCS7 Signals1 Input*

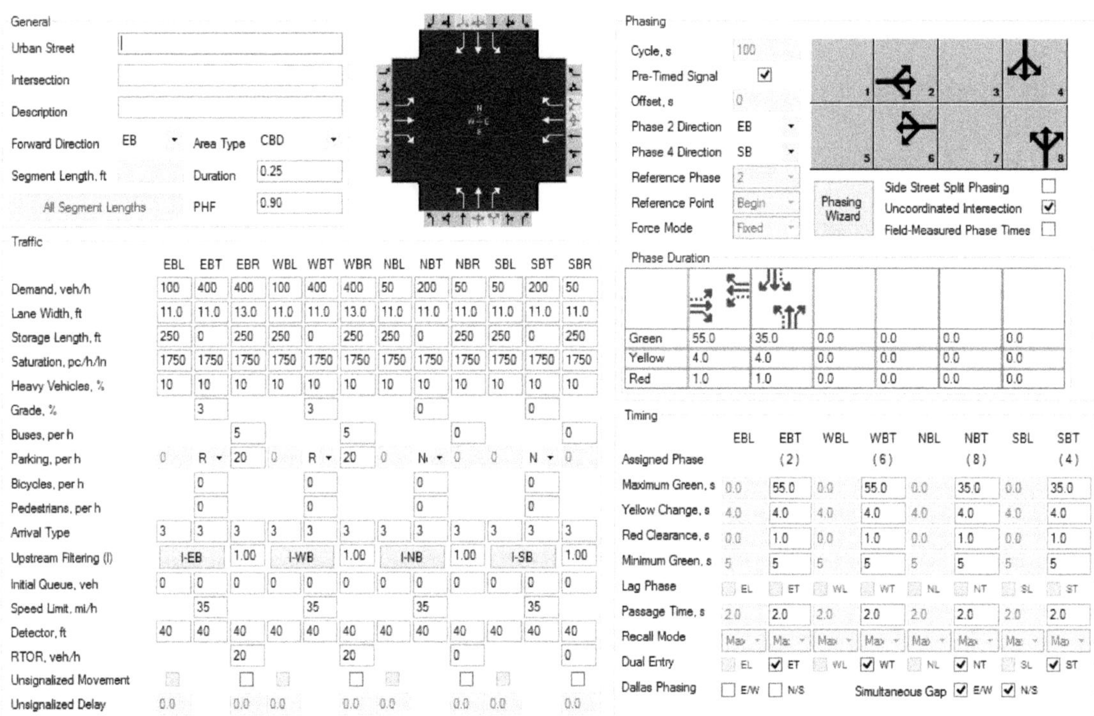

Figure 6.62 *HCS7 Signals1 Output*

General Information				Intersection Information	
Agency				Duration, h	0.25
Analyst		Analysis Date	October 23, 2010	Area Type	CBD
Jurisdiction		Time Period		PHF	0.90
Urban Street		Analysis Year	2010	Analysis Period	1> 7:00
Intersection		File Name	TextSignals1.xus		
Project Description					

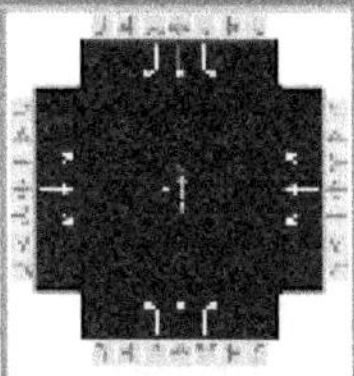

Demand Information	EB			WB			NB			SB		
Approach Movement	L	T	R	L	T	R	L	T	R	L	T	R
Demand (*v*), veh/h	100	400	400	100	400	400	50	200	50	50	200	50

Signal Information										
Cycle, s	100.0	Reference Phase	2							
Offset, s	0	Reference Point	Begin	Green	55.0	35.0	0.0	0.0	0.0	0.0
Uncoordinated	Yes	Simult. Gap E/W	On	Yellow	4.0	4.0	0.0	0.0	0.0	0.0
Force Mode	Fixed	Simult. Gap N/S	On	Red	1.0	1.0	0.0	0.0	0.0	0.0

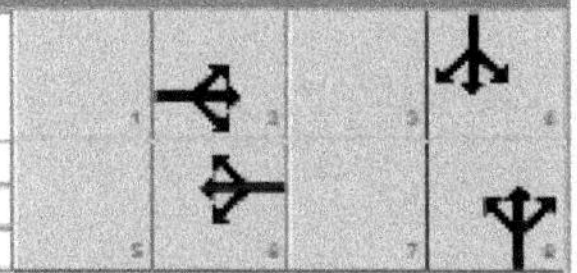

Timer Results	EBL	EBT	WBL	WBT	NBL	NBT	SBL	SBT
Assigned Phase		2		6		8		4
Case Number		5.0		5.0		5.0		5.0
Phase Duration, s		60.0		60.0		40.0		40.0
Change Period, ($Y+R_c$), s		5.0		5.0		5.0		5.0
Max Allow Headway (*MAH*), s		3.4		3.4		3.3		3.3
Queue Clearance Time (g_s), s		37.9		37.9		18.4		18.4
Green Extension Time (g_e), s		5.2		5.2		1.3		1.3
Phase Call Probability		1.00		1.00		1.00		1.00
Max Out Probability		0.20		0.20		0.00		0.00

Movement Group Results	EB			WB			NB			SB		
Approach Movement	L	T	R	L	T	R	L	T	R	L	T	R
Assigned Movement	5	2	12	1	6	16	3	8	18	7	4	14
Adjusted Flow Rate (*v*), veh/h	111	444	422	111	444	422	56	222	56	56	222	56
Adjusted Saturation Flow Rate (*s*), veh/h/ln	797	1408	952	797	1408	952	977	1452	1231	977	1452	1231
Queue Service Time (g_s), s	10.7	20.8	35.9	10.7	20.8	35.9	4.6	11.7	3.1	4.6	11.7	3.1
Cycle Queue Clearance Time (g_c), s	31.4	20.8	35.9	31.4	20.8	35.9	16.4	11.7	3.1	16.4	11.7	3.1
Green Ratio (*g/C*)	0.55	0.55	0.55	0.55	0.55	0.55	0.35	0.35	0.35	0.35	0.35	0.35
Capacity (*c*), veh/h	345	775	523	345	775	523	299	508	431	299	508	431
Volume-to-Capacity Ratio (*X*)	0.322	0.574	0.807	0.322	0.574	0.807	0.186	0.437	0.129	0.186	0.437	0.129
Back of Queue (*Q*), ft/ln (50 th percentile)	59.2	182.2	241.5	59.2	182.2	241.5	32	116.2	25.5	32	116.2	25.5
Back of Queue (*Q*), veh/ln (50 th percentile)	2.2	6.7	8.9	2.2	6.7	8.9	1.2	4.3	0.9	1.2	4.3	0.9
Queue Storage Ratio (*RQ*) (50 th percentile)	0.24	0.00	0.97	0.24	0.00	0.97	0.13	0.00	0.10	0.13	0.00	0.10
Uniform Delay (d_1), s/veh	25.1	14.8	18.2	25.1	14.8	18.2	31.2	24.9	22.1	31.2	24.9	22.1
Incremental Delay (d_2), s/veh	2.5	3.1	12.5	2.5	3.1	12.5	1.4	2.7	0.6	1.4	2.7	0.6
Initial Queue Delay (d_3), s/veh	0.0	0.0	0.0	0.0	0.0	0.0	0.0	0.0	0.0	0.0	0.0	0.0
Control Delay (*d*), s/veh	27.6	17.9	30.8	27.6	17.9	30.8	32.6	27.7	22.7	32.6	27.7	22.7
Level of Service (LOS)	C	B	C	C	B	C	C	C	C	C	C	C
Approach Delay, s/veh / LOS	24.5	C		24.5	C		27.7	C		27.7	C	
Intersection Delay, s/veh / LOS	25.3						C					

PROBLEM TWO

Building on Problem One, compute the delay and LOS for the EBR lane group after moving 10 seconds from the NS phases to the EW phases, maintaining the 100-second cycle length.

1. Compute the EBR flow rate.

Flow Rate: $n_{60} = 400 - 20(\text{RTOR}) = 380$ veh/h $v_{EBR} = (380)/0.90 = 422$ veh/h
EQUATION 19-4

2. Compute the EBR saturation flow rate.

Adjusted Saturation Flow Rate: s = (1750)(1.04)(0.89)(0.80)(0.96)(0.90)(0.85) = 952 pc/h/ln
EQUATION 19-8

3. Compute the EBR effective green time.

Phase Duration: Phase duration = 65 + 4 + 1 = 70 s
EQUATION 19-2

Effective Green Time: $l_1 = 2$ s $l_2 = 4 + 1 - 2 = 3$ s $g = 70 - 2 - 3 = 65$ s
EQUATION 19-3

4. Compute the EBR capacity.

Capacity: $g/C = 65/100 = 0.65$ $c = 952(0.65) = 618$ veh/h
EQUATION 19-16

5. Compute the EBR volume-to-capacity ratio.

Volume-to-Capacity Ratio: $v/c = 422/618 = 0.68$
EQUATION 19-16

6. Determine the EBR progression factor.

Platoon Ratio Isolated = Random Arrivals: $R_p = 1.0$
EXHIBIT 19-13

Proportion Arriving on Green: $P = 1.0(65/100) = 0.65$
EQUATION 19-15

Progression Factor: $PF = 1.0$
EQUATIONS 19-20 & 19-21

7. Compute EBR delay.

Uniform Delay: $d_1 = 1.0((0.5)(100)(1–0.65)^2)/(1–(0.68)(0.65)) = 11.0$ s/veh
EQUATION 19-19

Incremental Delay:

$d_2 = 900(0.25)((0.68 – 1) + ((0.68\text{-}1)^2 + 8(0.5)(1.0)(0.68)/618(0.25))^{0.5} = 6.0$ s/veh
EQUATION 19-26

Initial Queue Delay: $d_3 = 0$ s No initial queue
EQUATION 19-44

Control Delay: $d = 11.0 + 6.0 = 17.0$ s/veh
EQUATION 19-18

8. Determine the EBR level of service.

LOS: LOS = B (between 10 and 20)
EXHIBIT 19-8

HCS7

These screenshots show how Problem Two is coded into *HCS7* by shifting the 10 seconds from NS to EW, followed by the results as presented in the formatted report. All input data, intermediate computations, and final results are shown in logical order, leading to delay and level of service.

Figure 6.63 *HCS7 Signals2 Input*

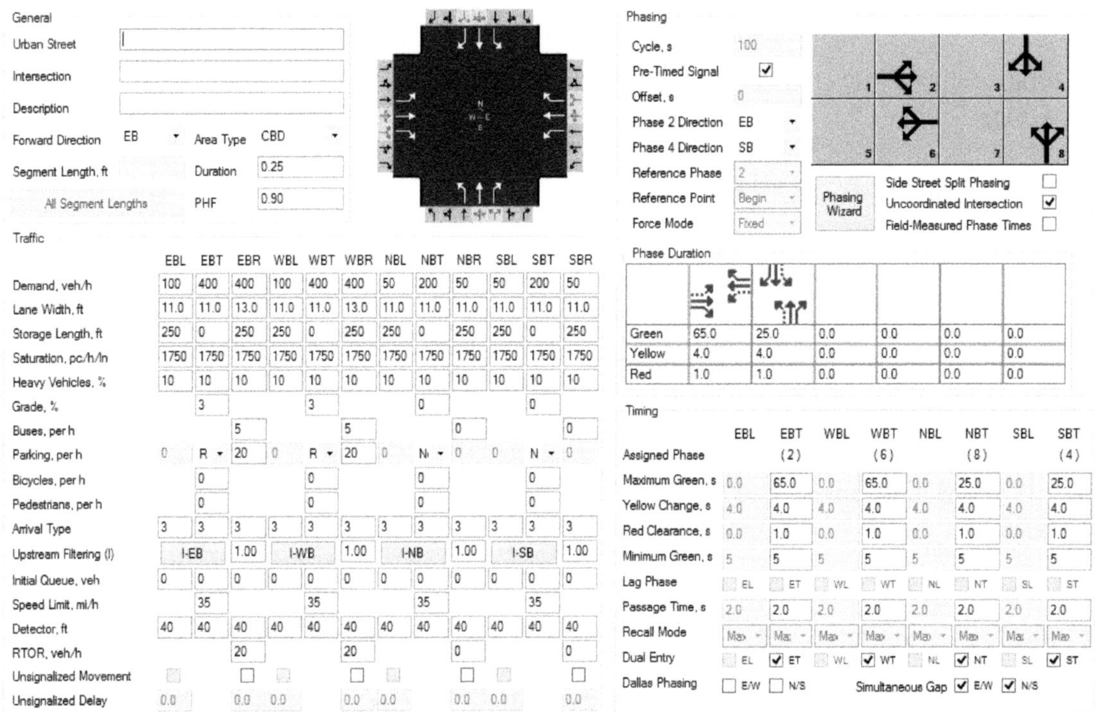

Figure 6.64 *HCS7 Signals2 Output*

General Information				Intersection Information	
Agency				Duration, h	0.25
Analyst		Analysis Date	October 23, 2010	Area Type	CBD
Jurisdiction		Time Period		PHF	0.90
Urban Street		Analysis Year	2010	Analysis Period	1> 7:00
Intersection		File Name	TextSignals2.xus		
Project Description					

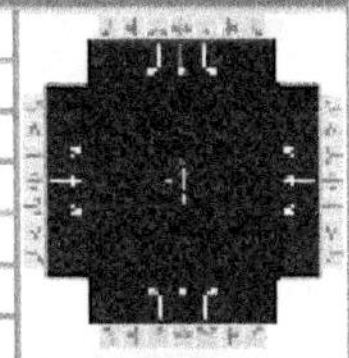

Demand Information	EB			WB			NB			SB		
Approach Movement	L	T	R	L	T	R	L	T	R	L	T	R
Demand (v), veh/h	100	400	400	100	400	400	50	200	50	50	200	50

Signal Information										
Cycle, s	100.0	Reference Phase	2							
Offset, s	0	Reference Point	Begin	Green	65.0	25.0	0.0	0.0	0.0	0.0
Uncoordinated	Yes	Simult. Gap E/W	On	Yellow	4.0	4.0	0.0	0.0	0.0	0.0
Force Mode	Fixed	Simult. Gap N/S	On	Red	1.0	1.0	0.0	0.0	0.0	0.0

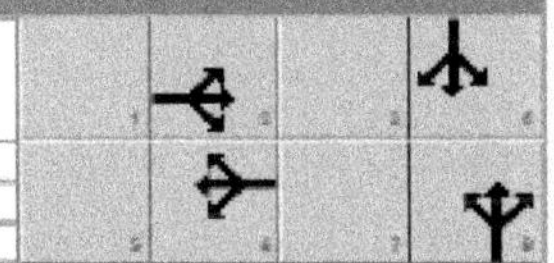

Timer Results	EBL	EBT	WBL	WBT	NBL	NBT	SBL	SBT
Assigned Phase		2		6		8		4
Case Number		5.0		5.0		5.0		5.0
Phase Duration, s		70.0		70.0		30.0		30.0
Change Period, ($Y+R_c$), s		5.0		5.0		5.0		5.0
Max Allow Headway (*MAH*), s		3.4		3.4		3.3		3.3
Queue Clearance Time (g_s), s		29.9		29.9		20.9		20.9
Green Extension Time (g_e), s		5.8		5.8		0.7		0.7
Phase Call Probability		1.00		1.00		1.00		1.00
Max Out Probability		0.02		0.02		0.68		0.68

Movement Group Results	EB			WB			NB			SB		
Approach Movement	L	T	R	L	T	R	L	T	R	L	T	R
Assigned Movement	5	2	12	1	6	16	3	8	18	7	4	14
Adjusted Flow Rate (v), veh/h	111	444	422	111	444	422	56	222	56	56	222	56
Adjusted Saturation Flow Rate (s), veh/h/ln	797	1408	952	797	1408	952	977	1452	1231	977	1452	1231
Queue Service Time (g_s), s	8.3	16.1	27.9	8.3	16.1	27.9	5.3	13.6	3.5	5.3	13.6	3.5
Cycle Queue Clearance Time (g_c), s	24.4	16.1	27.9	24.4	16.1	27.9	18.9	13.6	3.5	18.9	13.6	3.5
Green Ratio (g/C)	0.65	0.65	0.65	0.65	0.65	0.65	0.25	0.25	0.25	0.25	0.25	0.25
Capacity (c), veh/h	461	915	618	461	915	618	184	363	308	184	363	308
Volume-to-Capacity Ratio (X)	0.241	0.486	0.683	0.241	0.486	0.683	0.302	0.612	0.181	0.302	0.612	0.181
Back of Queue (Q), ft/ln (50 th percentile)	42.9	128.3	163.1	42.9	128.3	163.1	39.9	145.9	30.8	39.9	145.9	30.8
Back of Queue (Q), veh/ln (50 th percentile)	1.6	4.8	6.0	1.6	4.8	6.0	1.5	5.4	1.1	1.5	5.4	1.1
Queue Storage Ratio (*RQ*) (50 th percentile)	0.17	0.00	0.65	0.17	0.00	0.65	0.16	0.00	0.12	0.16	0.00	0.12
Uniform Delay (d_1), s/veh	15.2	8.9	11.0	15.2	8.9	11.0	41.6	33.2	29.5	41.6	33.2	29.5
Incremental Delay (d_2), s/veh	1.2	1.8	6.0	1.2	1.8	6.0	4.2	7.5	1.3	4.2	7.5	1.3
Initial Queue Delay (d_3), s/veh	0.0	0.0	0.0	0.0	0.0	0.0	0.0	0.0	0.0	0.0	0.0	0.0
Control Delay (d), s/veh	16.4	10.8	17.0	16.4	10.8	17.0	45.8	40.7	30.7	45.8	40.7	30.7
Level of Service (LOS)	B	B	B	B	B	B	D	D	C	D	D	C
Approach Delay, s/veh / LOS	14.1		B	14.1		B	39.9		D	39.9		D
Intersection Delay, s/veh / LOS	20.7						C					

CHAPTER 7 Streets

Urban Street models multiple signalized intersections to include the running time along the segments with through lane group delay to compute travel speed for level of service. The analysis of each signalized intersection (covered in Chapter Five) is used in this procedure, but with additional segment information that allows for more adjustments to saturation flow and a better computation for proportion arriving on green for a more appropriate calculation of through lane group delay. Procedures are defined for determining pedestrian, bicycle, and transit level of service as well. Procedures are defined for modeling travel time reliability with demand, weather, incidents, work zones, and special events varying for a distribution of results for hundreds of scenarios to produce travel time reliability indices at three different percentiles.

Required Data

Beyond those data required to model one signalized intersection, other information must be provided to define the segment. These parameters include: length of segment; speed limit; restricted median and curb percentages; percent on-street parking; access point density; mid-segment lanes and demand; and access point lanes and demands. Calibration for local conditions and mid-segment lane blockage can also be accommodated.

Flow Balancing

For these iterative models to work, the flows from intersection to intersection (including access points) must balance. The procedure assumes the departing flows are correct and adjusts the arriving flows downstream to match. Collecting field data on different days (which is common) will mean these values will essentially never match, so some adjustments will virtually always be required. Adjustments of up to 10% or so are acceptable with volumes varying by that much day to day, but beyond 10% may represent an issue—usually the failure to include access points in the analysis, where more traffic turns off of than on to the segment (or vice versa).

Running Time

Base Free-Flow Speed

Base free-flow speed is included in the running time computations, but also serves to stratify travel speed for the purposes of determining level of service. There are five components to HCM Equation 18-3, including four computational terms and one calibration term, with all computational adjustments coming from HCM Exhibit 18-11. The Speed Constant (S_o) is determined from the segment speed limit. The Adjustment for Cross-Section (f_{cs}) is a function of percent restrictive median and curb presence. The Adjustment for Access Points (f_A) is determined from the access point density and number of segment lanes. The Adjustment for Parking (f_P) uses the percentage of on-street parking. If calibration is desired, the increase or decrease in speed is provided directly.

Figure 7.1 *HCM Equation 18-3*

$$S_{fo} = S_{calib} + S_0 + f_{CS} + f_A + f_{pk}$$

where

S_{fo} = base free-flow speed (mi/h),
S_{calib} = base free-flow speed calibration factor (mi/h),
S_0 = speed constant (mi/h),
f_{CS} = adjustment for cross section (mi/h),
f_A = adjustment for access points (mi/h), and
f_{pk} = adjustment of on-street parking (mi/h).

Figure 7.2 *HCM Exhibit 18-11*

Speed Limit (mi/h)	Speed Constant S_0 (mi/h)[a]	Median Type	Percent with Restrictive Median (%)	Adjustment for Cross Section f_{CS}(mi/h)[b] No Curb	Curb
25	37.4	Restrictive	20	0.3	–0.9
30	39.7		40	0.6	–1.4
35	42.1		60	0.9	–1.8
40	44.4		80	1.2	–2.2
45	46.8		100	1.5	–2.7
50	49.1	Nonrestrictive	Not applicable	0.0	–0.5
55	51.5	No median	Not applicable	0.0	–0.5

Access Density D_a (points/mi)	Adjustment for Access Points f_A by Lanes N_{th} (mi/h)[c] 1 Lane	2 Lanes	3 Lanes	Percent with On-Street Parking (%)	Adjustment for Parking (mi/h)[d]
0	0.0	0.0	0.0	0	0.0
2	–0.2	–0.1	–0.1	20	–0.6
4	–0.3	–0.2	–0.1	40	–1.2
10	–0.8	–0.4	–0.3	60	–1.8
20	–1.6	–0.8	–0.5	80	–2.4
40	–3.1	–1.6	–1.0	100	–3.0
60	–4.7	–2.3	–1.6		

Free-Flow Speed

Free-flow speed (S_f) is computed from the base free-flow speed to account for the length of the segment using HCM Equation 18-4. The free-flow speed is not permitted to be lower than the posted speed limit, so this becomes the bottom of the range allowed in HCM Equation 18-5.

where

f_L = signal spacing adjustment factor,

S_{fo} = base free-flow speed (mi/h), and

L_S = distance between adjacent signalized intersections (ft).

where S_f is the free-flow speed (mi/h), S_{pl} is the posted speed limit.

Figure 7.3 *HCM Equation 18-4*

$$f_L = 1.02 - 4.7\frac{S_{fo} - 19.5}{\max(L_S, 400)} \leq 1.0$$

Figure 7.4 *HCM Equation 18-5*

$$S_f = S_{fo} f_L \geq S_{pl}$$

Think about your own driving:
With free-flow speed defined as the average speed under low-volume conditions, when was the last time you were driving along an arterial at midnight (essentially alone) that was signed for a 45 mi/h posted speed limit, and you were driving 43 mi/h? While the computations can yield a free-flow speed lower than the provided speed limit, this becomes unreasonable and is therefore prohibited.

Proximity Adjustment

The proximity adjustment factor (f_v) is calculated using HCM Equation 18-6 using the free-flow speed with the mid-segment number of lanes (N_{th}) and demand flow rate (v_m). This adjustment accounts for traffic density within the running time and travel speed computations—notably missing in the HCM 2000.

where

f_v = proximity adjustment factor,

v_m = midsegment demand flow rate (veh/h),

N_{th} = number of through lanes on the segment in the subject direction of travel (ln), and

S_f = free-flow speed (mi/h).

Figure 7.5 *HCM Equation 18-6*

$$f_v = \frac{2}{1 + \left(1 - \frac{v_m}{52.8\, N_{th}\, S_f}\right)^{0.21}}$$

Access Points

Vehicles turning into access points along the segment can introduce delay to the following through traffic. For right turns, the following vehicle must slow down for the vehicle in front to navigate the turn at a much slower speed. For left turns, the following vehicle may be faced with a queue as the vehicle in front waits for a gap in oncoming traffic to make the left turn, especially on undivided segments with no turn lanes. These delays are modeled in HCM Chapter 30, Section 4, which includes thirty-seven equations and requires individual access point data. They can also be estimated for planning purposes using HCM Exhibit 18-13 assuming 10% left and right

Figure 7.6 *HCM Exhibit 18-13*

Midsegment Volume (Veh/h/ln)	Through Vehicle Delay (s/veh/pt) by Number of Through Lanes		
	1 Lane	2 Lanes	3 Lanes
200	0.04	0.04	0.05
300	0.08	0.08	0.09
400	0.12	0.15	0.15
500	0.18	0.25	0.15
600	0.27	0.41	0.15
700	0.39	0.72	0.15

turns (that can be modified proportionally if the turn percentages are lower).

> ***Think about your own driving:***
> You are driving along an arterial at 45 mi/h about 100 feet behind the car in front of you within a moving platoon when that car decides to turn right into a gas station driveway. You have to slow down to about 10 mi/h because of that vehicle's speed making the right turn, causing you delay in getting back up to 45 mi/h. The more access points and activity to and from them, the higher the overall delay due to turns within the running time computations.

Running Time

Running time is the average time to travel from one signal to the next, not including the delay at the downstream signalized intersection. It can be computed by using the free-flow speed, proximity adjustment factor, and delay due to turns along with start-up lost time and segment length, following HCM Equation 18-7.

Figure 7.7 *HCM Equation 18-7*

$$t_R = \frac{6.0 - l_1}{0.0025\,L} f_x + \frac{3{,}600\,L}{5{,}280\,S_f} f_v + \sum_{i=1}^{N_{ap}} d_{ap,i} + d_{other}$$

with

$$f_x = \begin{cases} 1.00 & \text{(signalized or STOP-controlled through movement)} \\ 0.00 & \text{(uncontrolled through movement)} \\ \min\left[\frac{v_{th}}{c_{th}}, 1.00\right] & \text{(YIELD-controlled through movement)} \end{cases}$$

where

t_R = segment running time (s);

l_1 = start-up lost time = 2.0 if signalized, 2.5 if STOP- or YIELD-controlled (s);

L = segment length (ft);

f_x = control-type adjustment factor;

v_{th} = through-demand flow rate (veh/h);

c_{th} = through-movement capacity (veh/h);

$d_{ap,i}$ = delay due to left and right turns from the street into access point intersection i (s/veh);

N_{ap} = number of influential access point approaches along the segment = $N_{ap,s}$+ $P_{ap,lt}\, N_{ap,o}$ (points);

$N_{ap,s}$ = number of access point approaches on the right side in the subject direction of travel (points);

$N_{ap,o}$ = number of access point approaches on the right side in the opposing direction of travel (points);

$P_{ap,lt}$ = proportion of $N_{ap,o}$ that can be accessed by a left turn from the subject direction of travel;

d_{other} = delay due to other sources along the segment (e.g., curb parking or pedestrians) (s/veh); and

all other variables are as previously defined.

Travel Speed

Travel speed adds the downstream through movement delay to the running time. The major difference between this delay and the one computed for an isolated signalized intersection is the ability to compute the proportion arriving on green. This is very important, since the more vehicles that arrive on green with zero delay lowers the through movement delay average. Computing the flow profile from the upstream signalized intersection, including the effects on running speed and platoon integrity, is critical to achieving an accurate computation of this delay.

Figure 7.8 *HCM Exhibit 18-14*

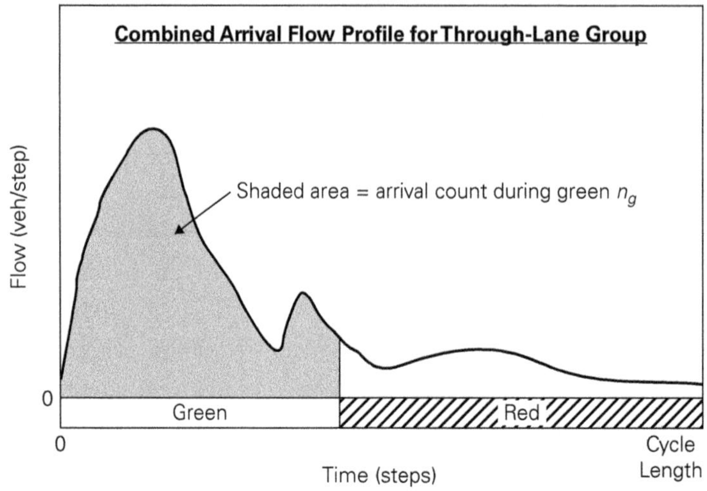

Flow Profile

The flow profile is computed following the process defined in HCM Chapter 30, Section 3 that comes from TRANSYT-7F, a well-established signal timing model. The platoon is modeled from the upstream signal contributing movements to the downstream approach that includes platoon dispersion (as vehicles get farther apart with increased speeds) and platoon decay (caused by access points increasing the distance between vehicles with the delay due to turns discussed earlier). The flow profile is illustrated in HCM Exhibit 18-14, and the proportion arriving on green is computed using HCM Equation 18-9.

Figure 7.9 *HCM Equation 18-9*

$$P = \frac{n_g}{q_d C}$$

where

P = proportion of vehicles arriving during the green indication,

n_g = arrival count during green (veh),

q_d = arrival flow rate for downstream lane group (veh/s), and

C = cycle length (s).

Saturation Flow

Another difference from analyzing isolated signalized intersections is the ability to model to segment situation by adjusting the saturation flow rate. Mid-segment blockage and sustained spillback are both modeled by iterating between the level of restriction in the segment by the blockage or spillback and the reduction created for the movements entering the segment. The results of these iterative methods generate the saturation flow rate adjustments for downstream lane blockage (f_{ms}) and sustained spillback (f_{sp}), following procedures in HCM Chapter 30, Section 3 and HCM Chapter 20, Section 3, respectively.

Travel Speed

Travel speed under prevailing demand and signal conditions is computed, following HCM Equation 18-15, which adds the running time to the downstream through movement delay (improved with proportion arriving on green from the flow profile and saturation flow rate adjustments for mid-segment blockage and sustained spillback), then converts this time to speed with the segment length.

Figure 7.10 *HCM Equation 18-15*

$$S_{T,seg} = \frac{3{,}600\,L}{5{,}280\,(t_R + d_t)}$$

where

$S_{T,seg}$ = travel speed of through vehicles for the segment (mi/h),
L = segment length (ft),
t_R = segment running time (s), and
d_t = through delay (s/veh).

Level of Service

Level of service is determined from HCM Exhibit 18-1 from the travel time stratified by base free-flow speed with the v/c ratio over 1.0 for the downstream through movement generating LOS F.

Facilities

Weighted averages of base free-flow speed and travel speed for all segments in an urban streets facility are computed following HCM Equations 16-2 and 16-3, respectively. Those averages are used to determine level of service following HCM Exhibit 16-3.

Figure 7.11 *HCM Equation 18-1*

LOS	Travel Speed Threshold by Base Free-Flow Speed (mi/h)							Volume-to-Capacity Ratio*
	55	50	45	40	35	30	25	
A	>44	>40	>36	>32	>28	>24	>20	≤ 1.0
B	>37	>34	>30	>27	>23	>20	>17	
C	>28	>25	>23	>20	>18	>15	>13	
D	>22	>20	>18	>16	>14	>12	>10	
E	>17	>15	>14	>12	>11	>9	>8	
F	≤17	≤15	≤14	≤12	≤11	≤9	≤8	
F				any				> 1.0

Figure 7.12 *HCM Equation 16-2*

$$S_{fo,F} = \frac{\sum_{i=1}^{m} L_i}{\sum_{i=1}^{m} \frac{L_i}{S_{fo,i}}}$$

where

$S_{fo,F}$ = base free-flow speed for the facility (mi/h),
L_i = length of segment i (ft),
m = number of segments on the facility, and
$S_{fo,i}$ = base free-flow speed for segment i (mi/h).

Figure 7.13 *HCM Equation 16-3*

$$S_{T,F} = \frac{\sum_{i=1}^{m} L_i}{\sum_{i=1}^{m} \frac{L_i}{S_{T,seg,i}}}$$

Travel Time Reliability

Most HCM analyses are based on fifteen-minute data from a typical peak period, but there may be a need to look at many more scenarios to cover the variations throughout the year. Modeling (potentially) hundreds of scenarios with varying parameters like demand, weather, incidents, work zones, and special events can generate a distribution of results to produce travel time reliability results like travel time indices and reliability ratings, following HCM Exhibit 17-2.

SCENARIOS

Demand data can normally be obtained from permanent counting stations that generate

Figure 7.14 *HCM Exhibit 16-3*

LOS	Travel Speed Threshold by Base Free-Flow Speed (mi/h)							Volume-to-Capacity Ratio*
	55	50	45	40	35	30	25	
A	>44	>40	>36	>32	>28	>24	>20	≤ 1.0
B	>37	>34	>30	>27	>23	>20	>17	
C	>28	>25	>23	>20	>18	>15	>13	
D	>22	>20	>18	>16	>14	>12	>10	
E	>17	>15	>14	>12	>11	>9	>8	
F	≤17	≤15	≤14	≤12	≤11	≤9	≤8	
F	any							> 1.0

adjustments by time of day, day of week, and month of year to create the distribution over time. Weather data are provided in a database for over one hundred cities to be able to select the one nearest the analysis facility. Incident data would generally come from crash reports for a given area, but can be in various formats with a variety of details. Work zone scenarios would be run independently with the work zone parameters in place for inserting during the appropriate time periods. Special events would also be analyzed separately with the typical increase in demand for inserting for those time periods.

Once the demand, weather, and incident databases are built and the analyses for any work

Figure 7.15 *HCM Exhibit 17-2*

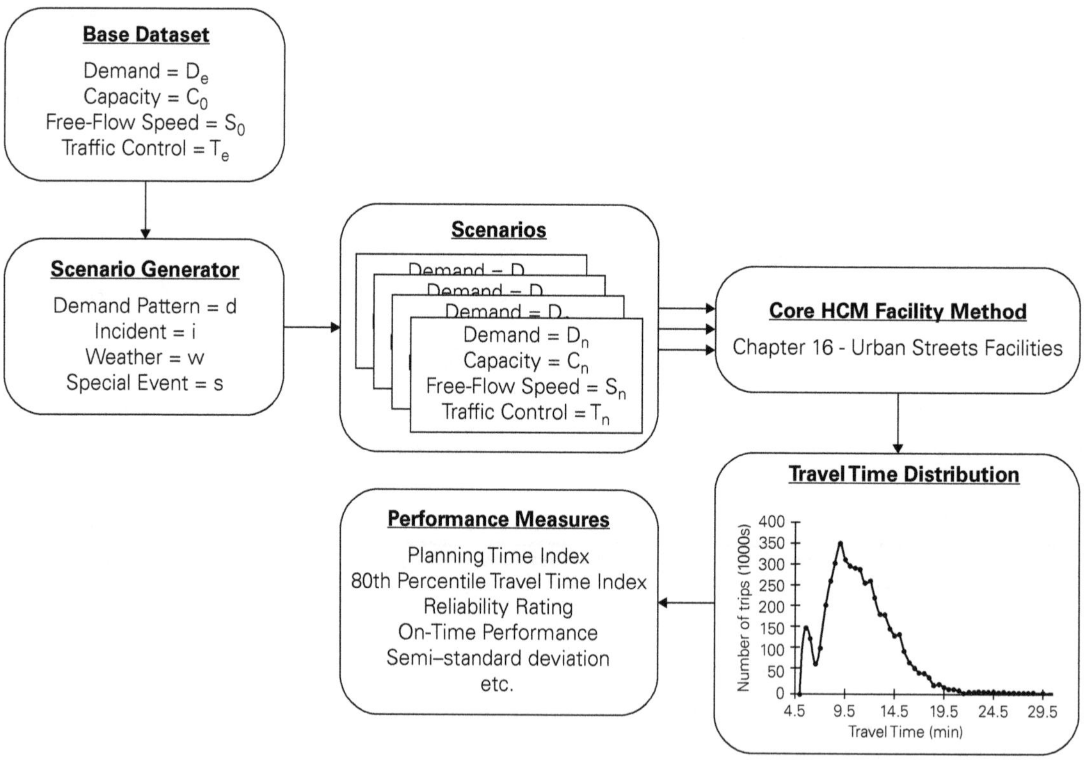

zones and/or special events are run, a base data set is provided for the typical fifteen-minute analysis for generating the additional scenarios for a specified set of days and times. While the work zone and special event timing is known, the demand, weather, and incident data must be randomized to populate the individual runs.

RESULTS

Results from all scenario analyses are compiled as a distribution to generate specific performance measures, including travel time, free-flow travel time, travel time index, percentile travel time index, and reliability rating. Actual travel time over the entire facility will vary with the conditions for each scenario. Free-flow travel time is simply the facility length divided by free-flow speed. Travel Time Index (TTI) is the ratio of the actual travel time to the free-flow travel time. Percentile travel time index is provided for the 50th (mean index), 80th (operational), and 95th (planning time index) percentiles. Reliability rating is the percentage of vehicle miles traveled with a travel time index under 1.33 that is considered reliable.

ATDM

Active Transportation Demand Management (ATDM) tactics include arterial monitoring; signal and speed control; geometric configuration; and demand modification. While all tactics are discussed, none has any analytical guidance with general topics shown in HCM Exhibit 17-1.

Optimization

Consistent with HCM Chapter 29, Section 4, one alternative tool that can be used to optimize cycle length, offsets, and phase splits is the Highway Capacity Software (HCS). The model uses a genetic algorithm process to iterate among variables using the HCM procedures directly, including multiple-period analysis, flow profile, and access point adjustments. The ability to specify weight to a given direction for a

Figure 7.16 *HCM Exhibit 17-1*

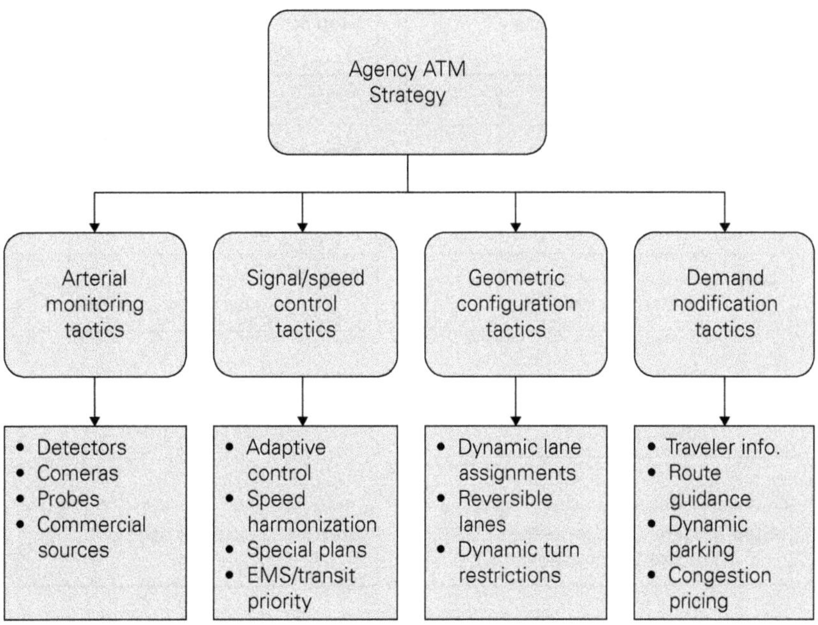

particular time-of-day timing plan is available as well to generate the optimum parameters for a given objective function among percent base free-flow speed (LOS), arterial delay, arterial stops, travel time, overall delay, and balanced delay.

Think about your own driving:

As you drive along an arterial, you want good coordination among the signals so you arrive on green and are not delayed at as many as possible. Most systems optimize the cycle, splits, and offsets to accomplish the best timing for different periods of the day since patterns change. For example, on a given arterial, most traffic may be going in one direction in the morning (to work) and the opposite direction in the afternoon (to home). Time-of-day plans account for this by changing throughout the day to accommodate these predictable patterns.

Pedestrians

Pedestrian level of service along an urban street segment includes the signalized intersection index score combined with an index score for the segment link as a function of average pedestrian space. The link score is based on many parameters, including vehicular lane widths and buffer; sidewalk width; on-street parking; and vehicular demand and speed, following the procedure illustrated in HCM Exhibit 18-17.

REQUIRED DATA

For the fairly comprehensive procedure, there are many parameters needed to compute the pedestrian link index score. For the pedestrian walkway, the data needed include: total width; fixed object width; shy distance; buffer width; and proportions adjacent to window displays,

Figure 7.17 *HCM Exhibit 18-17*

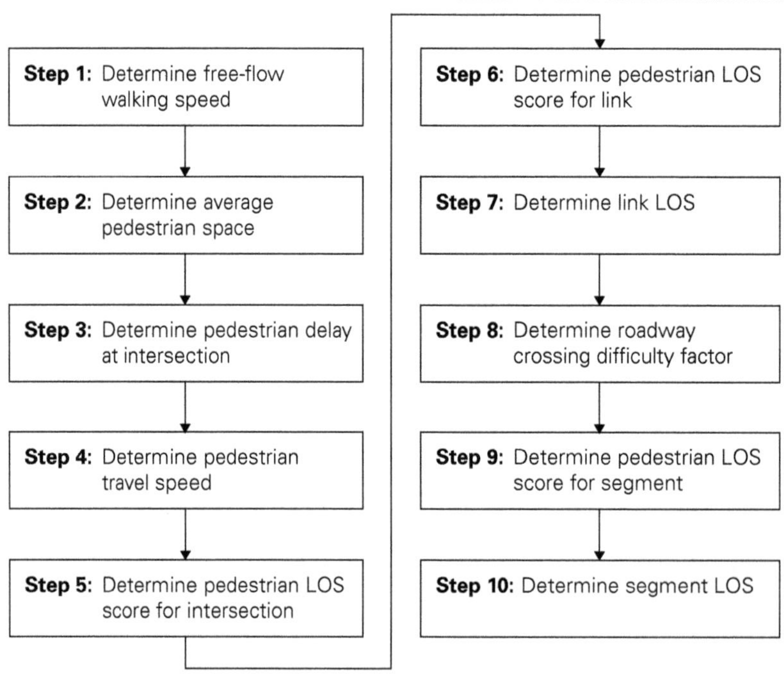

building face, and fence or wall. Pedestrian space computations require the pedestrian flow rate and free-flow walking speed (that can be influenced by pedestrian age and sidewalk grade).

For other index terms, intersection geometry and demand information are needed, including number of through traffic lanes; vehicular major-street demand; vehicular running speed; outside lane, bicycle lane, buffer, and sidewalk widths; and proportion on-street parking. To compute the pedestrian segment index score for level of service, the pedestrian crossing delay is needed as a function of pedestrian waiting and diversion delay that required the distance to the nearest signal. Results from the intersection pedestrian procedure are also needed for the intersection index score and waiting delay values.

PROCEDURE

The effective sidewalk width is computed following HCM Equations 18-23 through 18-27 that are used in HCM Equation 18-28 to calculate the pedestrian flow rate. The average walking speed used this flow rate to compute the pedestrian walking speed in HCM Equation 18-29 that is used in HCM Equation 18-30 to calculate the average pedestrian space.

Figure 7.18 *HCM Equation 18-23*

$$W_E = W_T - W_{O,i} - W_{O,o} - W_{s,i} - W_{s,o} \geq 0.0$$

Figure 7.19 *HCM Equation 18-24*

$$W_{s,i} = \max(W_{buf}, 1.5)$$

Figure 7.20 *HCM Equation 18-25*

$$W_{s,o} = 3.0\,p_{\text{window}} + 2.0p_{\text{building}} + 1.5p_{\text{fence}}$$

Figure 7.21 *HCM Equation 18-26*

$$W_{O,i} = w_{O,i} - W_{s,i} \geq 0.0$$

Figure 7.22 *HCM Equation 18-27*

$$W_{O,o} = w_{O,o} - W_{s,o} \geq 0.0$$

where

W_E = effective sidewalk width (ft),
W_T = total walkway width (ft),
$W_{O,i}$ = adjusted fixed-object effective width on inside of sidewalk (ft),
$W_{O,o}$ = adjusted fixed-object effective width on outside of sidewalk (ft),
$W_{s,i}$ = shy distance on inside (curb side) of sidewalk (ft),
$W_{s,o}$ = shy distance on outside of sidewalk (ft),
W_{buf} = buffer width between roadway and sidewalk (ft),
p_{window} = proportion of sidewalk length adjacent to a window display (decimal),
p_{building} = proportion of sidewalk length adjacent to a building face (decimal),
p_{fence} = proportion of sidewalk length adjacent to a fence or low wall (decimal),
$w_{O,i}$ = effective width of fixed objects on inside of sidewalk (ft), and
$w_{O,o}$ = effective width of fixed objects on outside of sidewalk (ft).

Figure 7.23 *HCM Equation 18-28*

$$v_p = \frac{v_{ped}}{60W_E}$$

where

v_p = pedestrian flow per unit width (p/ft/min),
v_{ped} = pedestrian flow rate in the subject sidewalk (walking in both directions) (p/h), and
W_E = effective sidewalk width (ft).

Figure 7.24 *HCM Equation 18-29*

$$S_p = \left(1 - 0.00078\, v_p^2\right) S_{pf} \geq 0.5\, S_{pf}$$

where S_p is the pedestrian walking speed (ft/s), s_{pf} is the free-flow pedestrian walking speed (ft/s), and v_p is the pedestrian flow per unit width (p/ft/min).

Figure 7.25 *HCM Equation 18-30*

$$A_p = 60\frac{S_p}{v_p}$$

where A_p is the pedestrian space (ft²/p) and all other variables are as previously defined.

The cross-section adjustment factor is computed using HCM Equation 18-33. The motorized vehicle volume adjustment factor is computed using HCM Equation 18-34. The motorized vehicle speed adjustment factor is computed using HCM Equation 18-35. These are combined in HCM Equation 18-32 to compute the pedestrian link index score and determine the pedestrian link level of service with HCM Exhibit 18-2.

Figure 7.26 *HCM Equation 18-33*

$$F_W = -1.2276\ln\left(W_v + 0.5W_l + 50\,p_{pk} + W_{buf}f_b + W_{aA}f_{SW}\right)$$

Figure 7.27 *HCM Equation 18-34*

$$F_v = 0.0091\frac{v_m}{4N_{th}}$$

Figure 7.28 *HCM Equation 18-35*

$$F_S = 4\left(\frac{S_R}{100}\right)^2$$

Figure 7.29 *HCM Equation 18-32*

$$I_{p,\text{link}} = 6.0468 + F_w + F_v + F_S$$

where

$I_{p,\text{link}}$ = pedestrian LOS score for link;

F_w = cross-section adjustment factor;

F_v = motorized vehicle volume adjustment factor;

F_S = motorized vehicle speed adjustment factor;

$\ln(x)$ = natural log of x;

W_v = effective total width of outside through lane, bicycle lane, and shoulder as a function of traffic volume (see Exhibit 18-19) (ft);

W_l = total width of shoulder, bicycle lane, and parking lane (see Exhibit 18-19) (ft);

p_{pk} = proportion of on-street parking occupied (decimal);

W_{buf} = buffer width between roadway and available sidewalk (= 0.0 if sidewalk does not exist) (ft);

f_b = buffer area coefficient = 5.37 for any continuous barrier at least 3 ft high that is located between the sidewalk and the outside edge of roadway; otherwise use 1.0;

W_A = available sidewalk width = 0.0 if sidewalk does not exist or $W_T - W_{buf}$ if sidewalk exists (ft);

W_{aA} = adjusted available sidewalk width = min(W_A, 10) (ft);

f_{sw} = sidewalk width coefficient = 6.0 – 0.3 W_{aA};

v_m = midsegment demand flow rate (direction nearest to the subject sidewalk) (veh/h);

N_{th} = number of through lanes on the segment in the subject direction of travel (ln); and

S_R = motorized vehicle running speed = (3,600 L)/(5,280 t_R) (mi/h).

The pedestrian diversion distance is computed using HCM Equation 18-36 for use in HCM Equation 18-37 to determine the pedestrian diversion delay. The roadway crossing difficulty factor is computed from HCM Equation 18-38.

PEDESTRIAN LOS

Use HCM Equation 18-39 to find the pedestrian index score for use in HCM Exhibit 18-2 to determine the pedestrian segment level of service.

Figure 7.30 *HCM Exhibit 18-2*

Pedestrian LOS Score	LOS by Average Pedestrian Space (ft²/p)					
	>60	>40–60	>24–40	>15–24	>8.0–15*	≤8.0*
≤2.00	A	B	C	D	E	F
>2.00–2.75	B	B	C	D	E	F
>2.75–3.50	C	C	C	D	E	F
>3.50–4.25	D	D	D	D	E	F
>4.25–5.00	E	E	E	E	E	F
>5.00	F	F	F	F	F	F

Figure 7.31 *HCM Equation 18-36*

$$D_d = 2D_c$$

where

D_d = diversion distance (ft), and

D_c = distance to nearest signal-controlled crossing (ft).

Figure 7.32 *HCM Equation 18-37*

$$d_{pd} = \frac{D_d}{S_p} + d_{pc}$$

where

d_{pd} = pedestrian diversion delay (s/p),

D_d = diversion distance (ft),

S_p = pedestrian walking speed (ft/s), and

d_{pc} = pedestrian delay when crossing the segment at the nearest signal-controlled crossing (s/p).

Figure 7.33 *HCM Equation 18-38*

$$F_{cd} = 1.0 + \frac{0.10\,d_{px} - (0.318\,I_{p,link} + 0.220 I_{p,int} + 1.606)}{7.5}$$

where

F_{cd} = roadway crossing difficulty factor,

d_{px} = crossing delay = min(d_{pd}, d_{pw}, 60) (s/p),

d_{pd} = pedestrian diversion delay (s/p),

d_{pw} = pedestrian waiting delay (s/p),

$I_{p,link}$ = pedestrian LOS score for link, and

$I_{p,int}$ = pedestrian LOS score for intersection.

Figure 7.34 *HCM Equation 18-39*

$$I_{p,seg} = 0.75\left[\frac{(F_{cd}\,I_{p,link}+1)^3\,\frac{L}{S_p} + (I_{p,int}+1)^3\,d_{pp}}{\frac{L}{S_p} + d_{pp}}\right]^{\frac{1}{3}} + 0.125$$

Figure 7.35 *HCM Exhibit 18-2*

Pedestrian LOS Score	LOS by Average Pedestrian Space (ft²/p)					
	>60	>40–60	>24–40	>15–24	>8.0–15*	≤8.0*
≤2.00	A	B	C	D	E	F
>2.00–2.75	B	B	C	D	E	F
>2.75–3.50	C	C	C	D	E	F
>3.50–4.25	D	D	D	D	E	F
>4.25–5.00	E	E	E	E	E	F
>5.00	F	F	F	F	F	F

Bicycles

Bicycle level of service along an urban street segment requires determining the bicycle delay at the boundary intersection and computing the bicycle travel speed along the segment. The intersection index score is combined with an index score for the segment that includes cross-section, motorized vehicle volume and speed, and pavement condition terms, following the procedure illustrated in HCM Exhibit 18-23.

Figure 7.36 *HCM Exhibit 18-23*

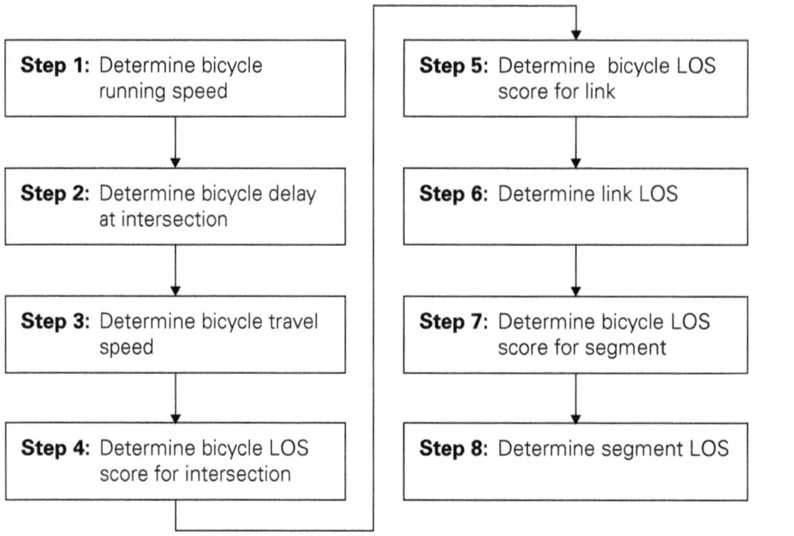

REQUIRED DATA

The data needed to compute speed include segment length and bicycle running speed along with the boundary intersection vehicular delay. The cross-section adjustment uses the widths of the outside through lane, bicycle lane, and paved shoulders with the width of the parking lane and its proportion occupied, as well as the vehicular volume, percentage heavy vehicles, and running speed. The volume adjustment factor also needs the number of through lanes, and pavement adjustment factor requires the pavement rating. Access points on the right side along the segment are used to compute the unsignalized conflicts factor.

PROCEDURE

The bicycle running speed is assumed to be 15 mi/h if no field data are available. Bicycle delay is equal to the through vehicle delay at the downstream boundary intersection from the signalized intersection procedure. The bicycle travel speed is computed using HCM Equation 18-40.

Figure 7.37 *HCM Equation 18-40*

$$S_{Tb,seg} = \frac{3{,}600L}{5{,}280(t_{Rb} + d_b)}$$

where

$S_{Tb,seg}$ = travel speed of through bicycles along the segment (mi/h),

L = segment length (ft),

t_{Rb} = segment running time of through bicycles = (3,600 L)/(5,280 S_b) (s),

s_b = bicycle running speed (mi/h), and

d_b = bicycle control delay (s/bicycle).

The cross-section adjustment factor is computed using HCM Equation 18-42 with HCM Exhibit 18-24. The motorized vehicle volume adjustment factor is computed using HCM Equation 18-43. The motorized vehicle speed adjustment factor is computed using HCM Equation 18-44. The pavement condition adjustment factor is calculated using HCM Equation 18-45 with the pavement condition rating from HCM Exhibit 18-22. These are combined in HCM Equation 18-41 to compute the pedestrian link index score and determine the pedestrian link level of service with HCM Exhibit 18-3.

Figure 7.38 *HCM Equation 18-42*

$$F_W = -0.005 W_e^2$$

Figure 7.39 *HCM Exhibit 18-24*

Condition	Variable When Condition Is Satisfied	Variable When Condition Is Not Satisfied
$p_{pk} = 0.0$	$W_t = W_{ol} + W_{bl} + W_{os}^* + W_{pk}$	$W_t = W_{ol} + W_{bl} + W_{os}^*$
$v_m > 160$ veh/h or street is divided	$W_v = W_t$	$W_v = W_t(2 - 0.005\ v_m)$
$W_l < 4.0$ ft	$W_e = W_v - 10\ p_{pk} \geq 0.0$	$W_e = W_v + W_l - 20\ p_{pk} \geq 0.0$
$V_m\ (1 - 0.01\ P_{HV}) < 200$ veh/h and $P_{HV} > 50\%$	$P_{HVa} = 50\%$	$P_{HVa} = P_{HV}$
$S_R < 21$	$S_{Ra} = 21$ mi/h	$S_{Ra} = S_R$
$v_m > 4\ N_{th}$	$v_{ma} = v_m$	$v_{ma} = 4\ N_{th}$

NOTES: W_t = total width of the outside through lane, bicycle lane, and paved shoulder (ft);
W_{ol} = width of outside through lane (ft);
W_{os}^* = adjusted width of paved outside shoulder; if curb is present $W_{os}^* + W_{os} - 1.5 \geq 0.0$, otherwise $W_{os}^* = W_{os}$ (ft);
W_{os} = width of paved outside shoulder (ft);
W_{bl} = width of bicycle lane = 0.0 if bicycle lane not provided (ft);
W_{pk} = width of striped parking lane (ft);
W_l = total width of shoulder, bicycle lane, and parking lane $W_l = W_{bi} + W_{os}^* + W_{pk}$(ft);
W_v = effective total width of outside through lane, bicycle lane, and shoulder as a function of traffic volume (ft);
p_{pk} = proportion of on-street parking occupied (decimal);
v_m = midsegment demand flow rate (veh/h);
P_{HV} = percent heavy vehicles in the midsegment demand flow rate (%); and
S_R = motorized vehicle running speed (mi/h).

Figure 7.40 *HCM Equation 18-43*

$$F_v = 0.507 \ln\left(\frac{v_{ma}}{4N_{th}}\right)$$

Figure 7.41 *HCM Equation 18-44*

$$F_S = 0.199[1.1199 \ln(S_{Ra} - 20) + 0.8103](1 + 0.1038\ P_{HV\,a})^2$$

Figure 7.42 *HCM Equation 18-45*

$$F_p \frac{7.066}{P_C^2}$$

Figure 7.43 *HCM Exhibit 18-22*

Pavement Condition Rating	Pavement Description	Motorized Vehicle Ride Quality and Traffic Speed
4.0 to 5.0	New or nearly new superior pavement. Free of cracks and patches.	Good ride.
3.0 to 4.0	Flexible pavements may begin to show evidence of rutting and fine cracks. Rigid pavements may begin to show evidence of minor cracking.	Good ride.
2.0 to 3.0	Flexible pavements may show rutting and extensive patching. Rigid pavements may have a few joint fractures, faulting, or cracking.	Acceptable ride for low-speed traffic but barely tolerable for high-speed traffic.
1.0 to 2.0	Distress occurs over 50% or more of the surface. Flexible pavement may have large potholes and deep cracks. Rigid pavement distress includes joint spalling, patching, and cracking.	Pavement deterioration affects the speed of free-flow traffic. Ride quality not acceptable.
0.0 to 1.0	Distress occurs over 75% or more of the surface. Large potholes and deep cracks exist.	Passable only at reduced speed and considerable rider discomfort.

Figure 7.44 *HCM Equation 18-41*

$$I_{b,link} = 0.760 + F_W + F_v + F_S + F_p$$

where

$I_{b,link}$ = bicycle LOS score for link,

F_w = cross-section adjustment factor,

F_v = motorized vehicle volume adjustment factor,

F_S = motorized vehicle speed adjustment factor,

F_P = pavement condition adjustment factor,

$\ln(x)$ = natural log of x,

W_e = effective width of outside through lane (see Exhibit 18-24) (ft),

v_{ma} = adjusted midsegment demand flow rate (see Exhibit 18-24) (veh/h),

N_{th} = number of through lanes on the segment in the subject direction of travel (ln),

S_{Ra} = adjusted motorized vehicle running speed (see Exhibit 18-24) (mi/h),

P_{Hva} = adjusted percent heavy vehicles in midsegment demand flow rate (see Exhibit 18-24) (%),and

P_c = pavement condition rating (see Exhibit 18-22).

Figure 7.45 *HCM Exhibit 18-3*

LOS	LOS Score
A	≤2.00
B	>2.00–2.75
C	>2.75–3.50
D	>3.50–4.25
E	>4.25–5.00
F	>5.00

BICYCLE LOS

The unsignalized conflicts factor is computed from the number of right-side access points on the segment from HCM Equation 18-47 for use in HCM Equation 18-46 to find the bicycle index score that determines the bicycle segment level of service defined in HCM Exhibit 18-3 to determine the pedestrian segment level of service.

Figure 7.46 *HCM Equation 18-47*

$$F_c = 0.035\left(\frac{5,280\ N_{ap,s}}{L} - 20\right)$$

Figure 7.47 *HCM Equation 18-46*

$$I_{b,seg} = 0.75\left[\frac{(F_c + I_{b,link} + 1)^3 t_{R,b} + (I_{b,int} + 1)^3 d_b}{t_{R,b} + d_b}\right]^{\frac{1}{3}} + 0.125$$

where

$I_{b,seg}$ = bicycle LOS score for segment;

$I_{b,link}$ = bicycle LOS score for link;

F_c = unsignalized conflicts factor;

$I_{b,int}$ = bicycle LOS score for intersection; and

$N_{ap,s}$ = number of access point approaches on the right side in the subject direction of travel (points).

Figure 7.48 *HCM Exhibit 18-3*

LOS	LOS Score
A	≤2.00
B	>2.00–2.75
C	>2.75–3.50
D	>3.50–4.25
E	>4.25–5.00
F	>5.00

Transit

Transit level of service along an urban street segment includes estimating the running speed as a function of vehicle running speed and number of transit stops. The delay from stops is computed using the deceleration and acceleration times along with the time to serve passengers. The computation of travel speed combined the transit running speed with the travel time loss. The wait-ride score using the headway and perceived travel time factors to be combined with the pedestrian link score to calculate the transit segment score to determine level of service, following HCM Exhibit 18-26.

Figure 7.49 *HCM Exhibit 18-26*

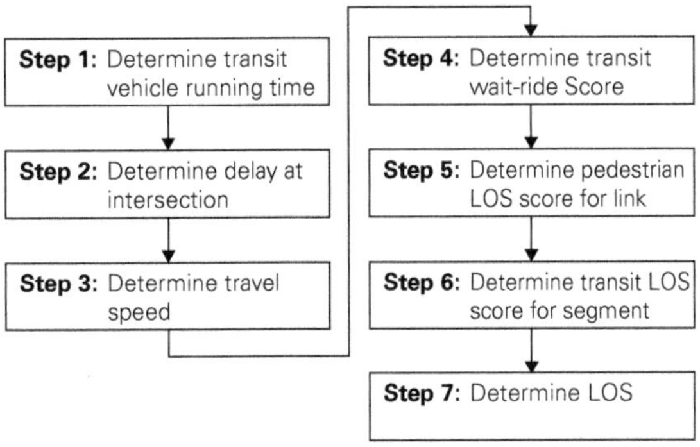

REQUIRED DATA

The segment length, vehicle running time, and number of transit stops are needed to compute the transit running speed. The green-to-cycle ratio for the downstream boundary signal is used for nearside stops. The average dwell time and its proportion during the effective green time is used for computing delay due to stops. The area type, the transit lane allocation, and traffic conditions are used with any signal timing priority to find the signal running time loss; then, the transit frequency is used to compute the headway factor. Passengers per seat and stops with benches and/or shelters are needed to calculate the perceived travel time factor. The pedestrian link index score is used to incorporate walking to the bus stop.

PROCEDURE

The transit running speed is computed using HCM Equation 18-48. Acceleration-deceleration delay is calculated following HCM Equations 18-49 and 18-50, with delay due to serving passengers calculated using HCM Equation 18-51 and delay due to stops computed from HCM Equation 18-52. Transit running time comes from HCM Equation 18-53 with intersection delay computed from HCM Equation 18-54, with HCM Exhibit 18-27 defining the running time loss. The running time and intersection delay are combined to compute travel speed in HCM Equation 18-55.

Figure 7.50 *HCM Equation 18-48*

$$S_{Rt} = \min\left(S_R, \frac{61}{1+e^{-1.00+(1{,}185 N_{ts}/L)}}\right)$$

where

S_{Rt} = transit vehicle running speed (mi/h),
L = segment length (ft),
N_{ts} = number of transit stops on the segment for the subject route (stops),
S_R = motorized vehicle running speed = (3,600 L)/(5,280 t_R) (mi/h), and
t_R = segment running time (s).

Figure 7.51 *HCM Equation 18-49*

$$d_{ad} = \frac{5{,}280}{3{,}600}\left(\frac{S_{Rt}}{2}\right)\left(\frac{1}{r_{at}} + \frac{1}{r_{dt}}\right) f_{ad}$$

where

d_{ad} = transit vehicle acceleration–deceleration delay due to a transit stop (s),
r_{at} = transit vehicle acceleration rate = 3.3 (ft/s^2),
r_{dt} = transit vehicle deceleration rate = 4.0 (ft/s^2),
f_{ad} = proportion of transit vehicle stop acceleration–deceleration delay not due to traffic control,
x = volume-to-capacity ratio of the link's rightmost lane on a roundabout approach,
g = effective green time (s), and
C = cycle length (s).

The headway factor is computed using HCM Equation 18-56, and the perceived travel time factor comes from HCM Equations 18-57 through 18-61. These are combined in HCM Equation 18-62 to compute the wait-ride score.

TRANSIT LOS

The wait-ride score is used with the pedestrian link index score to calculate the transit segment index score in HCM Equation 18-63 for use in HCM Exhibit 18-3 to determine level of service.

Figure 7.52 *HCM Equation 18-50*

$$f_{ad} = \begin{cases} 1.00 & \text{(stops not near side of a boundary intersection)} \\ 0.00 & \text{(near-side stops at all-way and major-street two-way STOP-controlled intersections)} \\ 1-x & \text{(near-side stops at roundabouts)} \\ g/c & \text{(near-side stops at traffic signals)} \end{cases}$$

Figure 7.53 *HCM Equation 18-51*

$$d_{ps} = t_d f_{dt}$$

where

d_{ps} = transit vehicle delay due to serving passengers (s),

t_d = average dwell time (s), and

f_{dt} = proportion of dwell time occurring during effective green (= g/C at near-side stops at signalized intersections and 1.00 otherwise, where g and C are as previously defined).

Figure 7.55 *HCM Equation 18-53*

$$t_{Rt} = \frac{3{,}600\,L}{5{,}280 S_{Rt}} + \sum_{i=1}^{N_{ts}} d_{ts,i}$$

where t_{Rt} is the segment running time of transit vehicle (s), $d_{ts,i}$ is the delay due to a transit vehicle stop for passenger pickup at stop *i* within the segment (s), and all other variables are as previously defined.

Figure 7.54 *HCM Equation 18-52*

$$d_{ts} = d_{ad} + d_{ps} + d_{re}$$

where d_{ts} is the delay due to a transit vehicle stop (s), d_{re} is the reentry delay (s),

Figure 7.56 *HCM Equation 18-54*

$$d_t = t_l 60\left(\frac{L}{5{,}280}\right)$$

where

d_t = through delay (s/veh),

t_l = transit vehicle running time loss (min/mi), and

L = segment length (ft).

Figure 7.57 *HCM Exhibit 18-27*

Area Type	Transit Lane Allocation	Traffic Condition	Running Time Loss by Signal Condition (min/mi): Typical	Signals Set for Transit	Signals More Frequent Than Transit Stops
Central business district	Exclusive	No right turns	1.2	0.6	1.5–2.0
		With right-turn delay	2.0	1.4	2.5–3.0
		Blocked by traffic	2.5–3.0	Not available	3.0–3.5
	Mixed traffic	Any	3.0	Not available	3.5–4.0
Other	Exclusive	Any	0.7 (0.5–1.0)	Not available	Not available
	Mixed traffic	Any	1.0 (0.7–1.5)	Not available	Not available

Figure 7.58 *HCM Equation 18-55*

$$S_{Tt,seg} = \frac{3{,}600\,L}{5{,}280\,(t_{Rt} + d_t)}$$

where $S_{Tt,seg}$ is the travel speed of transit vehicles along the segment (mi/h), t_{Rt} is the segment running time of transit vehicle (s), and all other variables are as previously defined.

Figure 7.59 *HCM Equation 18-56*

$$F_h = 4.00\,e^{-1.434/(v_s + 0.001)}$$

where

F_h = headway factor, and
v_s = transit frequency for the segment (veh/h).

Figure 7.60 *HCM Equation 18-57*

$$F_{tt} = \frac{(e-1)T_{btt} - (e+1)T_{ptt}}{(e-1)T_{ptt} - (e+1)T_{btt}}$$

Figure 7.61 *HCM Equation 18-58*

$$T_{ptt} = \left(a_1 \frac{60}{S_{Tt,seg}}\right) + (2\,T_{ex}) - T_{at}$$

Figure 7.62 *HCM Equation 18-59*

$$a_1 = \begin{cases} 1.00 & F_l \le 0.80 \\ 1 + \dfrac{4(F_l - 0.80)}{4.2} & 0.80 \le F_l \le 1.00 \\ 1 + \dfrac{4(F_l - 0.80) + (F_l - 1.00)(6.5 + 5[F_l - 1.00])}{4.2F_l} & F_l > 1.00 \end{cases}$$

Figure 7.63 *HCM Equation 18-60*

$$T_{at} = \frac{1.3p_{sh} + 0.2p_{be}}{L_{pt}}$$

where

F_{tt} = perceived travel time factor;
e = ridership elasticity with respect to changes in the travel time rate = –0.40;
T_{btt} = base travel time rate = 6.0 for the central business district of a metropolitan area with 5 million persons or more, otherwise = 4.0 (min/mi);
T_{ptt} = perceived travel time rate (min/mi);
T_{ex} = excess wait time rate due to late arrivals (min/mi) = t_{ex}/L_{pt};
t_{ex} = excess wait time due to late arrivals (min);
T_{at} = amenity time rate (min/mi);
a_1 = passenger load weighting factor;
$S_{Tt,seg}$ = travel speed of transit vehicles along the segment (mi/h);
F_l = average passenger load factor (passengers/seat);
L_{Pt} = average passenger trip length = 3.7 typical (mi);
P_{sh} = proportion of stops on segment with shelters (decimal); and
P_{be} = proportion of stops on segment with benches (decimal).

Figure 7.64 *HCM Equation 18-61*

$$t_{ex} = [t_{late}(1 - p_{ot})]^2$$

where

t_{ex} = excess wait time due to late arrivals (min),
t_{late} = threshold late time = 5.0 typical (min), and
p_{ot} = proportion of transit vehicles arriving within the threshold late time (default = 0.75) (decimal).

Figure 7.65 *HCM Equation 18-62*

$$s_{w-r} = F_h\,F_{tt}$$

where

s_{w-r} = transit wait–ride score,
F_h = headway factor, and
F_{tt} = perceived travel time factor.

Figure 7.66 *HCM Equation 18-63*

$$I_{t,seg} = 6.0 - 1.50s_{w-r} + 0.15I_{p,link}$$

Figure 7.67 *HCM Exhibit 18-3*

LOS	LOS Score
A	≤2.00
B	>2.00–2.75
C	>2.75–3.50
D	>3.50–4.25
E	>4.25–5.00
F	>5.00

PROBLEM ONE

Determine the directional LOS for an 1800-foot, two-lane segment with a mid-segment flow of 1000 veh/h and a speed limit of 40 mi/h. There are six access points on the segment, 70% restrictive median with curb, and on-street parking exists for 70% of the length. The delay due to turns has been computed at 0.4 s, and the delay for the downstream through movement is 20.4 s.

1. Compute the base free-flow speed.

EXHIBIT 18-11

Speed Constant: S_o = 40 mi/h Speed Limit = 44.4 mi/h
Cross Section: f_{CS} = 70% Restrictive Median with Curb = – 2.0 mi/h
Access Points: f_A = (5280/1800)6 = 17.6 points/mi = – 0.7 mi/h
Parking: f_{pk} = 70% Parking = – 2.1 mi/h
Base Free-Flow Speed: $S_{fo} = 44.4 - 2.0 - 0.7 - 2.1 = 39.6$ mi/h

2. Compute the free-flow speed.

Signal Spacing Adjustment: $f_L = 1.02 - 4.7[(39.6 - 19.5)/1800] = 0.97$
EQUATION 18-4

Free-Flow Speed: $S_f = 39.6(0.97) = 38.3 \text{ mi/h} < 40 \rightarrow 40$ mi/h
EQUATION 18-5

3. Compute the proximity adjustment factor.

Proximity Adjustment: $f_v = 2 / [1 + [1 - 1000/52.8(2)(40)]^{0.21} = 1.03$
EQUATION 18-6

4. Compute the running time.

Running Time: $[(6.0 - 2)/0.0025(1800)](1) + [3600(1800)/5280(40)]1.03 + 0.4 = 32.9$ s
EQUATION 18-7

5. Compute the travel speed.

Travel Speed: 3600(1800) / [5280(32.9 + 20.4)] = 23.0 mi/h

EQUATION 18-15

6. Determine the EBR level of service.

LOS: LOS = C (between 20 and 27)

EXHIBIT 18-1

HCS7

These screenshots show how Problem One is coded into *HCS7*, followed by the results as presented in the formatted report. All input data, intermediate computations, and final results are shown in logical order, leading to delay and level of service.

Figure 7.68 *HCS7 Streets1 Input*

Segment Name		
	EB	WB
Upstream Width, ft	50	50
Restrictive Median, ft	70	70
Right-Hand Curb, %	70	70
Right-Hand Access Points	6	6
Mid-Segment Delay, s/veh	0.0	0.0
Field Travel Speed, mi/h	23.27	23.27
Field Free-Flow Speed, mi/h	40.00	40.00
On-Street Parking, %	70	70
Speed Calibration Factor, mi/h	0.0	0.0

Figure 7.69 *HCS7 Streets1 Output*

Segment Output Data		EBL	EBT	EBR	WBL	WBT	WBR
Segment	Movement	5	2	12	1	6	16
1	Bay/Lane Spillback Time, h		never			never	
1	Shared Lane Spillback Time, h	never		never	never		never
1	Base Free-Flow Speed, mph		40.52			40.52	
1	Running Time, s		32.59			32.59	
1	Running Speed, mph		37.65			37.65	
1	Through Delay, s/veh		20.14			20.14	
1	Travel Time, s		52.73			52.73	
1	Travel Speed, mph		23.27			23.27	
1	Stop Rate, stops/veh		0.58			0.58	
1	Spatial Stop Rate, stops/mi		1.69			1.69	
1	Through vol/cap Ratio		0.56			0.56	
1	Percent of Base FFS		57.45			57.45	
1	Level of Service		C			C	

PROBLEM TWO

Building on Problem One, determine the segment LOS if the length is increased to 5280 feet.

1. Compute the base free-flow speed.

EXHIBIT 18-11

Speed Constant: S_o = 40 mi/h Speed Limit = 44.4 mi/h

Cross-Section: f_{CS} = 70% Restrictive Median with Curb = – 2.0 mi/h

Access Points: f_A = 6 Access Points per Mile = – 0.3 mi/h

Parking: f_{pk} = 70% Parking = – 2.1 mi/h

Base Free-Flow Speed: $S_{fo} = 44.4 - 2.0 - 0.3 - 2.1 = 40.0$ mi/h

2. Compute the free-flow speed.

Signal Spacing Adjustment: $f_L = 1.02 - 4.7[(40.0 - 19.5)/5280] = 1.00$

EQUATION 18-4

Free-Flow Speed: $S_f = 40.0(1.00) = 40.0$ mi/h

EQUATION 18-5

3. Compute the proximity adjustment factor.

Proximity Adjustment: $f_v = 2 / [1 + [1 - 1000/52.8(2)(40)]^{0.21} = 1.03$

EQUATION 18-6

4. Compute the running time.

Running Time: [(6.0 – 2)/0.0025(5280)](1) + [3600(5280)/5280(40)]1.03 + 0.4 = 93.4 s
EQUATION 18-7

5. Compute the travel speed.

Travel Speed: 3600(5280) / [5280(93.4 + 20.4)] = 33.5 mi/h
EQUATION 18-15

6. Determine the EBR level of service.

LOS: LOS = A (greater than 32)
EXHIBIT 18-1

HCS7

These screenshots show how Problem Two is coded into *HCS7* by changing the segment length from 1800 feet to 5280 feet, followed by the results as presented in the formatted report. All input data, intermediate computations, and final results are shown in logical order, leading to delay and level of service.

Figure 7.70 *HCS7 Streets2 Input*

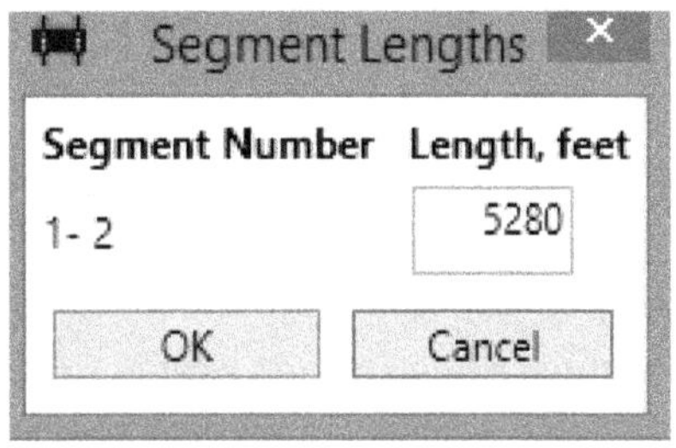

Figure 7.71 *HCS7 Streets2 Output*

Segment Output Data		EBL	EBT	EBR	WBL	WBT	WBR
Segment	Movement	5	2	12	1	6	16
1	Bay/Lane Spillback Time, h		never			never	
1	Shared Lane Spillback Time, h	never		never	never		never
1	Base Free-Flow Speed, mph		41.48			41.48	
1	Running Time, s		89.86			89.86	
1	Running Speed, mph		40.06			40.06	
1	Through Delay, s/veh		17.61			17.61	
1	Travel Time, s		107.47			107.47	
1	Travel Speed, mph		33.50			33.50	
1	Stop Rate, stops/veh		0.49			0.49	
1	Spatial Stop Rate, stops/mi		0.49			0.49	
1	Through vol/cap Ratio		0.56			0.56	
1	Percent of Base FFS		80.75			80.75	
1	Level of Service		A			A	

CHAPTER 8

Interchanges and Alternative Intersections

Interchange Ramp Terminals and Alternative Intersections are special cases of the Urban Streets analysis that models multiple signalized intersections along an arterial. Interchange Ramp Terminals models the (normally) two signals that control traffic to and from the freeway ramps at an interchange, including the Diverging Diamond Interchange design. Alternative Intersections models three distributed intersection designs, including Displaced Left Turn (DLT), Restricted Crossing U-Turn (RCUT), and Median U-Turn (MUT).

Ramp Terminals

In addition to analyzing typical signalized intersections, there is more focus on lane utilization to estimate the prepositioning at the first signal because of the intent at the second signal that occurs consistently at interchanges and results in more defined lane utilization adjustments to saturation flow. The saturation flow rate is also adjusted for traffic pressure to model driver aggression in congested conditions where headways are reduced. Beyond adjusting saturation flow rate for left and right turns, the turn radius is used for more precision. Additional lost time is also computed to model the effects of queue spillback and demand starvation between the ramp terminal signals. Level of service is computed for (potentially) fourteen origin-destination lane group pairs, but not for the interchange as a whole.

Required Data

Beyond the data required to model the two signalized intersections individually, only two addition values are needed to define the interchange. These parameters include the length of the segment between the two signalized intersections and the demand for interchange U-turns—vehicles exiting the freeway and turning onto the arterial in order to reenter the freeway traveling in the opposite direction.

Interchange Types

The ramps terminals procedures model eight conventional and one alternative interchange designs, including conventional diamond, six partial cloverleaf, single-point urban, and diverging diamond as shown in HCM Exhibits 23-15 through 23-18. The design being analyzed can affect several factors, including lane utilization, turn radius, additional lost time, and extra distance travel time.

Figure 8.1 *HCM Exhibit 23-15*

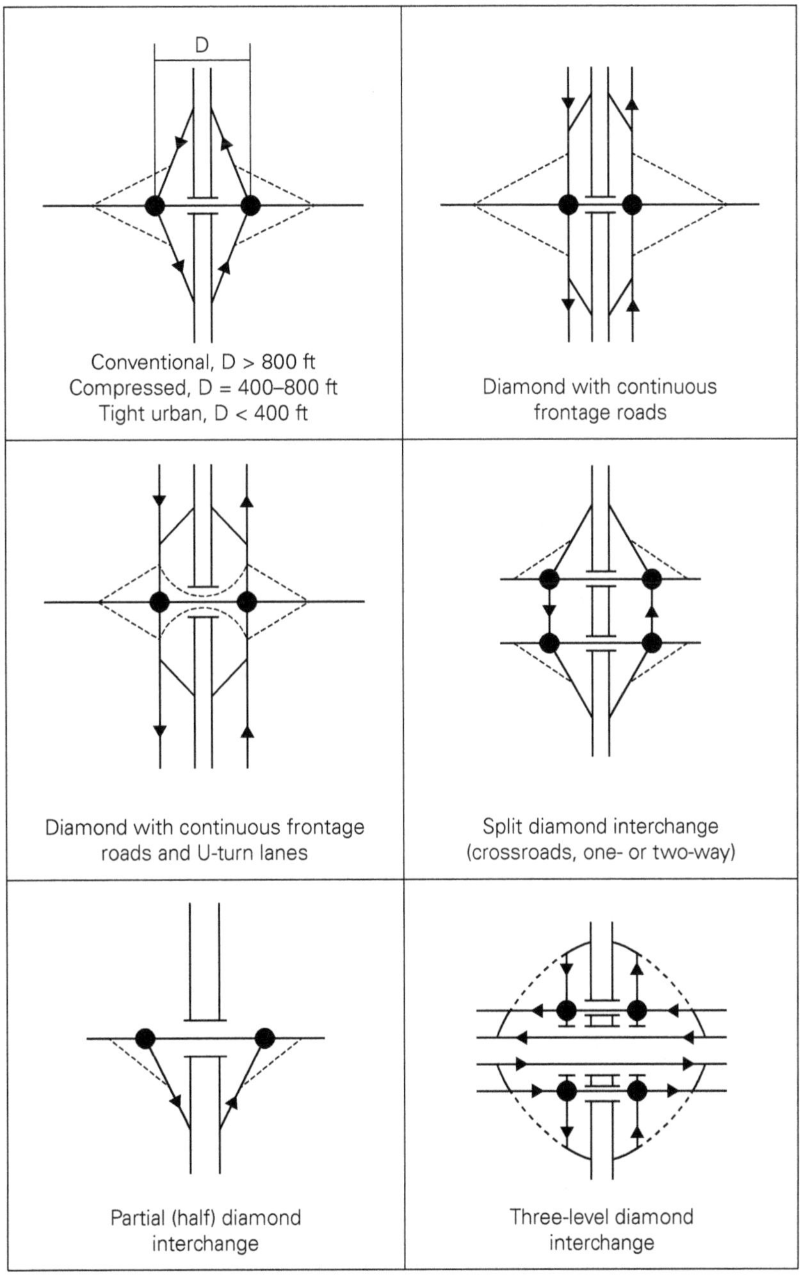

Figure 8.2 *HCM Exhibit 23-17*

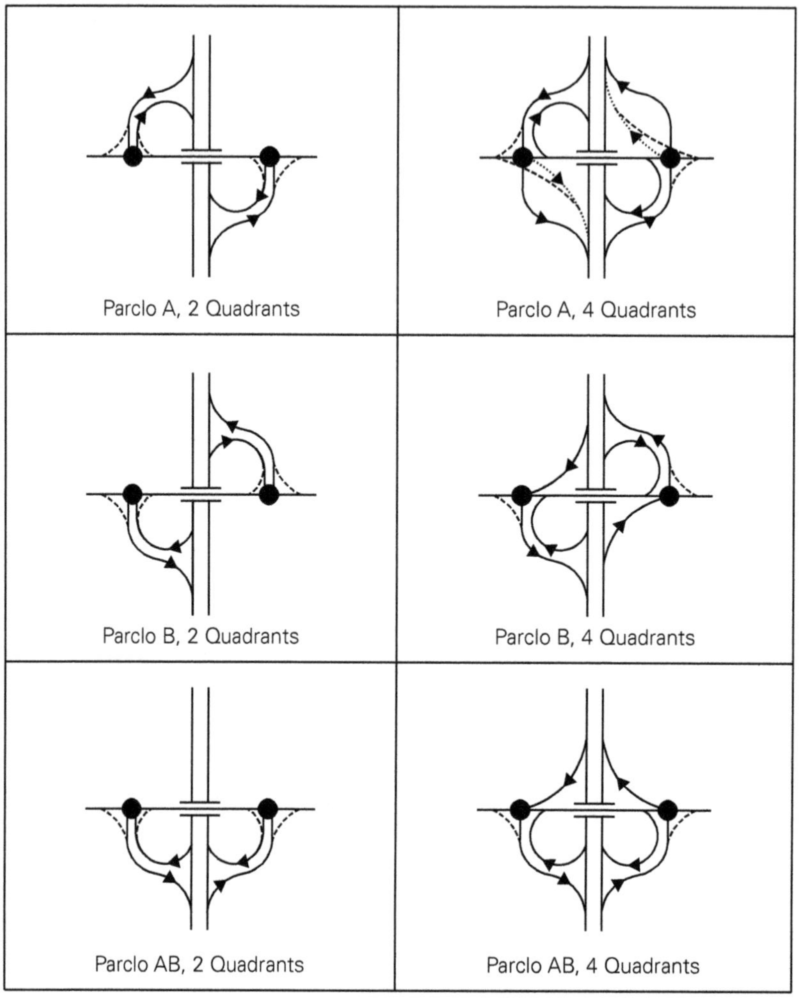

Figure 8.3 *HCM Exhibit 23-18*

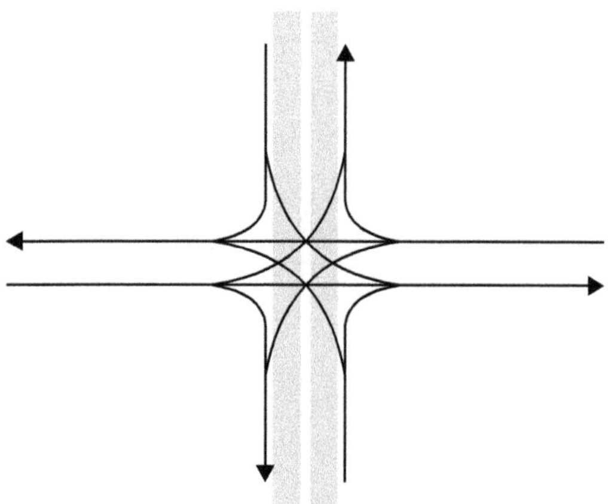

Figure 8.4 *HCM Exhibit 23-16*

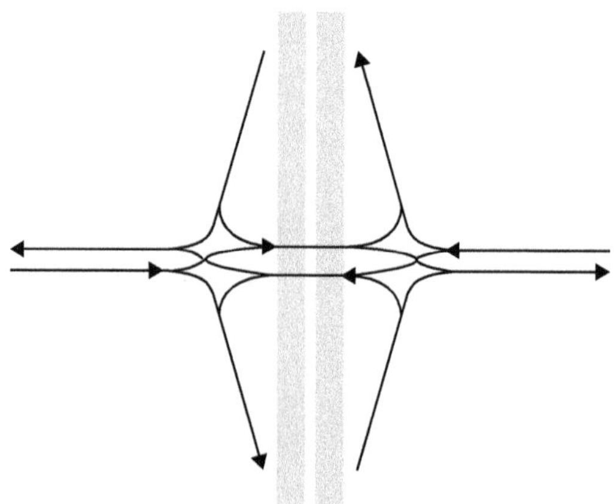

Saturation Flow

The adjusted saturation flow rate for movements within an interchange ramp terminal analysis follows the same procedure as for signalized intersections with three expansions (for lane utilization, left turns, and right turns) and one addition (for traffic pressure) as defined by HCM Equation 23-14.

Figure 8.5 *HCM Equation 23-14*

$$s = s_0 N f_W f_{HVg} f_p f_{bb} f_a f_{RT} f_{LT} f_{Lpb} f_{Rpb} f_v f_{LU} f_{DDI}$$

where

- s = adjusted saturation flow rate (veh/h),
- s_0 = base saturation flow rate per lane (from Exhibit 23-21),
- N = number of lanes in the lane group,
- f_w = adjustment factor for lane width (from Chapter 19),
- f_{HVg} = adjustment factor for heavy vehicles and grade (from Chapter 19),
- f_p = adjustment factor for existence of a parking lane and parking activity adjacent to the lane group (from Chapter 19),
- f_{bb} = adjustment factor for local bus blockage (from Chapter 19),
- f_a = adjustment factor for the area type (from Chapter 19),
- f_{RT} = adjustment for right-turning vehicle presence in the lane group (from Chapter 19, incorporating f_R; interchange saturation flow adjustment #4 from Equation 23-19),
- f_{LT} = adjustment for left-turning vehicle presence in the lane group (from Chapter 19, incorporating f_R; interchange saturation flow adjustment #4 from Equation 23-19),
- f_{Lpb} = pedestrian adjustment factor for left turns (from Chapter 19),
- f_{Rpb} = pedestrian-bicycle adjustment factor for right turns (from Chapter 19)
- f_v = adjustment for traffic pressure (interchange saturation flow rate adjustment #1 from Equation 23-15),
- f_{LU} = adjustment factor for lane utilization (interchange saturation flow rate adjustment #2 from Equation 23-16 and Equation 23-17), and
- f_{DDI} = adjustment for DDI crossover (interchange saturation flow rate adjustment #3 = 0.913 according to research (5).

Lane Utilization

The lane utilization adjustment to the saturation flow rate is expanded in this procedure to model situations where drivers select lanes at the upstream intersection based on intentions at the downstream intersection. With the distance between the terminals sometimes relatively short, this can significantly affect lane distribution in some situations as illustrated in HCM Exhibit 23-2.

HCM Equation 23-17 models this as a function of interchange design, number of lanes at each intersection, the segment length between the signals, and traffic turning movement proportions using the parameters defined in HCM Exhibit 23-24. This is also considered for closely spaced intersections.

Think about your own driving:
When you approach an interchange, you likely will put yourself in the leftmost lane at the first signal if your intent is to turn left at the second signal to get on the freeway, essentially prepositioning yourself to avoid not being able to change lanes within the short distance over or under the interchange bridge. Since others understand this need as well, the number of vehicles in one lane sometimes far outnumbers those in the adjacent lane for unequal lane distribution.

Figure 8.6 *HCM Exhibit 23-2*

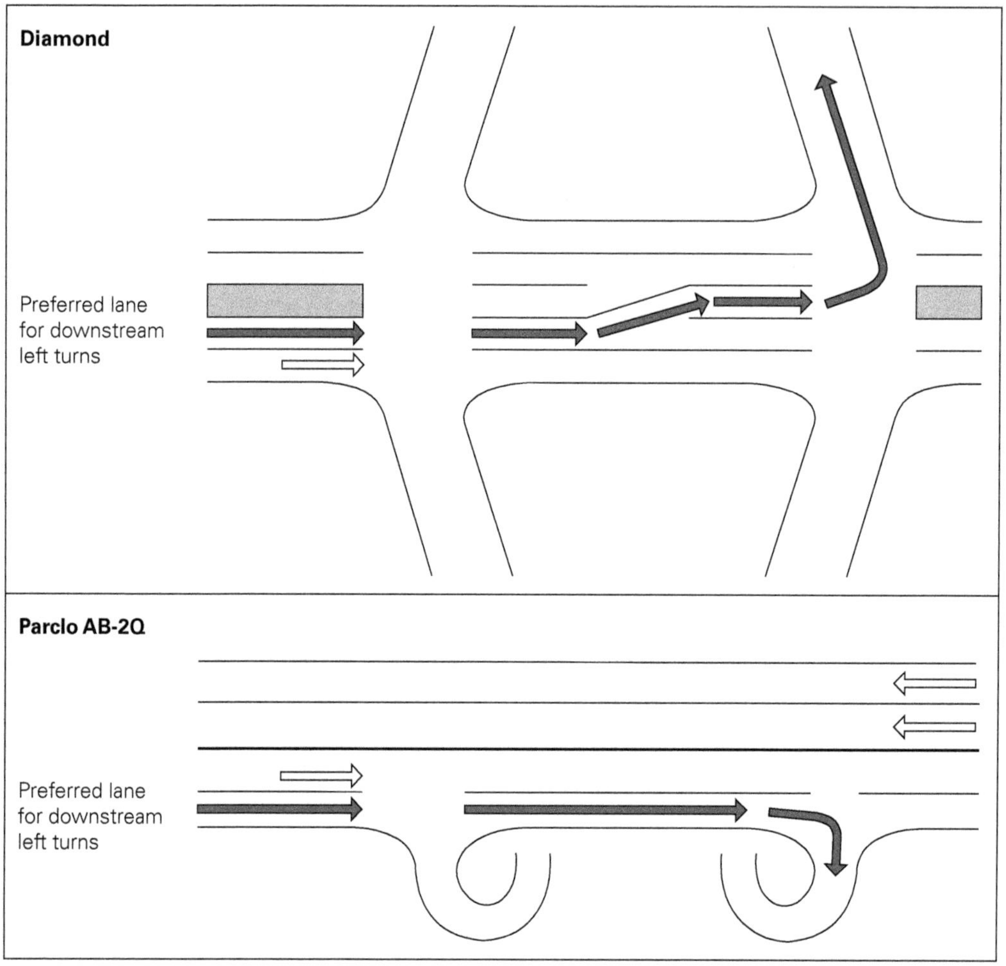

Figure 8.7 *HCM Equation 23-17*

$$\%V_{Li} = \frac{1}{n} + a_1\left(\frac{v_R}{v_L + v_R + v_T}\right) + a_2\left(\frac{v_L}{v_L + v_R + v_T}\right) + a_3\left(\frac{D \times v_L}{10^6}\right)$$

where

$\% V_{Li}$ = percent of traffic present in lane L_i, where L_1 represents the leftmost lane, L_2 represents the second lane from the left, and so forth;

n = number of lanes in the lane group;

a_i = coefficient for $i = 1$ through $i = 3$ (see Exhibit 23-24);

D = distance (ft) between the two intersections of the interchange (Equation 23-17 is valid for values of D below 800 ft);

v_R = O-D demand flow rate traveling through the first intersection and turning right at the second ($v_R = 0$ if there is an exclusive right-turn lane on the external approach);

v_L = O-D demand flow rate traveling through the first intersection and turning left at the second; and

v_T = O-D demand flow rate traveling through the first intersection and through the second.

Figure 8.8 *HCM Exhibit 23-24*

Interchange Type	Number of Lanes in Lane Group	Leftmost Lane (L_1)			Rightmost Lane (L_n)		
		a_1	a_2	a_3	a_1	a_2	a_3
Diamond	2	−0.154	0.187	−0.181	–	–	–
	3	−0.245	0.465	0	0.609	−0.326	0
	4	−0.328	0.684	0	0.640	−0.233	0
Parclo A-2Q	2	0	−0.527	0	–	–	–
	3	0	−0.363	0	0	0.605	0
	4	0	−0.257	0	0	0.747	0
Parclo B-2Q, B-4Q, AB-4Q (WB)	2	0.387	−0.344	0	–	–	–
	3	0.559	−0.218	0	−0.429	0.695	0
	4	0.643	−0.103	0	−0.359	0.794	0
Parclo A-4Q, AB-2Q (EB), AB-4Q (EB)	2	−0.306	−0.484	0	–	–	–
	3	−0.333	−0.289	0	0.579	0.428	0
	4	−0.233	−0.237	0	0.703	0.641	0
Parclo AB-2Q (WB)	2	0.468	0	0	–	–	–
	3	0.735	0	0	−0.308	0	0
	4	0.768	0	0	−0.202	0	0

Traffic Pressure

The saturation flow rate also adds an adjustment to account for shorter headways when the signalized approaches near capacity to account for the increase in driver aggression that accompanies congestion, as well as lowering rates for lower-than-normal demands. This relationship between the number of vehicles per cycle per lane and the saturation flow rate is defined for both left turns and through or right turns in HCM Equation 23-15 and illustrated for various demands in HCM Exhibit 23-23. Even though defined for interchange ramp terminal

Figure 8.9 *HCM Equation 23-15*

$$f_v = \begin{cases} \dfrac{1}{1.07 - 0.00672 \times \min(v_i', 30)} & \text{(left turn)} \\ \dfrac{1}{1.07 - 0.00486 \times \min(v_i', 30)} & \text{(through or right turn)} \end{cases}$$

where f_v is the adjustment factor for traffic pressure and v_i' is the demand flow rate per cycle per lane (veh/cycle/ln).

Figure 8.10 *HCM Exhibit 23-23*

Demand Flow Rate v_i' (veh/cycle/ln)	Movement Type	
	Left Turn	Through and Right Turn
3.0	0.953	0.947
6.0	0.971	0.961
9.0	0.991	0.974
12.0	1.011	0.988
15.0	1.032	1.003
18.0	1.054	1.018
21.0	1.077	1.033
24.0	1.100	1.049
30.0 or greater	1.152	1.082

intersections, this adjustment should be considered for any signalized intersection.

> ***Think about your own driving:***
> When you are in a situation where you did not get through a signalized intersection during the first cycle and had to wait for a second green light, you become anxious about a second cycle failure and tend to follow the vehicle in front of you a little closer to increase your chances of making it through, decreasing your headway and increasing the saturation flow rate.

Turn Radius

Turning vehicles must navigate the radius of the turn, which lowers saturation flow as compared with through vehicles. The assumed values for the equivalent passenger-car equivalents are 1.05 for left turns and 1.18 for right turns, resulting in adjustment factors of 0.95 and 0.85, respectively. These values assume a 112-foot radius for left turns and a 32-foot radius for right turns, but those can change for larger or skewed intersections. HCM Equations 23-19 through 23-23 define this adjustment as a function of turn radius as illustrated by HCM Exhibit 23-27. Even though defined for interchange ramp terminal intersections, this adjustment should be considered for any signalized intersection.

Figure 8.11 *HCM Equation 23-19*

$$f_R = \frac{1}{1+\frac{5.61}{R}}$$

where R is the radius of curvature of the left- or right-turning path (at the center of the path), in feet.

Figure 8.12 *HCM Equation 23-20*

For protected, exclusive left-tum lanes:

$$f_{LT} = f_R$$

Figure 8.13 *HCM Equation 23-21*

$$f_{LT} = \frac{1}{1+P_{LT}\left(\frac{1}{f_R}-1\right)}$$

where

f_{LT} = saturation flow adjustment factor for left turns;

P_{LT} = percentage of left turns in lane group; and

f_R = Interchange Saturation Flow Adjustment #4: Turn radius from Equation 23-19.

Figure 8.14 *HCM Equation 23-22*

For protected, exclusive right-tum lanes:

$$f_{RT} = f_R$$

Figure 8.15 *HCM Equation 23-23*

$$f_{RT} \frac{1}{1+P_{RT}\left(\frac{1}{f_R}-1\right)}$$

where

f_{RT} = saturation flow adjustment factor for right turns;

P_{RT} = percentage of right turns in lane group; and

f_R = Interchange Saturation Flow Adjustment #4: Turn radius from Equation 23-19.

Figure 8.16 *HCM Exhibit 23-27*

	Movement Type	
Radius of the Travel Path (ft)	**Left and Right Turn**	**Through**
25	0.817	1.00
50	0.899	1.00
100	0.947	1.00
150	0.964	1.00
200	0.973	1.00
250	0.978	1.00
300	0.982	1.00
350	0.984	1.00

Additional Lost Time

Additional lost time (beyond startup and clearance lost times defined for signalized intersections) may be incurred by upstream external approaches due to downstream queue or internal approaches due to demand starvation. If an internal queue limits access to the segment between the terminals, lost time occurs, while green time is displayed with vehicles restricted from moving as defined by HCM Equations 23-24 through 23-34 with HCM Exhibit 23-28. If the signal timing impedes the upstream movements from reaching the downstream signal while it is green, lost time occurs while green time is displayed with vehicles restricted from moving as defined by HCM Equations 23-38 and 23-39 with HCM Exhibit 23-33. This is also considered for closely spaced intersections, following HCM Equation 23-40 for downstream queues and HCM Equation 23-41 for demand starvation.

Figure 8.17 *HCM Equation 23-24*

$$t'_L = l_1 + L_{D-A} + Y - e \quad \text{(arterial)}$$

Figure 8.18 *HCM Equation 23-25*

$$t'_L = l_1 + L_{D-R} + L_{OL-DDI} + Y - e \quad \text{(ramp)}$$

where

t'_L = adjusted lost time (i.e., time when the signalized intersection is not used effectively by any movement) (s),

l_1 = start-up lost time (s),

L_{D-A} = lost time on external arterial approach due to presence of a downstream queue (lost time adjustment #1) (s),

L_{D-R} = external ramp lost time due to presence of downstream queue (lost time adjustment #1) (s),

L_{OL-DDI} = lost time on signalized external ramp approach at a DDI due to overlap phasing (lost time adjustment #2) (s),

Y = yellow-plus-all-red change-and-clearance interval (s), and

e = extension of effective green time into the clearance interval (s).

Figure 8.19 *HCM Equation 23-26*

$$t''_L = l_1 + L_{DS} + Y - e$$

where L_{DS} (lost time adjustment #3) is the additional lost time due to demand starvation (s), and other terms as previously defined.

Figure 8.20 *HCM Equation 23-27*

$$g' = G + Y - t_L'$$

where g' is the effective green time adjusted by presence of a downstream queue (s), G is the green time (s), and t_L' is adjusted lost time for external approaches (s).

Figure 8.21 *HCM Equation 23-28*

$$g' = G + Y - t_L''$$

where g' is the effective green time adjusted due to demand starvation (s), G is the green time (s), and t_L'' is the adjusted lost time for the internal approaches (s).

Figure 8.22 *HCM Equation 23-29*

Additional lost time on the external arterial approach:

$$L_{D-A} = G_A - 0.106DQ_A - 5.39\frac{CG_{UD}}{C}$$

Figure 8.23 *HCM Equation 23-30*

$$L_{D-R} = G_R - 0.106DQ_R - 5.39\frac{CG_{RD}}{C}$$

where

L_{D-A} = lost time on external arterial approach due to presence of downstream queue (s) (min=0),

L_{D-R} = lost time on external ramp approach due to presence of downstream queue (s) (min=0),

G_A = green interval for external arterial approach (s),

G_R = green interval for left-turning ramp movement (s),

DQ_A = distance to downstream queue at beginning of upstream arterial green (ft),

DQ_R = distance to downstream queue at beginning of upstream ramp green (ft),

CG_{UD} = common green time between upstream and downstream arterial through green (s),

CG_{RD} = common green time between upstream ramp green and downstream arterial through green (s), and

C = cycle length (s).

If Equation 23-29 or Equation 23-30 results in negative values, then the respective lost times L_{D-A} or L_{D-R} are zero. Furthermore, if DQ_A or DQ_R exceeds 200 ft, then the lost time will be zero.

DQ_A and DQ_R are calculated as follows:

$$DQ_A = D - Q_A$$

EQUATION 23-31

$$DQ_R = D - Q_R$$

EQUATION 23-32

where

D = distance corresponding to storage space between the two intersections of the interchange (ft),

Q_A = estimated average per lane queue length for through movement in downstream (internal) link at beginning of upstream arterial Phase A (ft), and

Q_R = estimated average per lane queue length for through movement in downstream (internal) link at beginning of upstream ramp Phase R (ft).

The downstream queue length (averaged across all through lanes) at the beginning of each upstream phase is estimated with the following equations:

Queue at the beginning of the upstream arterial Phase A:

$$Q_A = \left(0.0107\frac{v_R}{N_R} - 7.96\frac{G_D}{C} - 0.082CG_{UD} + 7.96\frac{GR}{C}\right)L_h$$

EQUATION 23-33

Queue at the beginning of the upstream ramp Phase R:

$$Q_R = \left(0.0107\frac{v_A}{N_A} - 7.96\frac{G_D}{C} - 0.082CG_{UD} + 7.96\frac{G_A}{C}\right)L_h$$

EQUATION 23-34

where

Q_A = queue at the beginning of upstream arterial Phase A (ft) (min=0);

Q_R = queue at the beginning of upstream ramp Phase R (ft) (min=0);

v_R = ramp flow feeding subject queue (veh/h);

v_A = arterial flow feeding subject queue (veh/h);

N_R = number of ramp lanes feeding the subject queue;

N_A = number of arterial lanes feeding the subject queue;

G_R = green interval for upstream left ramp movement (s);

G_A = green interval for upstream arterial through movement (s);

G_D = green interval for downstream arterial through movement (s);

Figure 8.24 *HCM Equation 23-31*

$$DQ_A = D - Q_A$$

Figure 8.25 *HCM Equation 23-32*

$$DQ_R = D - Q_R$$

where

D = distance corresponding to storage space between the two intersections of the interchange (ft),

Q_A = estimated average per lane queue length for through movement in downstream (internal) link at beginning of upstream arterial Phase A (ft), and

Q_R = estimated average per lane queue length for through movement in downstream (internal) link at beginning of upstream ramp Phase R (ft).

Figure 8.26 *HCM Equation 23-33*

Queue at the beginning of the upstream arterial Phase A:

$$Q_A = \left(0.0107\frac{v_R}{N_R} - 7.96\frac{G_D}{C} - 0.082CG_{UD} + 7.96\frac{G_R}{C}\right)L_h$$

Figure 8.27 *HCM Equation 23-34*

$$Q_R = \left(0.0107\frac{v_A}{N_A} - 7.96\frac{G_D}{C} - 0.082CG_{UD} + 7.96\frac{G_A}{C}\right)L_h$$

where

Q_A = queue at the beginning of upstream arterial Phase A (ft) (min=0);

Q_R = queue at the beginning of upstream ramp Phase R (ft) (min=0);

v_R = ramp flow feeding subject queue (veh/h);

v_A = arterial flow feeding subject queue (veh/h);

N_R = number of ramp lanes feeding the subject queue;

N_A = number of arterial lanes feeding the subject queue;

G_R = green interval for upstream left ramp movement (s);

G_A = green interval for upstream arterial through movement (s);

CG_{UD} = common green time between upstream arterial green and downstream through green (s);

CG_{RD} = common green time between upstream ramp green and downstream through green (s);

L_h = average queue spacing in a stationary queue, measured from front bumper to front bumper between successive vehicles (ft/veh); and

C = cycle length (s).

Figure 8.28 *HCM Exhibit 23-28*

green times between the upstream and downstream through phases (CG_{UD}) and between the upstream ramp and the downstream through phases (CG_{RD}).

EXHIBIT 23-28 ILLUSTRATION OF COMMON GREEN TIMES

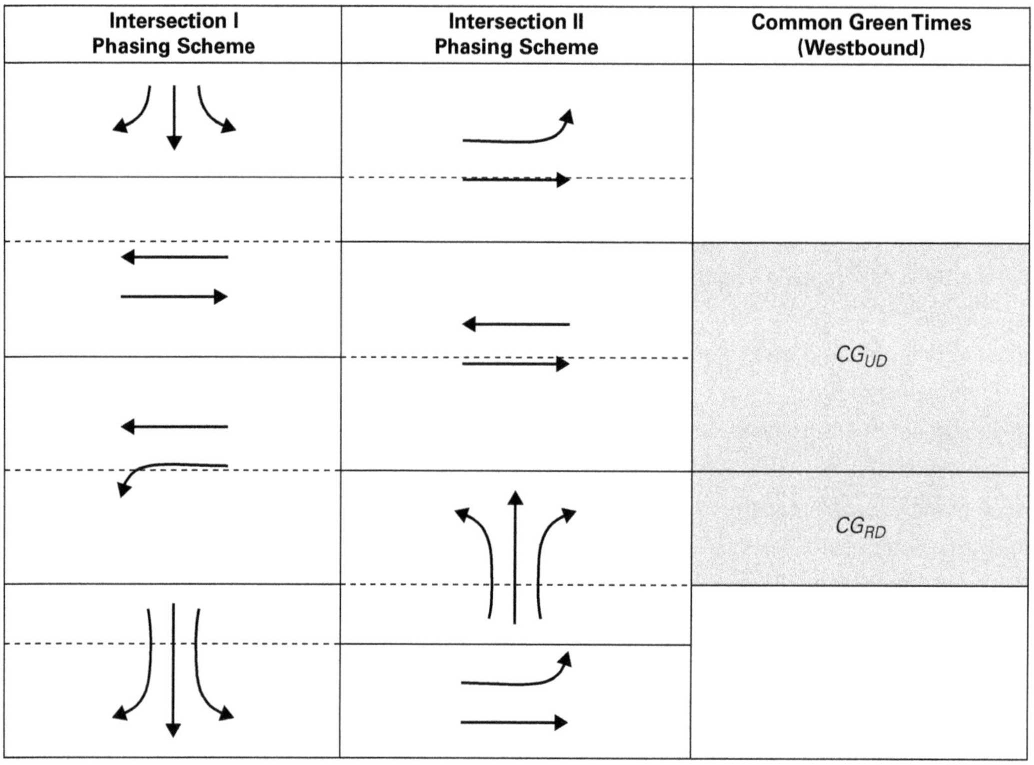

The additional lost time due to the presence of a downstream queue in the internal through movement is calculated for each of the upstream approaches with the following equations:

Additional lost time on the external arterial approach:

$$L_{D-A} = G_A - 0.106DQ_A - 5.39\frac{CG_{UD}}{C}$$

EQUATION 23-29

Additional lost time on the ramp approach:

$$L_{D-R} = G_R - 0.106DQ_R - 5.39\frac{CG_{RD}}{C}$$

EQUATION 23-30

where

$L_{D\text{-}A}$ = lost time on external arterial approach due to presence of downstream queue (s) (min=0),

$L_{D\text{-}R}$ = lost time on external ramp approach due to presence of downstream queue (s) (min=0),

G_A = green interval for external arterial approach (s),

G_R = green interval for left-turning ramp movement (s),

DQ_A = distance to downstream queue at beginning of upstream arterial green (ft),

DQ_R = distance to downstream queue at beginning of upstream ramp green (ft),

Figure 8.29 *HCM Equation 23-38*

$$L_{DS} = CG_{DS} - Q_{\text{Initial}} \times h_I$$

where

L_{DS} = additional lost time due to demand starvation (s);

CG_{DS} = common green time with demand starvation potential (s), as shown in Exhibit 23-33;

h_I = saturation headway for internal through approach (= 3,600/saturation flow per lane) (s); and

Q_{Initial} = length of queue stored at internal approach at beginning of interval during which this approach has demand starvation potential, calculated from Equation 23-39.

Figure 8.30 *HCM Equation 23-39*

$$Q_{\text{Initial}} = \left(\frac{v_{\text{Ramp}-L} \times C}{N_{\text{Ramp}-L} \times 3{,}600} - \frac{[CG_{RD} - t_L]}{h_I} \right) + \left(\frac{v_{\text{Arterial}} \times C}{N_{\text{Arterial}} \times 3{,}600} - \frac{[CG_{UD} - t_L]}{h_I} \right)$$

where

$v_{\text{Ramp-}L}$ = upstream ramp left-turning flow (v/h),

v_{Arterial} = upstream arterial through flow (v/h),

C = cycle length (s),

$N_{\text{Ramp-}L}$ = number of lanes for upstream ramp left-turning movement, ,

N_{Arterial} = number of lanes for upstream arterial through movement,

CG_{RD} = common green time between upstream ramp and downstream through green phase (s),

CG_{UD} = common green time between upstream through and downstream through green phase (s),

h_I = saturation headway for internal through approach (= 3,600/saturation flow per lane) (s), and

t_L = lost time per phase (s) from Equation 23-24 or Equation 23-25.

Figure 8.31 *HCM Exhibit 23-33*

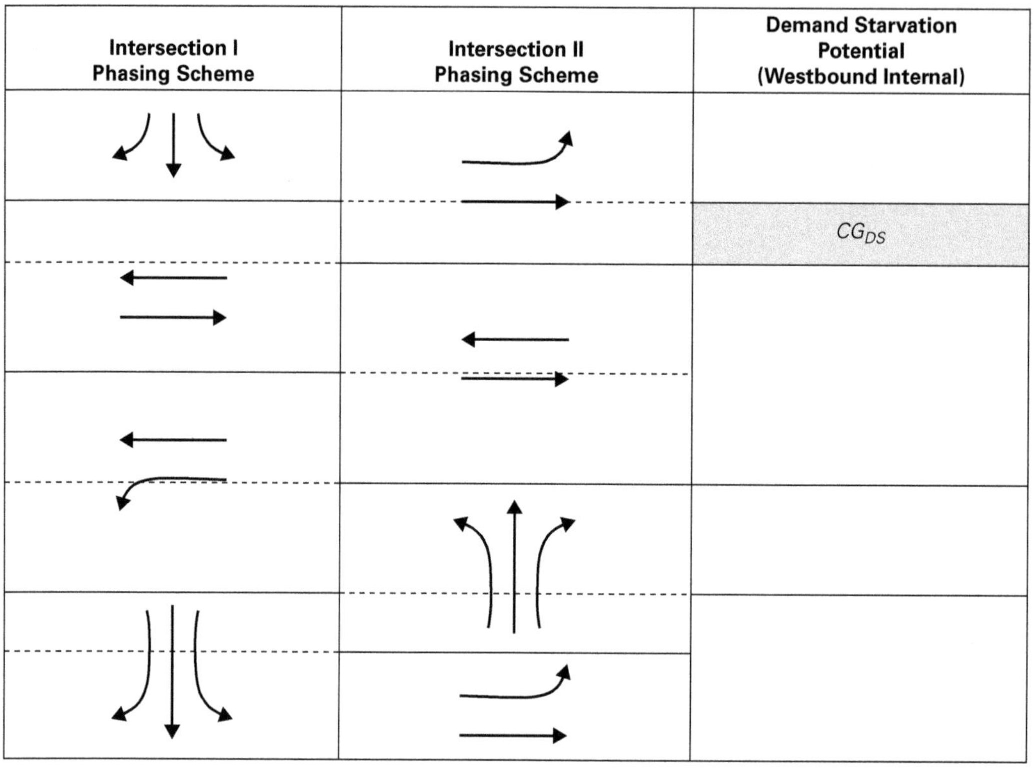

Figure 8.32 *HCM Equation 23-40*

$$L_{D-U_i} = G_{U_i} - 0.106DQ_i - 5.39\frac{CG_{U_iD}}{C}$$

where

L_{D-Ui} = lost time on upstream approach *i* due to presence of a downstream queue (s),

G_{U_i} = green interval for upstream approach *i* (s),

DQ_i = distance to downstream queue at beginning of upstream green for approach *i* (ft), and

CG_{U_iD} = common green time between upstream approach *i* and downstream through green (s).

Figure 8.33 *HCM Equation 23-41*

$$L_{DS} = CG_{DS} - Q_{\text{Initial}} \times h_l$$

where

L_{DS} = additional lost time due to demand starvation (s),

CG_{DS} = common green time with demand starvation potential (s),

h_l = saturation headway for through approach (= 3,600/saturation flow per lane) (s), and

Q_{Initial} = initial queue length at beginning of interval with demand starvation potential.

Origin Destination

The potentially fourteen origin-destination (O-D) movement pairs are illustrated in HCM Exhibit 23-20 for diamond and partial cloverleaf designs. The movement pairs are defined as A through N for each interchange type determining level of service by O-D individually. The

Figure 8.34 *HCM Exhibit 23-20*

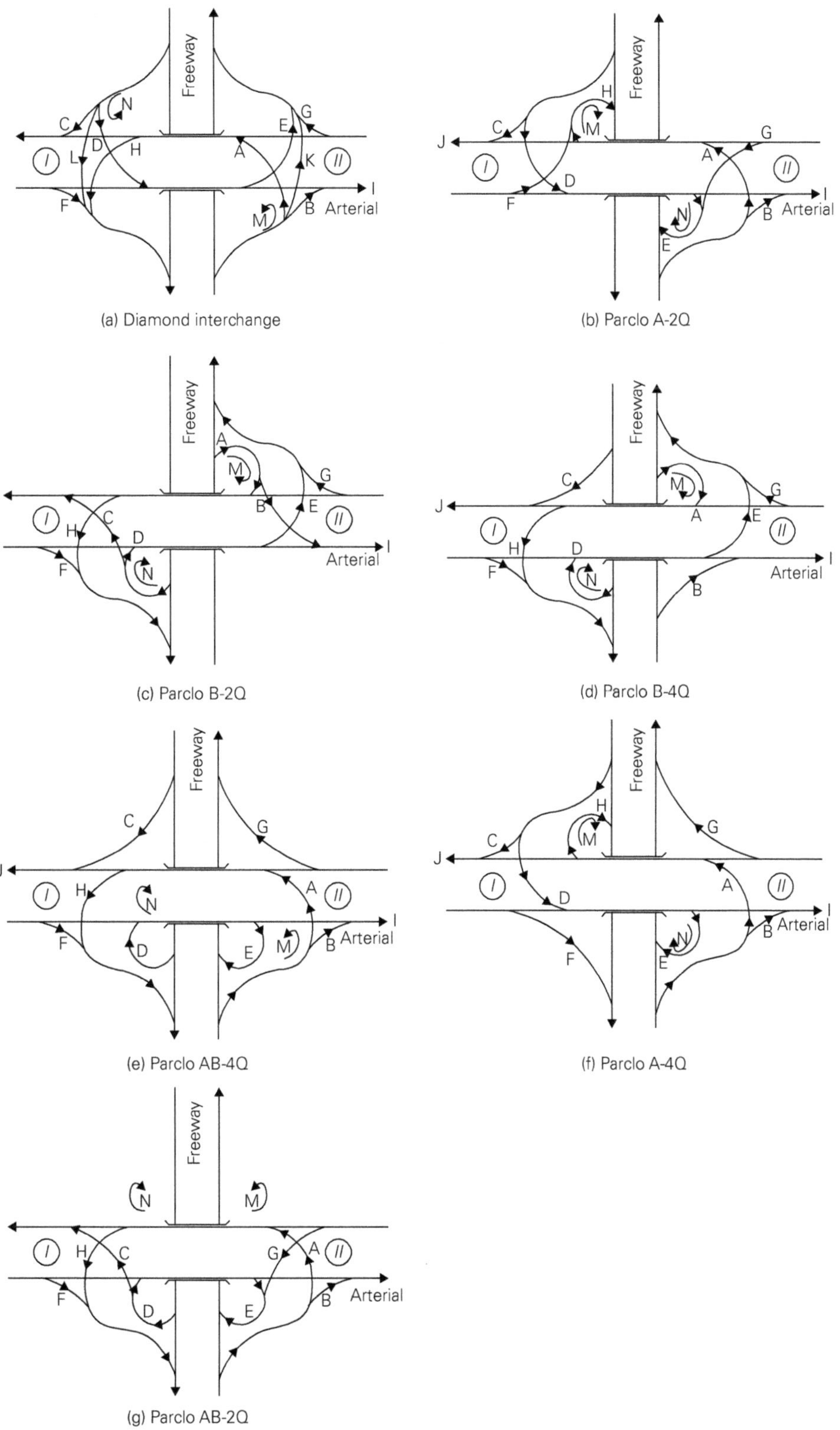

(a) Diamond interchange

(b) Parclo A-2Q

(c) Parclo B-2Q

(d) Parclo B-4Q

(e) Parclo AB-4Q

(f) Parclo A-4Q

(g) Parclo AB-2Q

Figure 8.35 *HCS7 Intchgs Graphic 1*

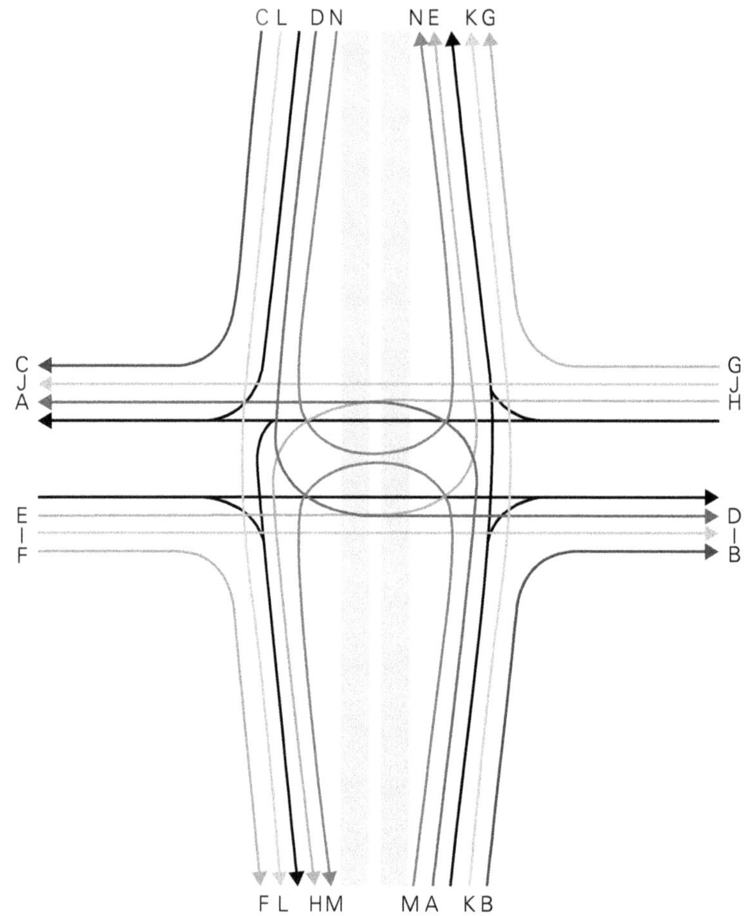

Highway Capacity Software (HCS) provides an alternative series of graphics for these O-D pairs to illustrate the implementation results as depicted in the diamond interchange example shown below.

Experienced Travel Time

Experienced Travel Time (*ETT*) is the sum of delays for each O-D movement through both intersections, then adding any Extra Distance Travel Time (*EDTT*) that may be incurred as defined for *ETT* in HCM Equation 23-50 and for *EDTT* in HCM Equation 23-51. These extra distances are illustrated for the diamond interchange in HCM Exhibits 23-7 and 23-8, with relationships shown in HCM Equations 23-3 through 23-10. For partial cloverleaf interchanges, HCM Exhibit 23-9 illustrates the extra distances with the relationships included.

Figure 8.36 *HCM Equation 23-50*

$$ETT = \sum d_i + \sum EDTT$$

Figure 8.37 *HCM Equation 23-51*

$$EDTT = \frac{D_t}{1.47 \times v_D} + a$$

where

$EDTT$ = extra distance travel time (s);

D_t = distance (ft) traveled along the loop ramp or diverted movement (ft);

v_D = design speed of the loop ramp or diverted movement (mi/h); and

a = delay due to deceleration into a turn and acceleration after the next turn (s); assumed to be 5 s for a loop ramp movement.

Figure 8.38 *HCM Exhibit 23-7*

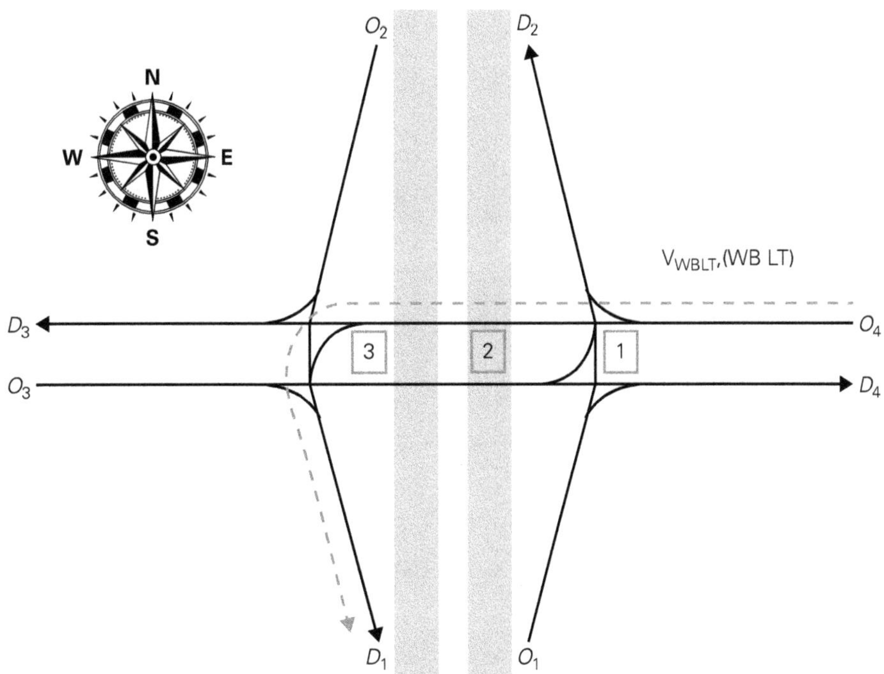

Figure 8.39 *HCM Exhibit 23-8*

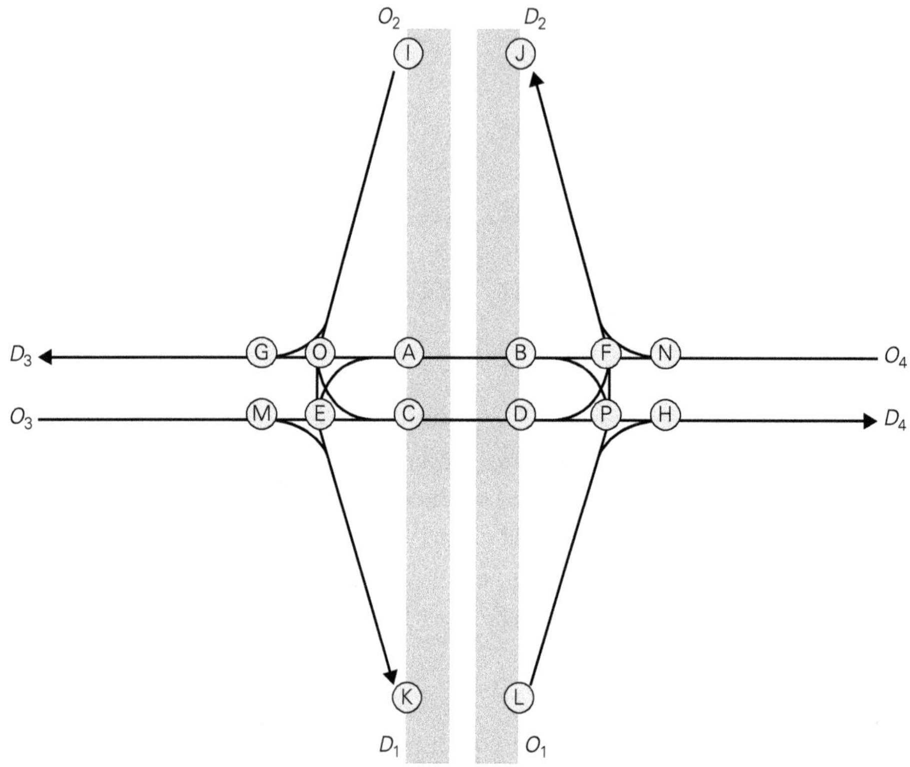

Figure 8.40 *HCM Equation 23-3*

$$EDTT_{13} = TT_{LPB} - TT_{LB}$$

Figure 8.41 *HCM Equation 23-4*

$$EDTT_{14} = TT_{LH} - TT_{LDH}$$

Figure 8.42 *HCM Equation 23-5*

$$EDTT_{24} = TT_{IOC} - TT_{IC}$$

Figure 8.43 *HCM Equation 23-6*

$$EDTT_{23} = TT_{IG} - TT_{IAG}$$

Figure 8.44 *HCM Equation 23-7*

$$EDTT_{32} = TT_{DFJ} - TT_{DJ}$$

Figure 8.45 *HCM Equation 23-8*

$$EDTT_{31} = TT_{MK} - TT_{MCK}$$

Figure 8.46 *HCM Equation 23-9*

$$EDTT_{41} = TT_{AEK} - TT_{AK}$$

Figure 8.47 *HCM Equation 23-10*

$$EDTT_{42} = TT_{NJ} - TT_{NBJ}$$

Figure 8.48 *HCM Exhibit 23-9*

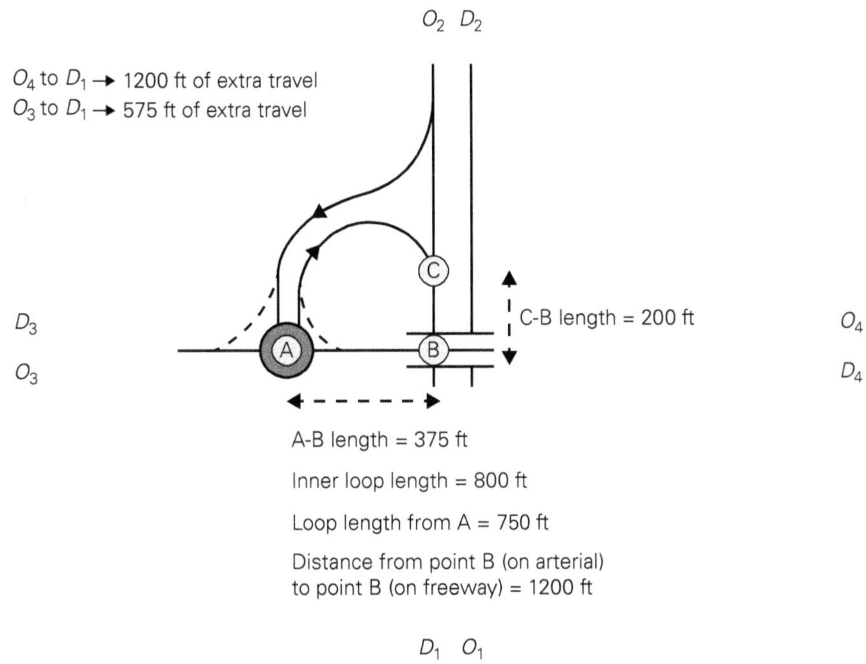

Level of Service

Level of service is defined for each O-D movement according to HCM Exhibit 23-10 with the volume-to-capacity ratio or the queue-storage ratio exceeding 1.0 for any approach dictating LOS F for that approach. The 120-second scale is higher than the signalized intersection scale to compensate for O-D movements that must travel through two intersections. Though no interchange level of service is defined, a weighted average of experienced travel time for all O-D pairs by flow rate could be considered.

Diverging Diamond Interchange

The diverging diamond interchange (DDI) is an alternative design being implemented widely because of its unique ability to significantly increase capacity without adding lanes that typically require rebuilding the bridge—saving millions. The crossing over of the arterial lanes at the ramp junctions allow for all interaction with the freeway to be free-flowing merges that avoid the conflicting left-turn phasing as illustrated in HCM Exhibit 23-30. The DDI procedures are consistent with other interchanges with a few modifications to the saturation flow adjustments for the crossover movement, lane utilization, and additional lost time for overlap phasing, as well as a gap acceptance procedure for yield-controlled movements.

SATURATION FLOW

The crossover movement includes an adjustment to the saturation flow rate, similar to that for left or right turns at any signalized intersection, with this calibrated value for the crossover shown to equal 0.913.

Figure 8.49 *HCM Exhibit 23-10*

ETT (s/veh)	Condition $v/c \leq 1$ and $R_Q \leq 1$ for Every Lane Group	 $v/c > 1$ for Any Lane Group	 $R_Q > 1$ for Any Lane Group
≤15	A	F	F
>15–30	B	F	F
>30–55	C	F	F
>55–85	D	F	F
>85–120	E	F	F
>120	F	F	F

Figure 8.50 *HCM Exhibit 23-30*

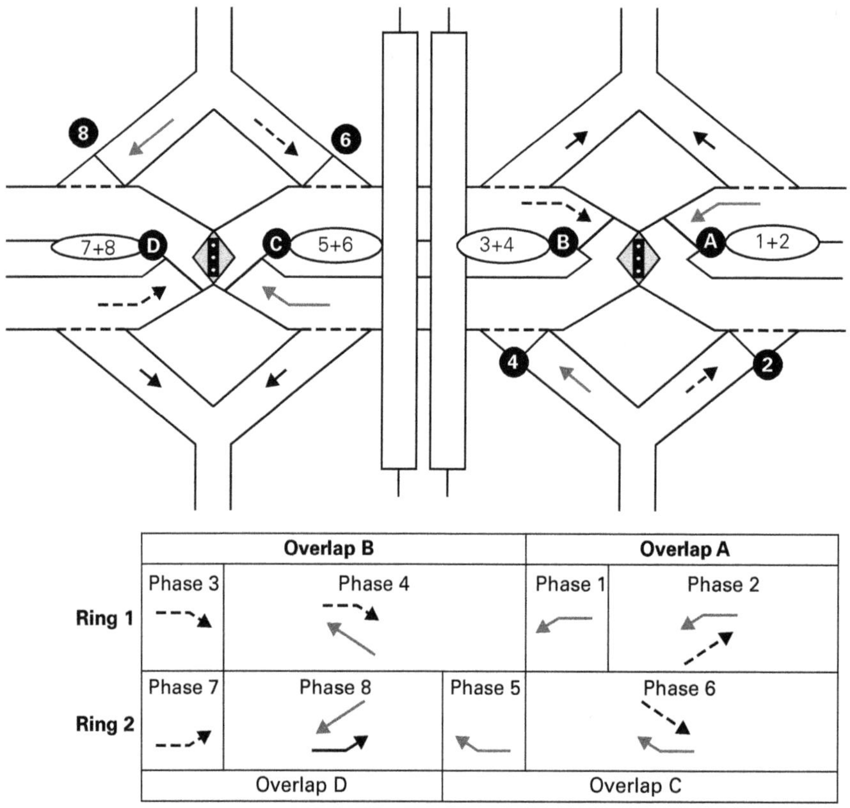

Think about your own driving:

When you are going through a DDI, crossing over the opposing lanes to essentially drive on the wrong side of the road through the interchange is still relatively new to most drivers. So, the slight turn to navigate the crossover becomes more cautious than the radius warrants, resulting in the nearly 9% reduction in capacity, with a left turn only a 5% reduction. This will likely improve over time as drivers get more comfortable toward a reduction more in line with this minimal radius.

LANE UTILIZATION

There are five models to estimate the lane utilization for different DDI lane configurations shown in HCM Exhibit 23-25. The coefficients for each model are defined in HCM Exhibit 23-26 and applied in HCM Equation 23-18 for the appropriate adjustment to the saturation flow rate.

ADDITIONAL LOST TIME

The typical additional lost time values for an interchange analysis are for internal queue and demand starvation, but cannot be applied to a DDI for internal queue and is assumed to be zero for demand starvation—both due to the unique signal phasing. There is a lost time component applied to the DDI for the overlap phasing required to account for the extra distance from the ramp to the crossover as a function of the geometric values L, D, and W defined in HCM Exhibit 23-31 applied in HCM Equation 23-37.

Figure 8.51 *HCM Exhibit 23-25*

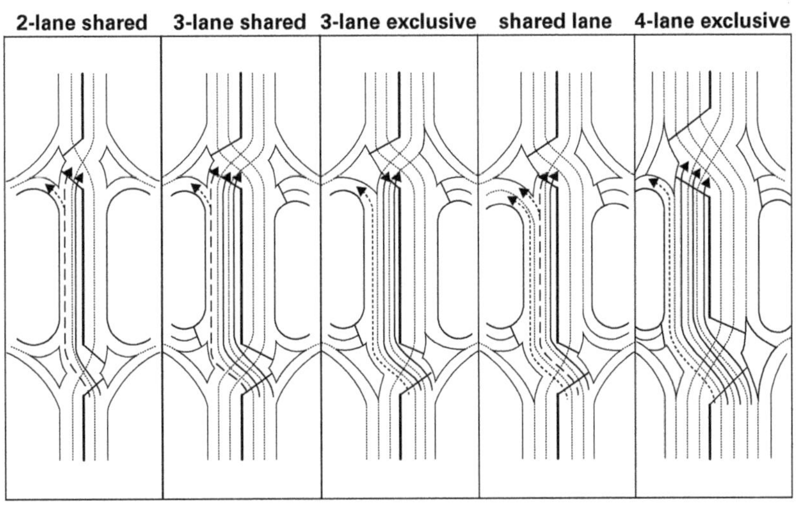

Figure 8.52 *HCM Exhibit 23-26*

Lane Configuration	Regime	Lane	a_1	a_2
2-lane shared	I(LTDR ≤ 0.35)	Left	0.2129	0.5250
	II (LTDR > 0.35)	Left	0.5386	0.4110
3-lane shared	I-1 (LTDR ≤ 0.13)	Middle	−0.1831	0.3863
	I-2 (0.13 < LTDR ≤ 0.43)	Leftmost	0.2245	0.3336
	II (LTDR > 0.43)	Leftmost	0.6460	0.1523
3-lane exclusive	I(LTDR ≤ 0.33)	Middle	−0.5983	0.5237
	II (LTDR > 0.33)	Leftmost	0.9695	0.0096
3-lane exclusive with middle shared lane	I(LTDR ≤ 0.50)	Middle	−0.2884	0.5626
	II (LTDR > 0.50)	Leftmost	0.4903	0.1761
4-lane Exclusive	I(LTDR ≤ 0.35)	Center-left	−0.5432	0.5095
	II (LTDR > 0.35)	Leftmost	0.9286	−0.0071

Figure 8.53 *HCM Equation 23-18*

$$\%V_{Li,DDI} = a_1 \times LTDR + a_2$$

where

$\%V_{Li,DDI}$ = percent of traffic present in lane L_i for a DDI, where L_1 represents the leftmost lane, L_2 represents the second lane from the left, and so forth;

a_i = coefficient for i = 1 and i = 2 (see Exhibit 23-26); and

$LTDR$ = left-turn demand ratio (decimal), calculated as left-turn demand at external crossover, divided by total approach volume.

Figure 8.54 *HCM Exhibit 23-31*

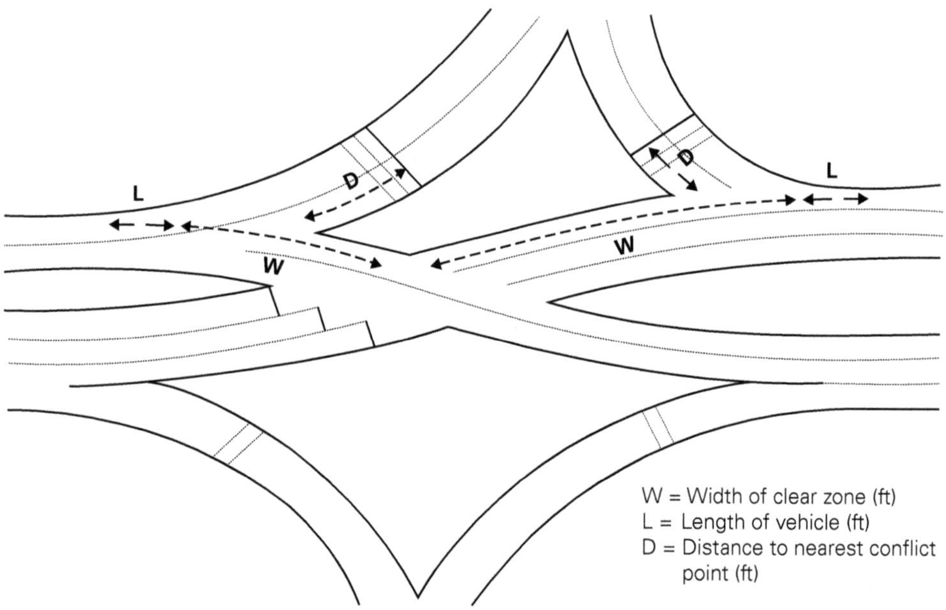

Figure 8.55 *HCM Equation 23-37*

$$L_{OL-DDI} = \frac{W + L - D}{1.467 \times S_f}$$

where

W = width of the clear zone for the longest vehicle path (ft), measured along the centerline of the outside lane, which is the closest conflicting vehicle path to the ramp;

L = design vehicle length (ft), typically 20 ft;

D = distance from the ramp movement stop bar to the conflict point (ft) measured along the centerline of the off-ramp approach; and

S_f = free-flow speed of the vehicle (mi/h).

YIELD CONTROL

Yield-controlled left or right turns are modeled in three regimes: 1) blocked by conflicting platoon serving the queue being served by the crossover signal; 2) gap acceptance in conflicting traffic during the crossover green after the initial queue; and 3) no conflicting flow with the crossover signal red, as illustrated in HCM Exhibit 23-34.

Regime 1 is estimated as the time required to serve the conflicting queue plus the clearance time. Regime 2 is computed using the roundabout procedure as an application of yield control, following HCM Equations 23-43 and 23-44 using the DDI parameters defined in HCM Exhibit 23-36.

Origin Destination

The origin-destination (O-D) movement pairs are illustrated in the Highway Capacity Software (HCS) graphic implementation results shown below.

Experienced Travel Time

Experience Travel Time (*ETT*) is the sum of delays for each O-D movement through both intersections, then adding any Extra Distance Travel Time (*EDTT*) that may be incurred as defined for *ETT* in HCM Equation 23-50 and for *EDTT* in HCM Equation 23-51. These extra distances are illustrated for a DDI in HCM Exhibit 23-11.

Level of Service

Level of service is defined for each O-D movement according to HCM Exhibit 23-10, with the volume-to-capacity ratio or the queue-storage ratio exceeding 1.0 for any approach dictating LOS F for that approach. The 120-second scale is higher than the signalized intersection scale to compensate for O-D movements that must travel through two intersections as illustrated in HCM Exhibit 23-11. Though no interchange level of service is defined, a weighted average of experienced travel time for all O-D pairs by flow rate could be considered.

Figure 8.56 *HCM Exhibit 23-34*

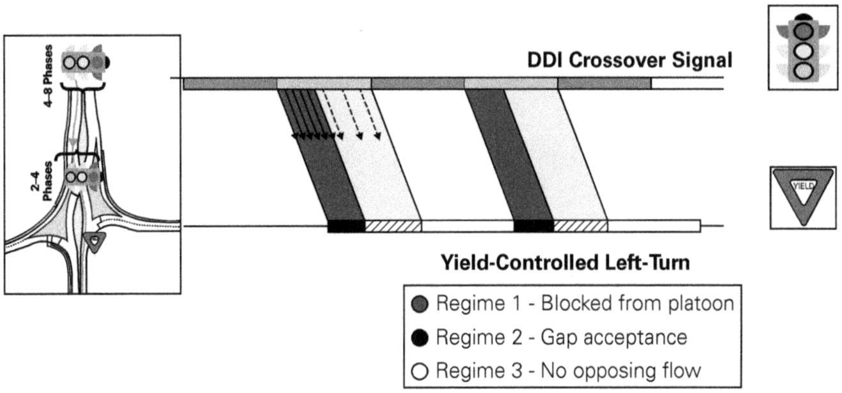

Figure 8.57 *HCM Equation 23-43*

$$C_{GA} = \frac{3{,}600}{t_f} \exp\left(-\frac{t_c - \frac{t_f}{2}}{3{,}600} \times q_c \right)$$

where

C_{GA} = the capacity during the gap-acceptance regime (veh/h),
q_c = conflicting flow rate (veh/h),
t_c = critical headway (s), and
t_f = follow-up headway (s).

Figure 8.58 *HCM Equation 23-44*

$$p_{GA,x} = \frac{g - (t_{CQ} + t_{clear})}{C}$$

where

$P_{GA,x}$ = proportion of time of gap-acceptance regime (decimal),
t_{CQ} = time to clear conflicting queue (s),
t_{clear} = time for last queued vehicle to clear distance from stop bar to yield point (s),
g = effective green time of the DDI crossover movement (s), and
C = cycle length of the DDI crossover signal (s).

Figure 8.59 *HCM Exhibit 23-36*

Parameter and Units	**Left Turns**	**Right Turns**
Critical headway, t_c (s)	3.9	1.8
Follow-up headway, t_f (s)	2.6	2.4
Capacity intercept (veh/h)	1,399	1,481
Capacity slope (veh/h)	–0.00073	–0.00016

Figure 8.60 *HCS7 Interchanges Graphic 2*

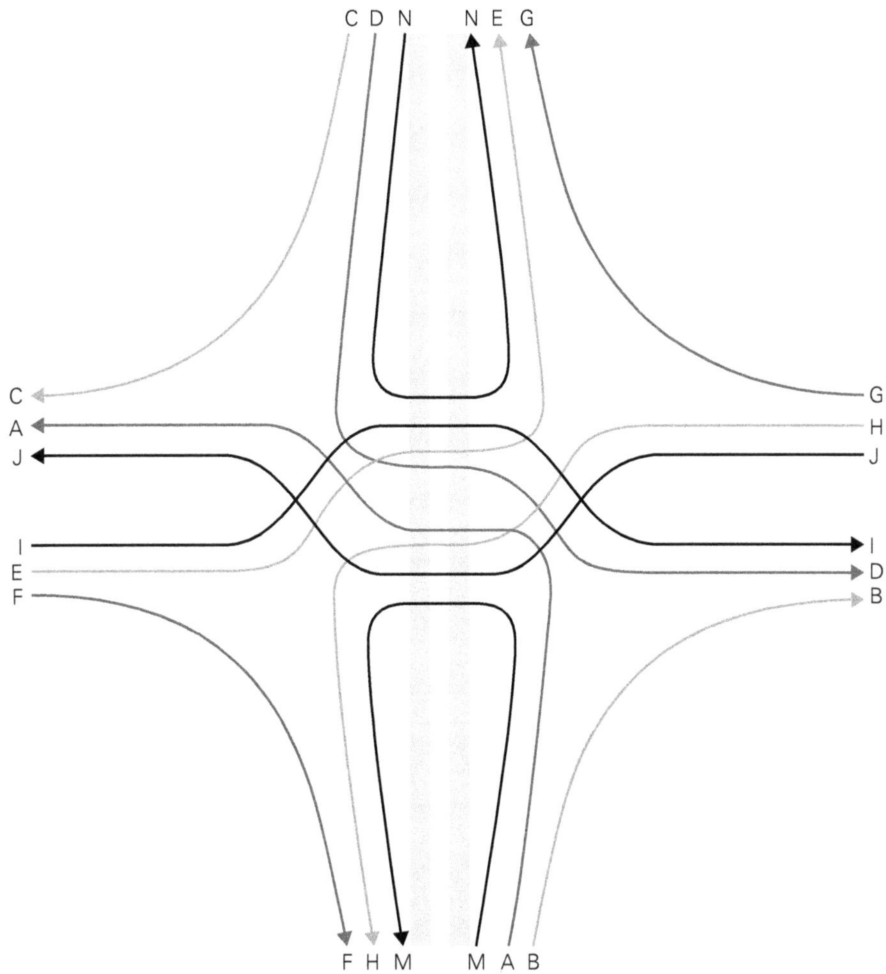

Figure 8.61 *HCM Equation 23-50*

$$ETT = \sum d_i + \sum EDTT$$

Figure 8.62 *HCM Equation 23-51*

$$EDTT = \frac{D_t}{1.47 \times v_D} + a$$

where

$EDTT$ = extra distance travel time (s);

D_t = distance (ft) traveled along the loop ramp or diverted movement (ft);

v_D = design speed of the loop ramp or diverted movement (mi/h); and

a = delay due to deceleration into a turn and acceleration after the next turn (s); assumed to be 5 s for a loop ramp movement.

Figure 8.63 *HCM Exhibit 23-11*

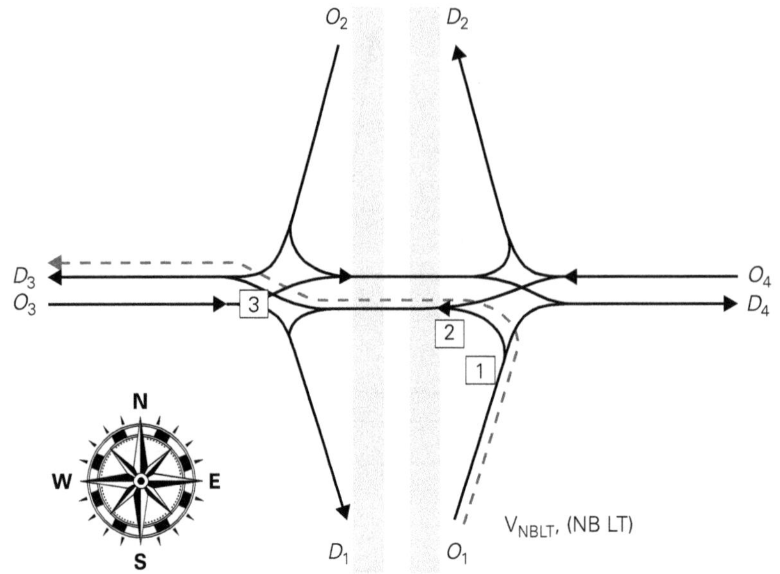

Figure 8.64 *HCM Exhibit 23-10*

	Condition		
ETT (s/veh)	**$v/c \leq 1$ and $R_a \leq 1$ for Every Lane Group**	**$v/c > 1$ for Any Lane Group**	**$R_a > 1$ for Any Lane Group**
≤15	A	F	F
>15–30	B	F	F
>30–55	C	F	F
>55–85	D	F	F
>85–120	E	F	F
>120	F	F	F

Problem One

Since the procedures for this analysis do not lend themselves to performing manually, the HCM Example Problem for a diamond interchange is illustrated here. The lane configurations, movement demands, grades, and segment lengths are shown in HCM Exhibit 34-3 with signal phasing and timing shown in HCM Exhibit 34-4. Other data include: turning radius as 50 feet for right turns and 75 feet for left turns; grade at 2% for ramp approaches; extra distance as 100 feet for ramp movements; heavy vehicles at 6.1%; and peak hour factor as 0.90.

The origin-destination (O-D) movement flow rates are shown in HCM Exhibit 34-5, and the final results for control delay, extra distance travel time, experienced travel time, and level of service are compiled in HCM Exhibit 34-16. Notice there is no weighted average generated to determine an interchange level of service.

Figure 8.65 *HCM Exhibit 34-3*

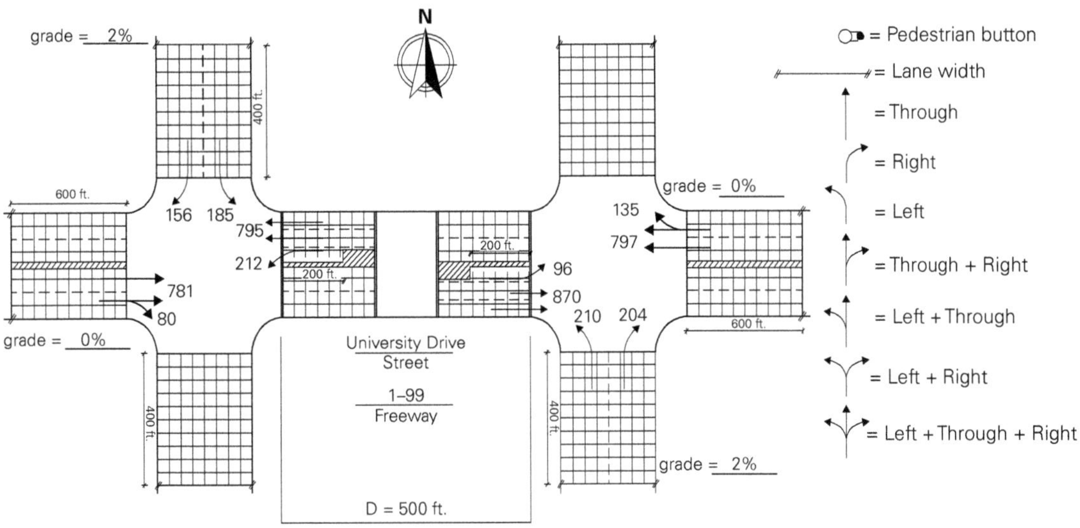

Figure 8.66 *HCM Exhibit 34-4*

	Intersection I			Intersection II		
Phase	**1**	**2**	**3**	**1**	**2**	**3**
NEMA	Φ (2+6)	Φ (1+6)	Φ (4+7)	Φ (2+6)	Φ (3+8)	Φ (2+5)
Green time	63	43	39	63	53	29
Yellow + all red	5	5	5	5	5	5
Offset		19			9	

Figure 8.67 *HCM Exhibit 34-5*

O-D Movement	Demand (veh/h)	PHF-Adjusted Demand (veh/h)
A	210	233
B	204	227
C	156	173
D	185	206
E	96	107
F	80	89
G	135	150
H	212	236
I	685	761
J	585	650
K	0	0
L	0	0
M	0	0
N	0	0

Figure 8.68 *HCM Exhibit 34-16*

O-D Movement	Ctrl. Delay (s)	EDTT (s)	ETT (s)	$v/c > 1$?	$R_Q > 1$?	LOS
A	45.6	1.9	47.5	No	No	C
B	43.7	–1.9	41.8	No	No	C
C	54.6	–1.9	52.7	No	No	C
D	63.6	1.9	65.5	No	No	D
E	99.2	1.9	101.1	No	No	E
F	44.2	–1.9	42.3	No	No	C
G	37.5	–1.9	35.6	No	No	C
H	82.7	1.9	84.6	No	No	D
I	52.0	0.0	52.0	No	No	C
J	39.8	0.0	39.8	No	No	C
K	–	–	–	–	–	–
L	–	–	–	–	–	–
M	88.5	3.8	92.3	No	No	E
N	110.8	3.8	114.8	No	No	E

HCS7

These screenshots show how Problem One is coded into *HCS7*, followed by the results as presented in the formatted report. All input data, intermediate computations, and final results are shown in logical order, leading to density and level of service. Note some delays differ due to flow profile and share-lane models in *HCS7*.

Figure 8.69 *HCS7 Intchgs1 Input*

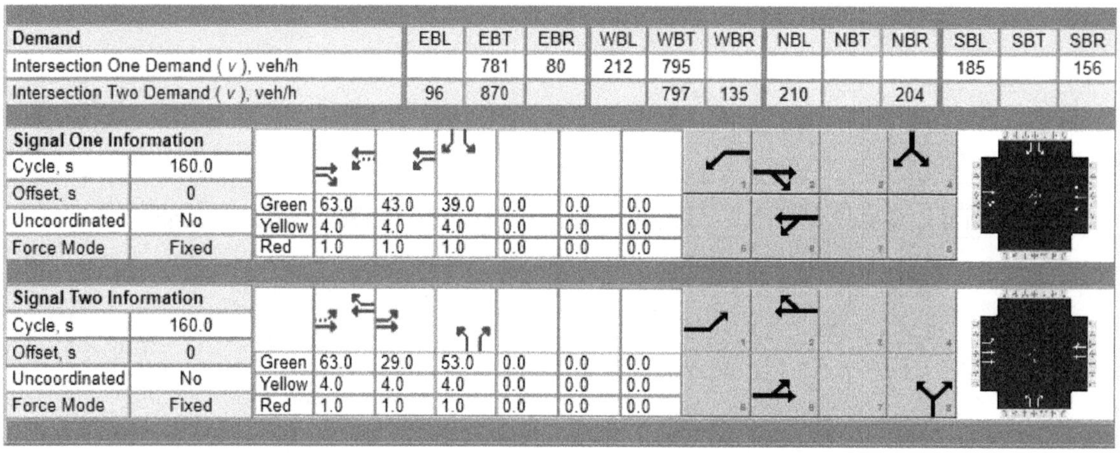

Demand	EBL	EBT	EBR	WBL	WBT	WBR	NBL	NBT	NBR	SBL	SBT	SBR
Intersection One Demand (v), veh/h		781	80	212	795					185		156
Intersection Two Demand (v), veh/h	96	870			797	135	210		204			

Signal One Information	
Cycle, s	160.0
Offset, s	0
Uncoordinated	No
Force Mode	Fixed

Green	63.0	43.0	39.0	0.0	0.0	0.0
Yellow	4.0	4.0	4.0	0.0	0.0	0.0
Red	1.0	1.0	1.0	0.0	0.0	0.0

Signal Two Information	
Cycle, s	160.0
Offset, s	0
Uncoordinated	No
Force Mode	Fixed

Green	63.0	29.0	53.0	0.0	0.0	0.0
Yellow	4.0	4.0	4.0	0.0	0.0	0.0
Red	1.0	1.0	1.0	0.0	0.0	0.0

Figure 8.70 *HCS7 Intchgs1 Output*

Interchange Results

O-D	Demand (veh/h)	Delay (s)	EDTT	ETT	v/c > 1 ?	RQ > 1 ?	LOS
A	233	64.1	1.9	66.0	No	No	D
B	227	43.7	-1.9	41.8	No	No	C
C	173	54.6	-1.9	52.7	No	No	C
D	206	77.2	1.9	79.1	No	No	D
E	107	86.0	1.9	87.9	No	No	E
F	89	34.8	-1.9	32.9	No	No	C
G	150	37.3	-1.9	35.4	No	No	C
H	236	86.0	1.9	87.9	No	No	F
I	761	56.5	0.0	56.5	No	No	D
J	650	58.1	0.0	58.1	No	No	D
K	0	-	0.0	-	-	-	-
L	0	-	0.0	-	-	-	-
M	0	-	0.0	-	-	-	-
N	0	-	0.0	-	-	-	-

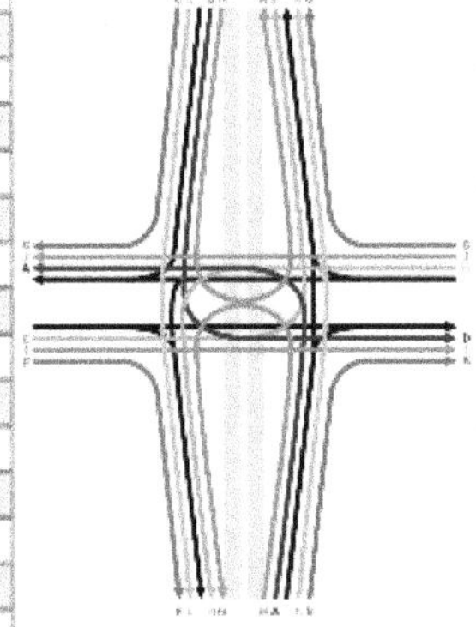

Problem Two

Since the procedures for this analysis do not lend themselves to performing manually, the HCM Example Problem for a signalized diverging diamond interchange is illustrated here. The lane configurations and movement demands are shown in HCM Exhibit 34-58 with signal timing, turn radius, clear zone distances, and volumes shown in HCM Exhibit 34-59. Other data include: turning radius as 75 feet for right turns and 150 feet for left turns; heavy vehicles at 6.1%; and peak hour factor as 0.95.

Figure 8.71 *HCM Exhibit 34-58*

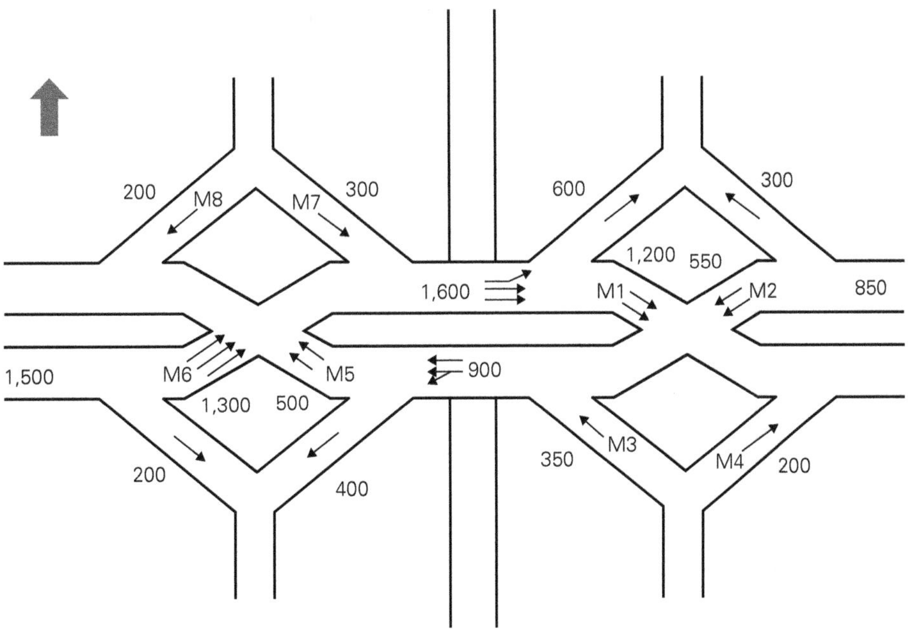

Figure 8.72 *HCM Exhibit 34-59*

Movement	West Crossover				East Crossover			
	M5	M6	M7	M8	M1	M2	M3	M4
Green time (s)	25	35	25	35	35	25	35	25
Yellow time (s)	4	4	4	4	4	4	4	4
All-red time (s)	1	1	1	1	1	1	1	1
Phase split (s)	30	40	30	40	40	30	40	30
Turn radius (ft)			150	75			150	75
Width of clear zone (ft)			200	100			200	100
Shortest distance, stop bar to conflict point (ft)			20	60			20	60
Volume (veh/h)	500	1,300	300	200	1,000	550	350	200

Figure 8.73 *HCM Exhibit 34-60*

O-D Movement	Demand (veh/h)	PHF-Adjusted Demand (veh/h)
A	350	368
B	200	211
C	200	211
D	300	316
E	600	632
F	200	211
G	300	316
H	400	421
I	700	737
J	150	158
K	0	0
L	0	0
M	0	0
N	0	0

Figure 8.74 *HCM Exhibit 34-65*

O-D	PHF-Adjusted Demand (veh/h)	Movement	Control Delay Components	Total Control Delay (s)	Extra Distance (ft)	EDTT (s)	ETT (s)	LOS
A	368	NB L	M3 + M5	39.5	100	1.9	41.4	C
B	211	NB R	M4	22.9	−100	−1.9	21.0	B
C	211	SB R	M8	13.3	−100	−1.9	11.4	A
D	316	SB L	M7 + M1	72.0	100	1.9	73.9	D
E	632	EB L	M6	19.4	100	1.9	21.3	B
F	211	EB R	N/A	0.0	0	0.0	0.0	A
G	316	WBR	N/A	0.0	0	0.0	0.0	A
H	421	WBL	M2	40.3	100	1.9	42.2	C
I	737	EB T	M6 + M1	40.8	40	0.8	41.6	C
J	158	WBT	M2 + M5	59.0	40	0.8	59.8	D

The origin-destination (O-D) movement flow rates are shown in HCM Exhibit 34-60, and the final results for control delay, extra distance travel time, experienced travel time, and level of service are compiled in HCM Exhibit 34-65. Notice there is no weighted average generated to determine an interchange level of service.

HCS7

These screenshots show how Problem Two is coded into *HCS7*, followed by the results as presented in the formatted report. All input data, intermediate computations, and final results are shown in logical order, leading to density and level of service. Note some delays differ due to flow profile and share-lane models in *HCS7*.

Figure 8.75 *HCS7 Intchgs2 Input*

Demand	EBL	EBT	EBR	WBL	WBT	WBR	NBL	NBT	NBR	SBL	SBT	SBR
Intersection One Demand (*v*), veh/h	1	1300	200	300	500			0	300		0	200
Intersection Two Demand (*v*), veh/h	600	1000		1	450	300		0	200		0	350

Signal One Information	
Cycle, s	70.0
Offset, s	0
Uncoordinated	No
Force Mode	Fixed

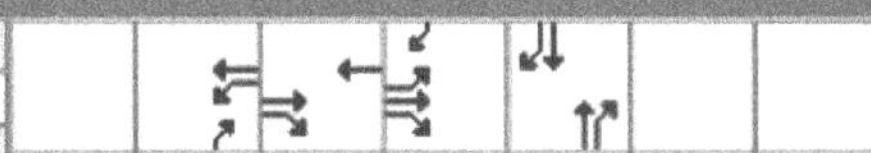

Green	25.0	0.0	34.0	1.0	0.0	0.0
Yellow	4.0	0.0	4.0	0.0	0.0	0.0
Red	1.0	0.0	1.0	0.0	0.0	0.0

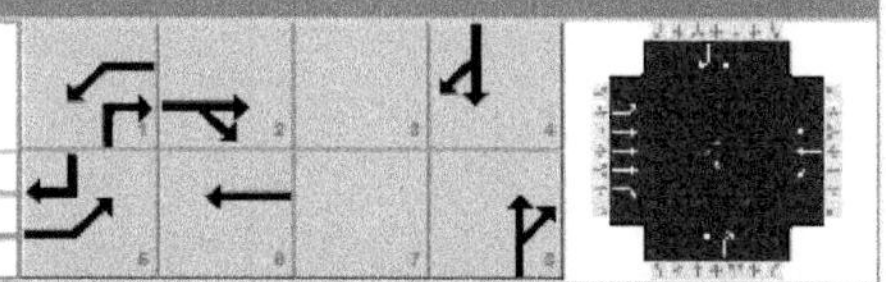

Signal Two Information	
Cycle, s	70.0
Offset, s	0
Uncoordinated	No
Force Mode	Fixed

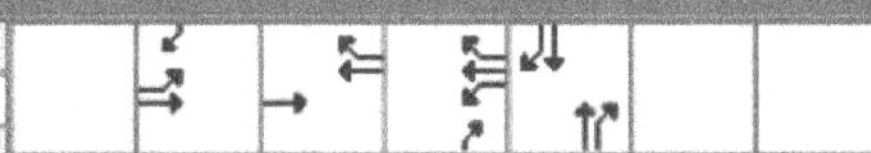

Green	34.0	0.0	25.0	1.0	0.0	0.0
Yellow	4.0	0.0	4.0	0.0	0.0	0.0
Red	1.0	0.0	1.0	0.0	0.0	0.0

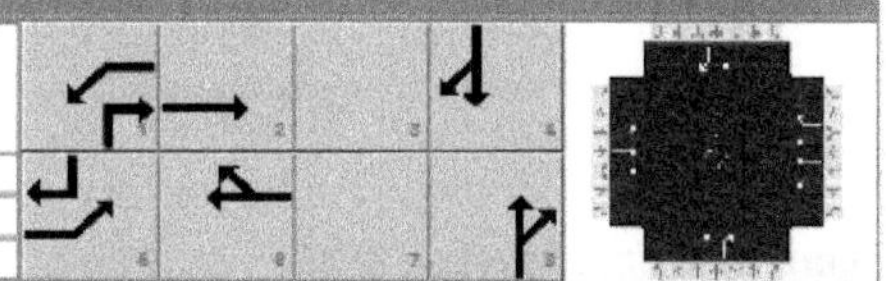

Figure 8.76 *HCS7 Intchgs2 Output*

Interchange Results

O-D	PHF-Adjusted Demand (veh/h)	Movement	Control Delay Components	Total Control Delay (s/veh)	Extra Distance (ft)	EDTT (s/veh)	ETT (s/veh)	LOS
A	368	NBL	M3 + M5	39.6	100	1.9	41.6	C
B	211	NBR	M4	23.7	-100	-1.9	21.8	B
C	211	SBR	M8	15.0	-100	-1.9	13.0	A
D	316	SBL	M7 + M1	61.5	100	1.9	63.5	D
E	632	EBL	M6	26.5	100	1.9	28.4	B
F	211	EBR	N/A	0.0	0	0.0	0.0	A
G	316	WBR	N/A	0.0	0	0.0	0.0	A
H	316	WBL	M2	29.7	100	1.9	31.6	C
I	737	EBT	M6 + M1	38.1	40	0.8	38.8	C
J	158	WBT	M2 + M5	47.8	40	0.8	48.6	C
M	--	--	--	--	--	--	--	--
N	--	--	--	--	--	--	--	--

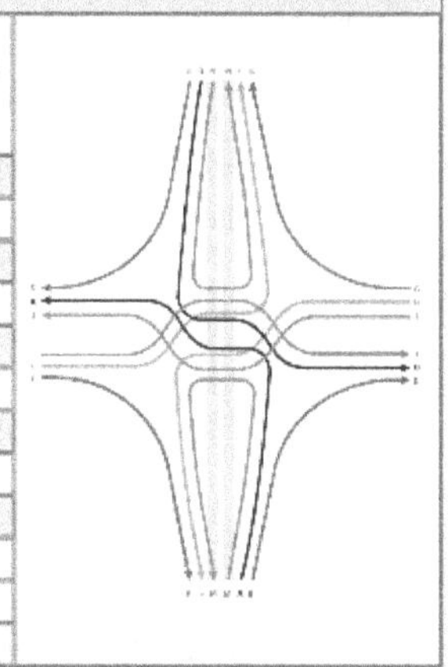

Alternative Intersections

Three alternative intersection configurations are modeled, including Displaced Left Turn (DLT) with signalized crossover intersections; Restricted Crossing U-Turn (RCUT) with signalized, stop-controlled, or merge cross-over intersections; and Median U-Turn (MUT) with signalized or unsignalized crossover intersections.

Figure 8.77 *HCM Exhibit 23-45*

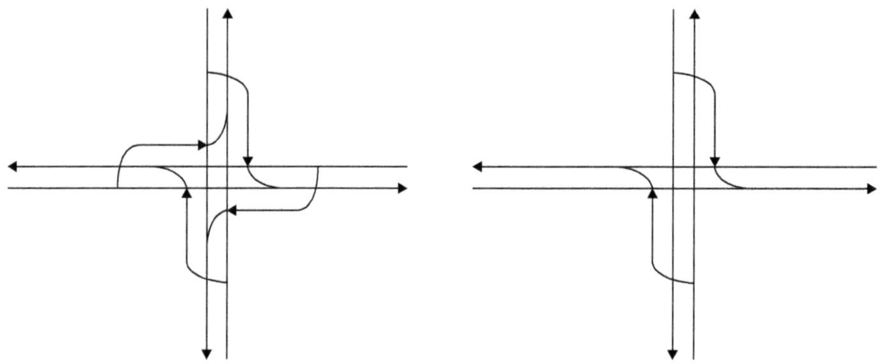

(a) Full DLT

(b) Partial DLT

Figure 8.78 *HCM Equation 23-60*

$$ETT_A = \frac{\Sigma(ETT_j \times v_j)}{\Sigma v_j}$$

where

ETT_A = approach experienced travel time (s/veh),
ETT_j = experienced travel time for movement j (s/veh),
v_j = demand flow rate for movement j (veh/h), and
$j \in$ all movements on the approach of interest.

Figure 8.79 *HCM Equation 23-61*

$$ETT_I = \frac{\Sigma(ETT_k \times v_k)}{\Sigma v_k}$$

where

ETT_I = intersection experienced travel time (s/veh),
ETT_k = experienced travel time for movement k (s/veh),
v_k = demand flow rate for movement k (veh/h), and
$k \in$ all movements at the intersection.

Figure 8.80 *HCM Exhibit 23-13*

ETT (s/veh)	Condition		
	$v/c \le 1$ and $R_Q \le 1$ for Every Lane Group	$v/c > 1$ for Any Lane Group	$R_Q > 1$ for Any Lane Group
≤10	A	F	F
>10–20	B	F	F
>20–35	C	F	F
>35–55	D	F	F
>55–80	E	F	F
>80	F	F	F

Figure 8.81 *HCM Exhibit 23-41*

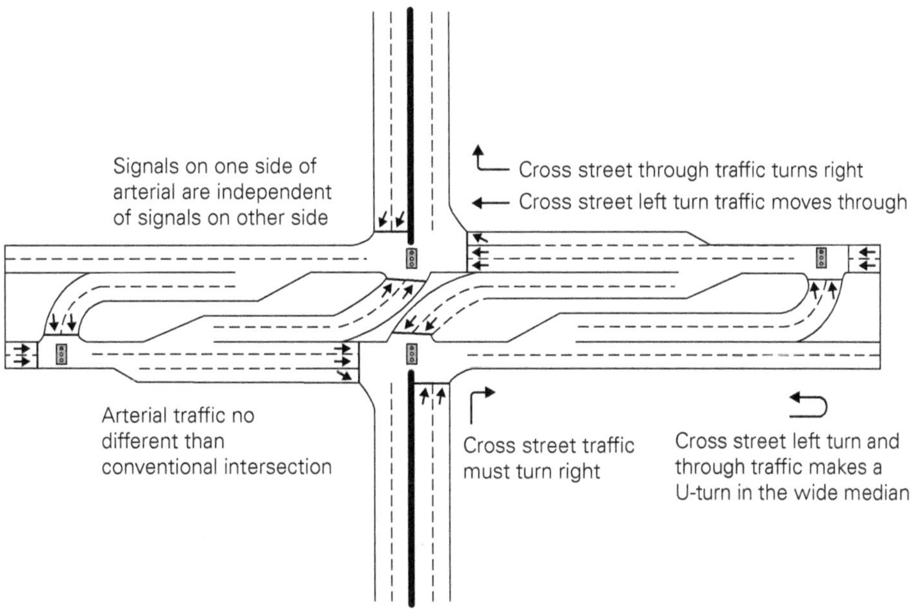

Figure 8.82 *HCM Exhibit 23-42*

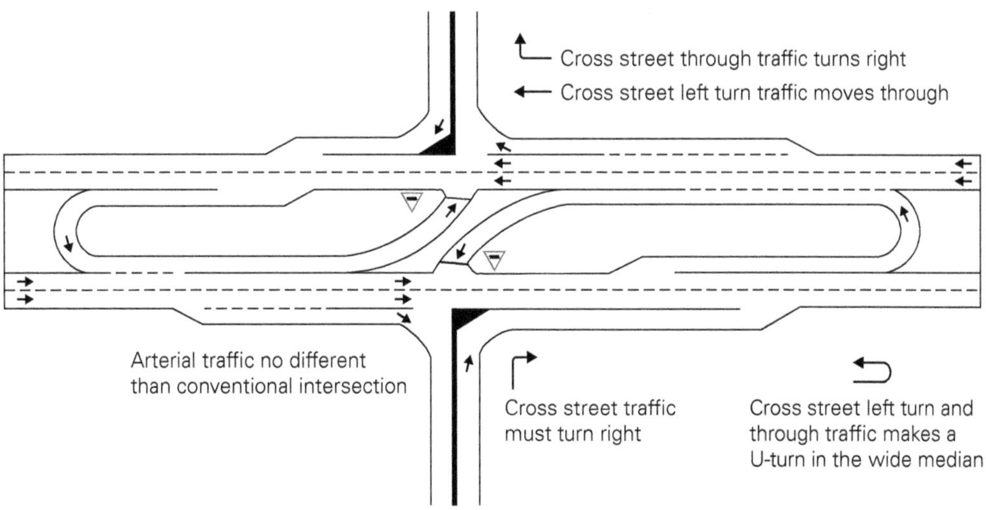

Required Data

The initial data are those required to model one conventional signalized intersection, with other information provided to define the distributed alternative intersection parameters. The turning movement data for the conventional intersection are redistributed to the crossover intersections during the conversion process. The additional information will include the distances and free-flow speeds to the crossover intersections, as well as the TWSC analyses for any unsignalized crossover movements and the turn radii for any signalized crossover intersections.

Displaced Left Turn (DLT)

A DLT consists of three signalized intersections, including the main intersection and two adjacent crossover intersections. The major-street left turns cross over in advance of the main intersection to merge with the cross street. This allows for the major-street left turn phase and opposing through movement phase to run together for increased efficiency. This design can apply to the major street only (Partial DLT) or to all four approaches (Full DLT) as illustrated in HCM Exhibit 23-45 (see Figure 8.77).

PROCEDURE

After the conventional turning movements are modified for the main and supplemental intersections, the three intersections are modeled, following the Signalized Intersection procedures covered in Chapter Six with the

Figure 8.83 *HCM Exhibit 23-53*

Median Width	Saturation Flow Rate Adjustment Factor
Narrow (<35 ft)	0.80
Typical (35–80 ft)	0.85
Very wide (>80 ft)	0.95

Figure 8.84 *HCM Exhibit 23-51*

		Default Arrival Type		
Intersection Type	**Movement**	**2nd Signal**	**3rd Signal**	**4th Signal**
Four-legged RCUT	Left from major street	3	None	None
	Major street through	5	None	None
	Right from major street	5	None	None
	Left from minor street	3	5	None
	Minor street through	3	5	None
Three-legged RCUT	Left from major street	3	None	None
	Left from minor street	3	None	None
Four-legged MUT with signals at U-turn crossovers	Left from major street	5	2	5
	Major street through	5	None	None
	Right from major street	5	None	None
	Left from minor street	3	5	None
Four-legged MUT with stop signs at U-turn crossovers	Left from major street	3	None	None
	Left from minor street	3	None	None

Figure 8.85 *HCM Exhibit 23-52*

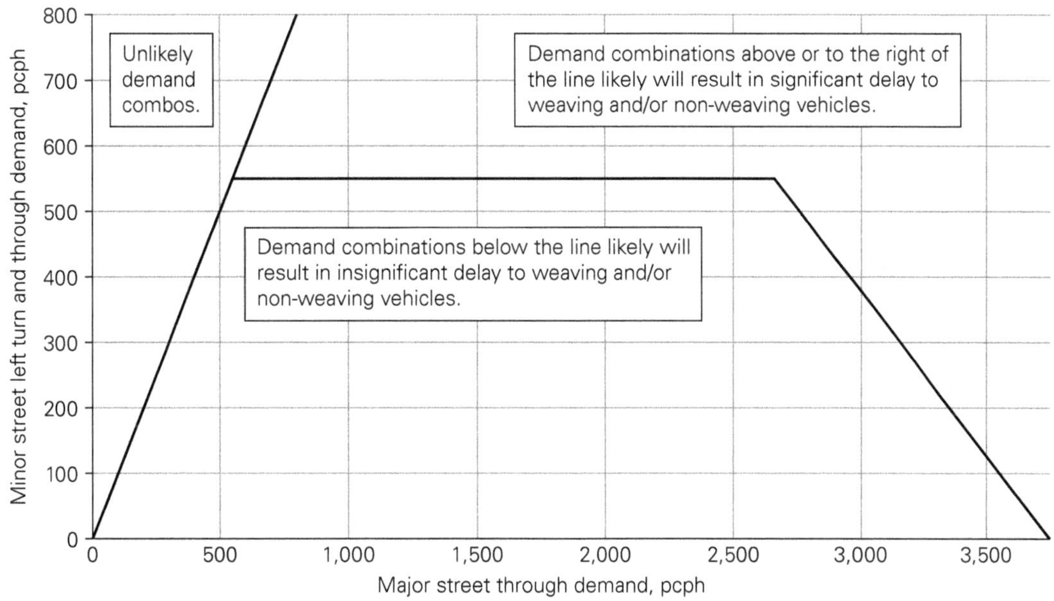

Figure 8.86 *HCM Equation 23-57*

$$EDTT = \frac{D_t + D_f}{1.47 \times S_f} + a$$

where

$EDTT$ = extra distance travel time (s),

D_t = distance from the main junction to the U-turn crossover (ft),

D_f = distance from the U-turn crossover to the main junction (ft),

1.47 = conversion factor from mi/h to ft/s,

S_f = major-street free-flow speed (mi/h), and

a = delay associated with deceleration into a turn and acceleration from the turn (s).

Figure 8.87 *HCM Equation 23-58*

$$EDTT = \frac{D_t + D_f}{1.47 \times S_f}$$

Figure 8.88 *HCM Equation 23-59*

$$ETT = \sum d_i + \sum EDTT$$

progression adjustments from the Urban Streets procedures covered in Chapter Seven. The downstream left-turn merge onto the cross street uses the right-turn adjustment for saturation flow because of the tight radius. Since there is assumed to be no extra distance traveled, Experienced Travel Time (*ETT*) is a sum of the delays for each O-D movement pair, following HCM Equation 23-60 (Figure 8.78) and for the intersection using HCM Equation 23-61 (Figure 8.79). Level of service is defined by HCM Exhibit 23-13 (8.80) with any lane group with a v/c ratio or queue-storage ratio greater than 1.0 dictating LOS F.

Restricted Crossing U-Turn (RCUT)

An RCUT can consist of three signalized intersections with the U-turn crossovers signalized, one signalized main intersection and two stop-controlled U-turn crossover intersections, or one yield-controlled intersection with all other movements merging as illustrated in

Figure 8.89 *HCM Exhibit 23-44*

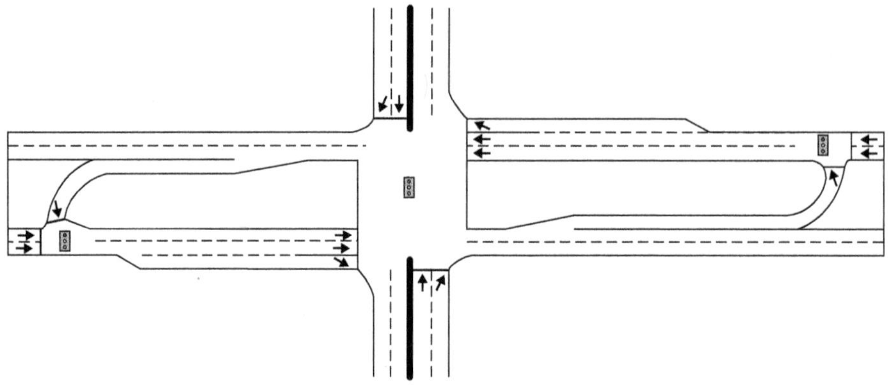

HCM Exhibits 23-41 (Figure 8.81) and 23-42 (Figure 8.82). All minor-street movements must turn right, then make the U-turn to return to the main intersection to complete the left-turn or through movements.

PROCEDURE

After the conventional turning movements are modified for the main and supplemental intersections, the three intersections are analyzed. With signalized U-turn crossovers, the saturation flow rate is adjusted as a function of the radius of the turn, following HCM Exhibit 23-53 (Figure 8.83). The progression is computed using the flow profile model or can be estimated using default arrival types listed in HCM Exhibit 23-51 (Figure 8.84). With stop-controlled U-turn crossovers, the default values of 4.4 seconds for critical headway and 2.6 seconds for follow-up headway can be used if field data are not available. With merges for all movements (except the major-street left turns and the main intersections that are yield controlled), control delay is assumed to be zero seconds if confirmed using HCM Exhibit 23-52 (Figure 8.85). Extra Distance Travel Time (*EDTT*) is defined by HCM Equation 23-57 (Figure 8.86)for merges (to account for acceleration time) and HCM Equation 23-58 (Figure 8.87) for signalized or stop-controlled U-turn crossovers. Experienced Travel Time (*ETT*) is a sum of the delays plus the *EDTT* for each O-D movement pair, following HCM Equation 23-59 (Figure 8.88) and for the intersection using HCM Equation 23-61. Level of service is defined by HCM Exhibit 23-13 with any lane group with a v/c ratio or queue-storage ratio greater than 1.0 dictating LOS F.

Median U-Turn (MUT)

An MUT can consist of three signalized intersections with the U-turn crossovers signalized (as illustrated in HCM Exhibit 23-44—Figure 8.89) or one signalized main intersection with two stop-controlled U-turn crossover intersections. Minor-street left-turn movements must turn right, then make the U-turn to return to the main intersection to complete the left-turn by going through there.

PROCEDURE

After the conventional turning movements are modified for the main and supplemental intersections, the three intersections are analyzed. With signalized U-turn crossovers, the saturation flow rate is adjusted as a function of the radius of the turn, following HCM Exhibit 23-53. The progression is computed using the flow profile model or can be estimated using default arrival types listed in HCM Exhibit 23-51. With stop-controlled U-turn crossovers, the default values of 4.4 seconds for critical headway and 2.6 seconds for follow-up headway can be used if field data are not available. Extra Distance Travel Time (*EDTT*) is defined by HCM Equation 23-58 for both signalized and stop-controlled U-turn crossovers. Experienced Travel Time (*ETT*) is a sum of the delays plus the *EDTT* for each O-D movement pair, following HCM Equation 23-59 and for the intersection using HCM Equation 23-61. Level of service is defined by HCM Exhibit 23-13 with any lane group with a v/c ratio or queue-storage ratio greater than 1.0 dictating LOS F.

PROBLEM ONE

Since the procedures for this analysis do not lend themselves to performing manually, the HCM Example Problem for a Displaced Left-Turn (DLT) intersection is illustrated here. For the conventional intersection, the lane configurations and movement demands are shown in HCM Exhibit 34-139 (Figure 8.90) with fully actuated signal phasing and phase durations shown in HCM Exhibit 34-140 (Figure 8.91) with a peak hour factor of 0.92 used for all movements.

The distributed movement flow rates are shown in HCM Exhibit 34-141 (Figure 8.92). The main and supplemental intersection lane configurations are shown in HCM Exhibit 34-142 (Figure 8.93) with the resulting phase durations shown in HCM Exhibit 34-143 (Figure 8.94). The final delay results are compiled in HCM Exhibit 34-145 (Figure 8.95). Notice there is a weighted average of delay of 28.5 seconds per vehicle to determine the alternative intersection level of service at LOS C (which contrasts with the initial conventional intersection delay of 64.1 seconds per vehicle and LOS E).

Figure 8.90 *HCM Exhibit 34-139*

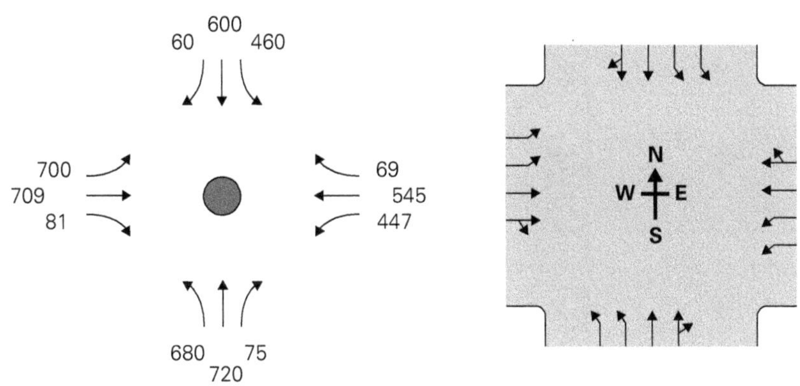

(a) Peak 15-min volumes (veh/h)

(b) Lane channelization

Figure 8.91 *HCM Exhibit 34-140*

Green	20.9	5.9	23.0	21.6	4.4	26.0
Yellow	4.0	4.0	4.0	4.0	4.0	4.0
Red	1.0	1.0	1.0	1.0	1.0	1.0

Figure 8.92 *HCM Exhibit 34-141*

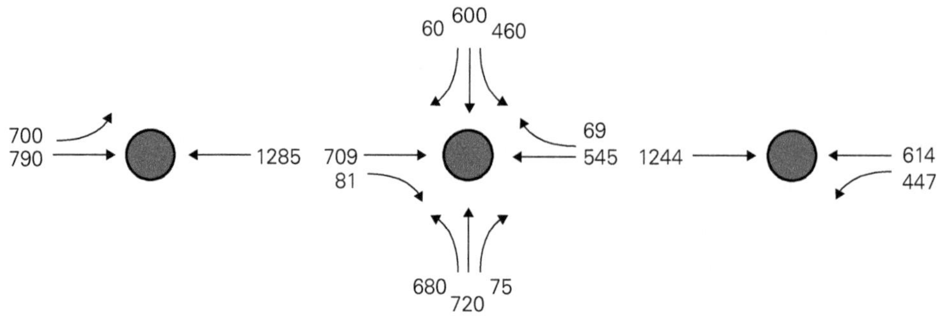

Figure 8.93 *HCM Exhibit 34-142*

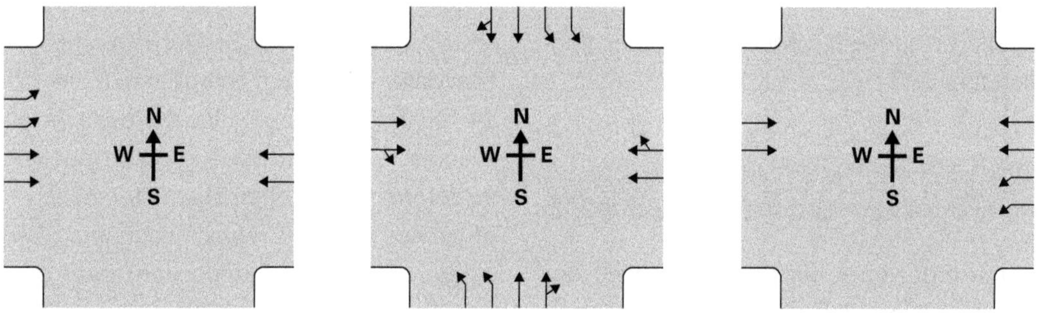

Figure 8.94 *HCM Exhibit 34-143*

Supplemental Intersection Timing Plans

Green	18.7	36.3
Yellow	4.0	4.0
Red	1.0	1.0

17.0	11.8	4.8	16.4
4.0	4.0	0.0	4.0
1.0	1.0	0.0	1.0

12.7	42.3
4.0	4.0
1.0	1.0

Main intersection timing plan

Figure 8.95 *HCM Exhibit 34-145*

Move-ment	Flows				Delays			Products		
	Orig.	Int 1	Int 2	Int 3	Int 1	Int 2	Int 3	Int 1	Int 2	Int 3
EB L	761	761			22.5			17,123	0	0
EB TH	437	859	437	1,352	0.4	41.9	2.5	344	18,310	3,380
EB R	422		422			42.5		0	17,935	0
WB L	486			486			25.7	0	0	12,490
WB TH	340	1,397	340	667	4.0	29.3	0.4	5,588	9,962	267
WB R	328		328			29.7		0	9,742	0
NB L	739		739			23.7		0	17,514	0
NB TH	439		439			19.8		0	8,692	0
NB R	425		425			19.8		0	8,415	0
SB L	500		500			26.2		0	13,100	0
SB TH	364		364			23.4		0	8,518	0
SB R	353		353			23.5		0	8,296	0
Total	5,594								159,675	
Avg.									28.5	

HCS7

These screenshots (Figures 8.96 and 8.97) show how Problem One is coded into *HCS7*, followed by the results as presented in the formatted report. All input data, intermediate computations, and final results are shown in logical order, leading to density and level of service.

Figure 8.96 *HCS7 AltInts1 Input*

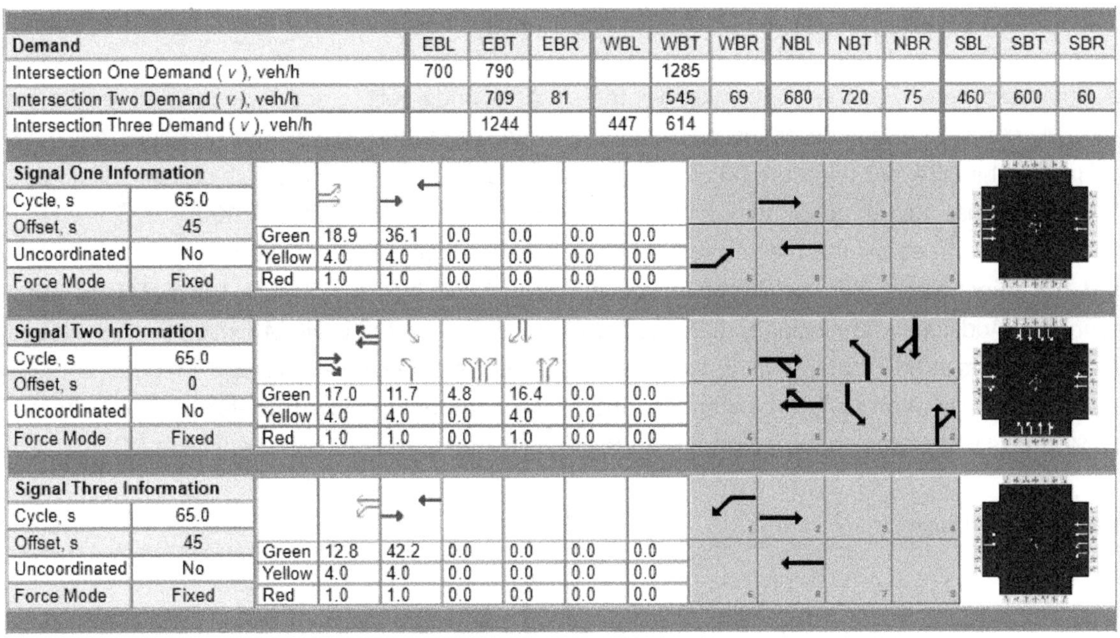

Demand	EBL	EBT	EBR	WBL	WBT	WBR	NBL	NBT	NBR	SBL	SBT	SBR
Intersection One Demand (v), veh/h	700	790			1285							
Intersection Two Demand (v), veh/h		709	81		545	69	680	720	75	460	600	60
Intersection Three Demand (v), veh/h		1244		447	614							

Signal One Information								
Cycle, s	65.0							
Offset, s	45	Green	18.9	36.1	0.0	0.0	0.0	0.0
Uncoordinated	No	Yellow	4.0	4.0	0.0	0.0	0.0	0.0
Force Mode	Fixed	Red	1.0	1.0	0.0	0.0	0.0	0.0

Signal Two Information								
Cycle, s	65.0							
Offset, s	0	Green	17.0	11.7	4.8	16.4	0.0	0.0
Uncoordinated	No	Yellow	4.0	4.0	0.0	4.0	0.0	0.0
Force Mode	Fixed	Red	1.0	1.0	0.0	1.0	0.0	0.0

Signal Three Information								
Cycle, s	65.0							
Offset, s	45	Green	12.8	42.2	0.0	0.0	0.0	0.0
Uncoordinated	No	Yellow	4.0	4.0	0.0	0.0	0.0	0.0
Force Mode	Fixed	Red	1.0	1.0	0.0	0.0	0.0	0.0

Figure 8.97 *HCS7 AltInts1 Output*

Alternative Intesection Results					
O-D	O-D Movements	Flow Rate (veh/h)	ETT Movements	ETT (s)	LOS
EBL	EBL (1)	761	EBL (1)	22.6	C
EBT	EBT (1) + EBT (2) + EBT (3)	437	EBT (2) + EBT (3)	44.3	D
EBR	EBT (1) + EBR (2)	422	EBR (2)	42.3	D
WBL	WBL (3)	486	WBL (3)	25.7	C
WBT	WBT (3) + WBT (2) + WBT (1)	340	WBT (2) + WBT (1)	33.4	C
WBR	WBT (3) + WBR (2)	328	WBR (2)	29.6	C
NBL	NBL (2) + WBT (1)	739	NBL (2) + WBT (1)	27.9	C
NBT	NBT (2)	439	NBT (2)	19.7	B
NBR	NBR (2) + EBT (3)	425	NBR (2) + EBT (3)	22.4	C
SBL	SBL (2) + EBT (3)	500	SBL (2) + EBT (3)	29.0	C
SBT	SBT (2)	364	SBT (2)	23.4	C
SBR	SBR (2) + WBT (1)	353	SBR (2) + WBT (1)	27.7	C

Overall Results	EB1	WB1	EB2	WB2	NB2	SB2	EB3	WB3
Approach Delay, s/veh	10.8	4.2	42.0	29.4	21.6	24.6	2.7	11.0
Approach LOS	B	A	D	C	C	C	A	B
Intersection ETT, s/veh / LOS	28.6				C			

PROBLEM TWO

Since the procedures for this analysis do not lend themselves to performing manually, the HCM Example Problem for a signalized Restricted Crossing U-Turn (RCUT) intersection is illustrated here. For the conventional intersection, the movement demands are shown in HCM Exhibit 34-129 (Figure 8.98). The main intersection has two through lanes with exclusive right-turn lanes on the major approaches, and the minor street has two lanes on each approach. Additional data include: distances to the crossovers at 800 feet with a 40-foot median and a 50-mile-per-hour free-flow speed; and 3.7% trucks with a peak hour factor of 0.93 for all movements. The signals are pretimed with a southbound 90-second cycle and 60 seconds for major street greens and 20 seconds for crossover greens, and a northbound 60-second cycle with 25 seconds for major street greens and 25 seconds of crossover greens—all having 4 seconds yellow and 1 second all-red times. The distributed movement demands and flow rates are shown in HCM Exhibits 34-130 (Figure 8.99) and 34-131 (Figure 8.100).

The control delay results are compiled in HCM Exhibit 34-132 with the *EDTT* included for *ETT* and LOS in HCM Exhibit 34-133. Notice there is a weighted average of delay of 22.8 seconds per vehicle to determine the alternative intersection level of service at LOS C.

Figure 8.98 *HCM Exhibit 34-129*

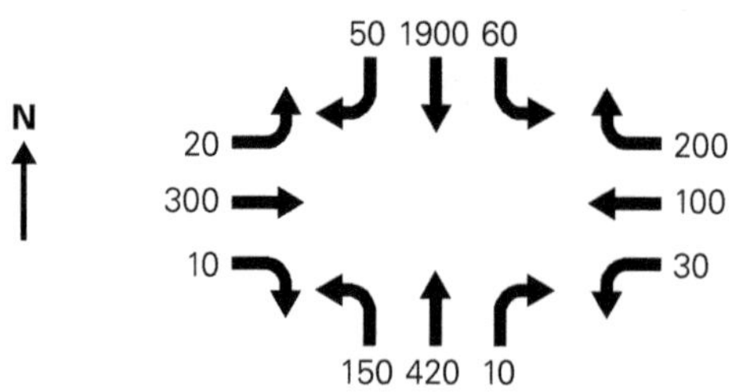

Figure 8.99 *HCM Exhibit 34-130*

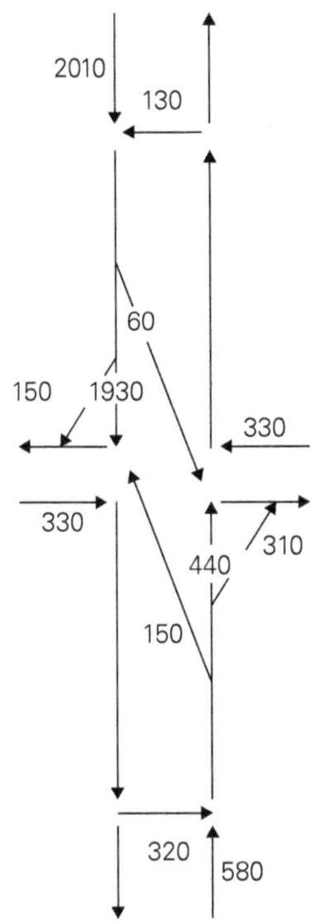

Figure 8.100 *HCM Exhibit 34-131*

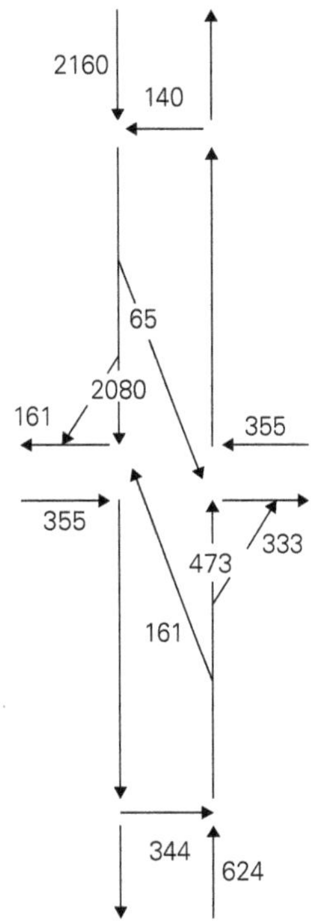

Figure 8.101 *HCM Exhibit 34-132*

Junction	Movement	*v/c*	95% Queue Length (veh)	Control Delay (s/veh)
North crossover	SB through	0.92	4.4	7.6
	WB crossover	0.40	5.0	33.3
West main intersection	SB through	0.89	3.2	5.4
	SB right turn	0.16	0.2	0.3
	EB right turn	0.58	6.4	35.1
	NB left turn	0.41	5.7	33.2
South crossover	NB through	0.43	1.4	4.1
	EB crossover	0.53	5.9	16.1
East main intersection	NB through	0.32	1.7	6.4
	NB right turn	0.51	3.1	9.1
	WB right turn	0.31	2.4	12.4
	SB left turn	0.09	0.8	10.8

Figure 8.102 *HCM Exhibit 34-133*

Movement	Control Delay (s/veh) by Traffic Control Device: First	Second	Third	*EDTT* (s/veh)	*ETT* (s/veh)	LOS
NB left	4.1	33.2	None	0	37.3	D
SB left	7.6	10.8	None	0	18.4	B
NB through	4.1	6.4	None	0	10.5	B
SB through	7.6	5.4	None	0	13.0	B
NB right	4.1	9.1	None	0	13.2	B
SB right	7.6	0.3	None	0	7.9	A
EB left	35.1	16.1	6.4	21.8	79.4	E
WB left	12.4	33.3	5.4	21.8	72.9	E
EB through	35.1	16.1	9.1	21.8	82.1	F
WB through	12.4	33.3	0.3	21.8	67.8	E
EB right	35.1	None	None	0	35.1	D
WB right	12.4	None	None	0	12.4	B

HCS7

These screenshots (Figures 8.103 and 8.104) show how Problem Two is coded into *HCS7*, followed by the results as presented in the formatted report. All input data, intermediate computations, and final results are shown in logical order, leading to density and level of service.

Figure 8.103 *HCS7 AltInts2 Input*

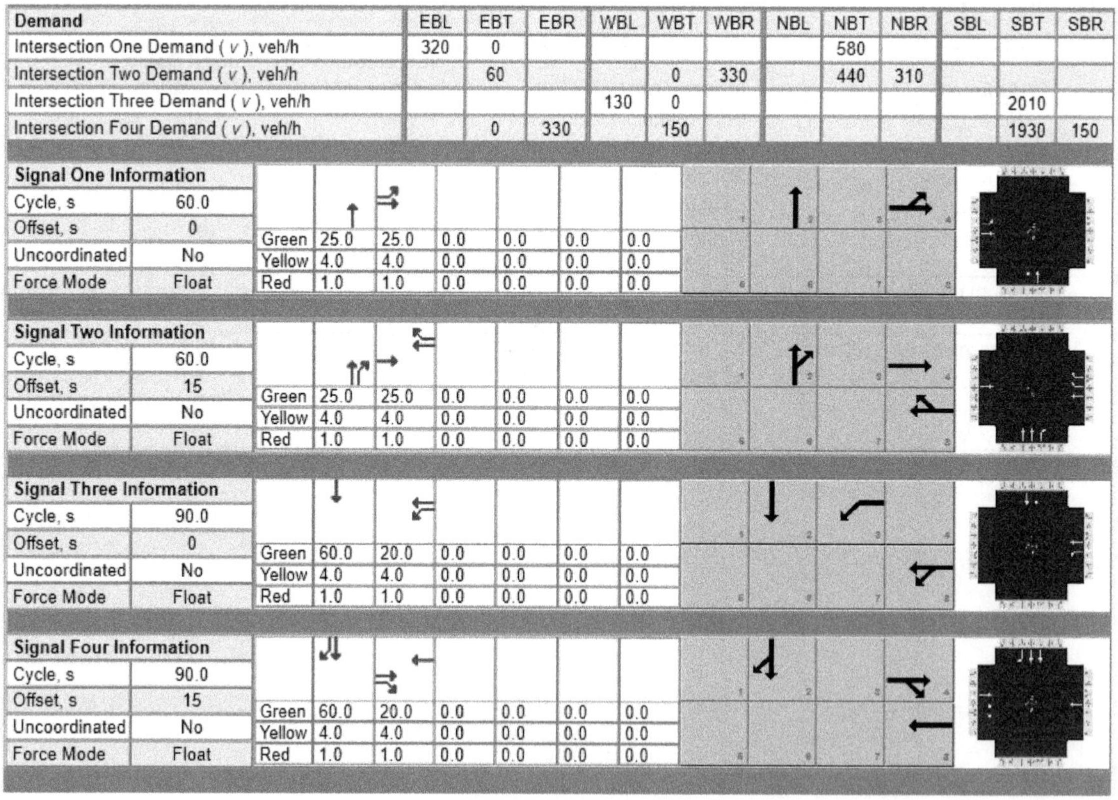

Figure 8.104 *HCS7 AltInts2 Output*

Alternative Intesection Results						
O-D	O-D Movements	Flow Rate (veh/h)	Control Delay (s/veh)	EDTT (s/veh)	ETT (s/veh)	LOS
EBL	EBR(4) + SBU(1) + NBT(2)	22	63.2	21.8	85.0	F
EBT	EBR(4) + SBU(1) + NBR(2)	323	63.2	21.8	85.0	F
EBR	EBR(4)	11	35.2	--	35.2	D
WBL	WBR(2) + NBU(3) + SBT(4)	32	51.2	21.8	73.0	E
WBT	WBR(2) + NBU(3) + SBR(4)	108	46.8	21.8	68.6	E
WBR	WBR(2)	215	12.4	--	12.4	B
NBL	NBT(1) + NBL(2)	161	36.7	--	36.7	D
NBT	NBT(1) + NBT(2)	568	16.0	--	16.0	B
NBR	NBT(1) + NBR(2)	400	16.0	--	16.0	B
SBL	SBT(3) + SBL(4)	65	18.5	--	18.5	B
SBT	SBT(3) + SBT(4)	2135	13.2	--	13.2	B
SBR	SBT(3) + SBR(4)	166	8.8	--	8.8	A
Overall Results						
Intersection ETT, s/veh | LOS		24.0		C		

PROBLEM THREE

Since the procedures for this analysis do not lend themselves to performing manually, the HCM Example Problem for a Median U-Turn (MUT) intersection with stop-controlled crossovers is illustrated here. For the conventional intersection, the movement demands are shown in HCM Exhibit 34-134. The main intersection has two through lanes with shared right-turn lanes on the major approaches, and the minor street has one through lane and one right-turn lane on each approach. Additional data include: distances to the crossovers at 500 feet and a 50-mile-per-hour free-flow speed; and 2.6% trucks with a peak hour factor of 0.95 for all movements. The signal is fully actuated with maximum green times of 50 seconds for the major-street phases and 30 seconds for the minor-street phases direction, with both having 4 seconds yellow and 1 second all-red times to generate phase durations, also shown in HCM Exhibit 34-134. The distributed movement demands and flow rates are shown in HCM Exhibits 34-135 and 34-136.

Figure 8.105 *HCM Exhibit 34-134*

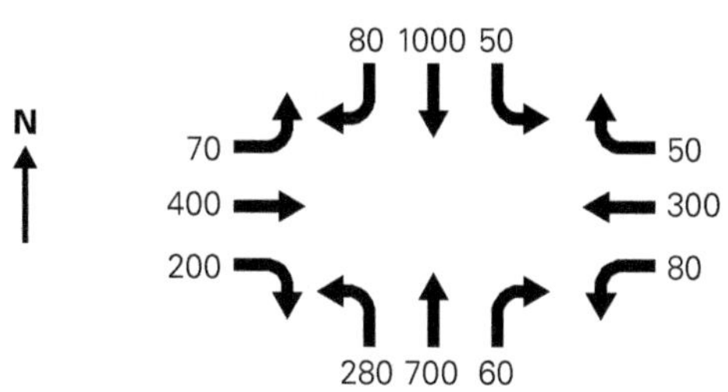

Green	39.8	18.9
Yellow	4.0	4.0
Red	1.0	1.0

Figure 8.106 *HCM Exhibit 34-135*

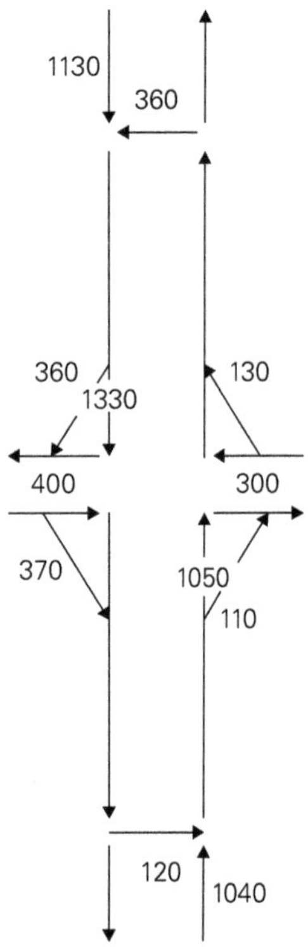

Figure 8.107 *HCM Exhibit 34-136*

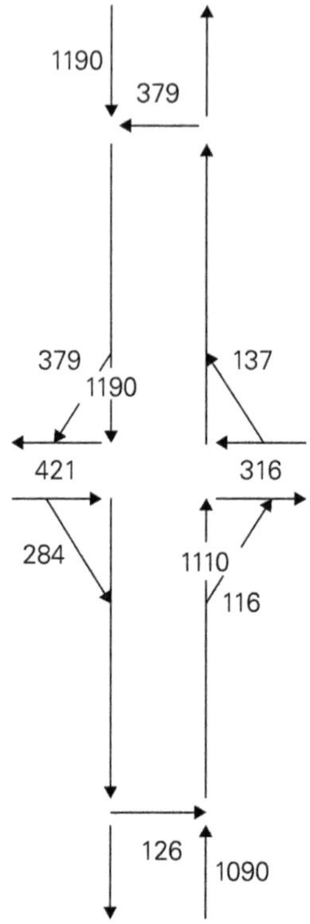

The control delay results are compiled in HCM Exhibit 34-137 with the *EDTT* included for *ETT* and LOS in HCM Exhibit 34-138. Notice there is a weighted average of delay of 21.8 seconds per vehicle (which contrasts with the initial conventional intersection delay of 33.7 seconds per vehicle) to determine the alternative intersection level of service at LOS C.

Figure 8.108 *HCM Exhibit 34-137*

Junction	Movement	*v*/*c*	95% Queue Length (veh)	Control Delay (s/veh)
North crossover	WB crossover	0.78	7.1	34.6
Main intersection	EB through	0.82	10.2	25.1
	EB right turn	0.74	7.1	23.7
	WB through	0.62	7.5	22.2
	WB right turn	0.35	3.0	20.2
	NB through	0.58	8.3	9.3
	NB right turn	0.58	8.0	9.4
	SB through	0.76	12.2	12.3
	SB right turn	0.80	12.0	13.7
South crossover	EB crossover	0.24	0.9	14.0

Figure 8.109 *HCM Exhibit 34-138*

Movement	Control Delay (s/veh) by Traffic Control Device			EDTT (s/veh)	ETT (s/veh)	LOS
	First	Second	Third			
NB left	9.3	34.6	13.7	20.4	78.0	E
SB left	12.3	14.0	9.4	20.4	56.1	E
NB through	9.3	None	None	0	9.3	A
SB through	12.3	None	None	0	12.3	B
NB right	9.4	None	None	0	9.4	A
SB right	13.7	None	None	0	13.7	B
EB left	23.7	14.0	9.3	20.4	67.4	E
WB left	20.2	34.6	12.3	20.4	87.5	F
EB through	25.1	None	None	0	25.1	C
WB through	22.2	None	None	0	22.2	C
EB right	23.7	None	None	0	23.7	C
WB right	20.2	None	None	0	20.2	C

HCS7

These screenshots show how Problem Three is coded into *HCS7*, followed by the results as presented in the formatted report. All input data, intermediate computations, and final results are shown in logical order, leading to density and level of service.

Figure 8.110 *HCS7 AltInts3 Input*

Demand	EBU	EBT	EBR	WBU	WBT	WBR	NBU	NBT	NBR	SBU	SBT	SBR
Intersection One Demand (*v*), veh/h							360	820			1130	
Intersection Two Demand (*v*), veh/h		400	270		300	130		1050	110		1130	360
Intersection Three Demand (*v*), veh/h								1040		120	1280	

Free-Flow Speed	
Segment One, mph	50.0
Segment Two, mph	50.0

Main Signal Information								
Cycle, s	68.4							
Offset, s	0	Green	39.7	18.8	0.0	0.0	0.0	0.0
Uncoordinated	Yes	Yellow	4.0	4.0	0.0	0.0	0.0	0.0
Force Mode	Fixed	Red	1.0	1.0	0.0	0.0	0.0	0.0

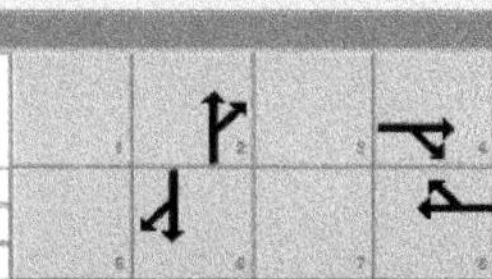

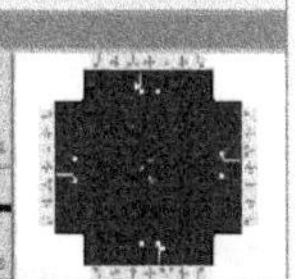

Figure 8.111 *HCS7 AltInts3 Output*

Alternative Intesection Results						
O-D	O-D Movements	Flow Rate (veh/h)	Control Delay (s/veh)	EDTT (s/veh)	ETT (s/veh)	LOS
EBL	EBR(2) + SBU(3) + NBT(2) + NBT(1)	73	47.2	21.8	69.0	E
EBT	EBT(2)	421	24.9	--	24.9	C
EBR	EBR(2) + SBT(3)	284	23.6	--	23.6	C
WBL	WBR(2) + NBU(1) + SBT(2) + SBT(3)	84	69.8	21.8	91.6	F
WBT	WBT(2)	316	22.1	--	22.1	C
WBR	WBR(2) + NBT(1)	137	20.1	--	20.1	C
NBL	NBT(3) + NBT(2) + NBU(1) + SBR(2)	294	60.4	21.8	82.2	F
NBT	NBT(3) + NBT(2) + NBT(1)	1105	9.3	--	9.3	A
NBR	NBT(3) + NBR(2)	116	9.3	--	9.3	A
SBL	SBT(1) + SBT(2) + SBU(3) + NBR(2)	52	35.7	21.8	57.5	E
SBT	SBT(1) + SBT(2) + SBT(3)	1189	12.1	--	12.1	B
SBR	SBT(1) + SBR(2)	379	13.5	--	13.5	B

Overall Results	EB	WB	NB	SB
Approach ETT, s/veh	28.6	32.5	23.4	13.9
Approach LOS	C	C	C	B
Intersection ETT, s/veh / LOS	21.9		C	

PART III
Application Extensions

This section deals with the use of these procedures and models in widely applied situations like signal timing optimization and traffic impact analysis. In practice, results from these methods are commonly used for signal phasing and timing parameters, as well as intersection and roadway design decisions. The HCS Streets module provides for optimizing signal timing for several objective functions. Traffic impact analysis was shown to be the most widely used application from an HCM survey.

Signal Timing Optimization

Signal timing parameters like cycle length, phase splits, and intersection offsets can be optimized using the Streets module in HCS that employs a genetic algorithm process to meet the defined objective function goals. A range of cycle lengths, as well as weighting for the dominant direction, can be specified. Objective functions include travel speed for LOS, arterial delay for platoon movement, overall delay for individual intersections with automobile emissions, and crash rates planned.

Streets

The Streets module in HCS uses genetic algorithm logic to generate scenarios toward converging on the optimum solution for the data provided and the goals selected. This process is HCM-based to include multiple-period analysis for congested corridors, the phase duration model for actuated movement, flow profile to model platooning along segments, and the effects of access points along the segments. A dynamic time-space diagram is generated to illustrate the progression of green bands, and a flow profile diagram is produced for visualizing the upstream departure and downstream arrival rates.

Arterials

Coordinated-actuated (and pretimed) operations for arterials require the optimization of several (sometimes competing) parameters, typically also by time of day based on relative movement demands and traffic patterns. This involves generating analyses for each period to be coordinated with reasonable settings, typically morning and afternoon peaks with off-peak and noon periods sometimes included. The objective function can vary with arterial or overall delay most used, followed by travel time and/or arterial stops. Normally, cycle length, phase splits, and intersection offsets are included, with forward weighting used to give priority to the major direction when patterns are heavily skewed during morning or afternoon rush periods.

Intersections

For isolated intersections normally operating in fully actuated or semi-actuated mode, the process is simplified with the goal essentially to optimize maximum green times by time of day based on varying demands. A signalized intersection analysis is run for each time period to be optimized with reasonable maximum green times—setting the objective function as overall delay to generate the maximum green times that yield to lowest overall intersection delay.

Traffic Impact Analysis

This procedure is defined to explain how existing, future, and mitigated analyses are used to establish improvements necessary for a development to avoid negatively impacting the adjacent traffic system. The process includes estimating additional trips generated by the development, assigning those trips with growth to the nearby streets and intersections, identifying the deficiencies created, and defining the mitigating improvements required to offset the additional traffic. There is also a comprehensive project described suitable for a semester-ending assignment.

Existing

The definition of the analysis of existing conditions to include the appropriate affected area is discussed. The size of the proposed development dictates the area that will be affected, so care must be taken to prescribe all intersections and roadways that should be included in the impact and study area. All intersections and roadways are analyzed using these procedures to determine the existing delays, queues, and LOS.

Future

The size of all aspects of the development must be known in order to quantify the trips that will be generated using the *ITE Trip Generation Manual*, which is the accepted standard for this process. The trips are then distributed appropriately to the various roadway segments with turning movement estimates at all affected intersections as well. The capacity analysis is then repeated with the future demand values to determine the delays, queues, and LOS values with additional development demand in place.

Mitigation

Improvements to mitigate any increased delays and queues, as well as any deterioration in LOS, are identified with cost estimates included. The local city, county, or state reviews and negotiates the proposed improvements to finalize a mitigation strategy that must be accepted by the developer for the project to move forward. Many times as part of this process, an unsignalized intersection will meet the criteria for signalization as determined by the *MUTCD* signal warrant procedure, so this can become part of the overall process as well.

CHAPTER 9 Signal Timing Optimization

Signal timing parameters like cycle length, phase splits, and intersection offsets can be optimized using the Streets module in HCS that employs a genetic algorithm (GA) process to meet the defined objective function goals. This process is HCM based to include multiple-period analysis for congested corridors, the phase duration model for actuated movement, flow profile to model platooning along segments, and the effects of access points along the segments.

Streets

The Streets module in HCS generates scenarios toward converging on the optimum solution for the data provided and the goals selected. The genetic algorithm search technique is specifically designed to locate the global optimum solution, provided that the original dataset already has an effective timing plan, the number of generations is sufficiently high, and the detailed settings are reasonable for the given arterial.

Optimization

A range of cycle lengths as well as weighting for the dominant direction can be specified in the Optimization dashboard shown below. Objective functions include Percent Base Free-Flow Speed for level of service; Arterial Delay for platoon movement; Overall Delay for individual intersections with Vehicle Emissions; and Crash Rates planned. The number of intersections and time periods in the initial analysis will dictate the time required for each optimization generation. If the arterial has many intersections and/or the analysis is for several time periods, each generation will take longer, and an optimization can require a significant amount of time.

When the optimization process detects that little additional improvement is possible, termination will indicate the convergence threshold has been reached. If the convergence has not been reached when the user-specified maximum number of generations occurs, the process will terminate—but before it has reached convergence, indicating more generations

Figure 9.1 *HCS7 Optimization Dashboard*

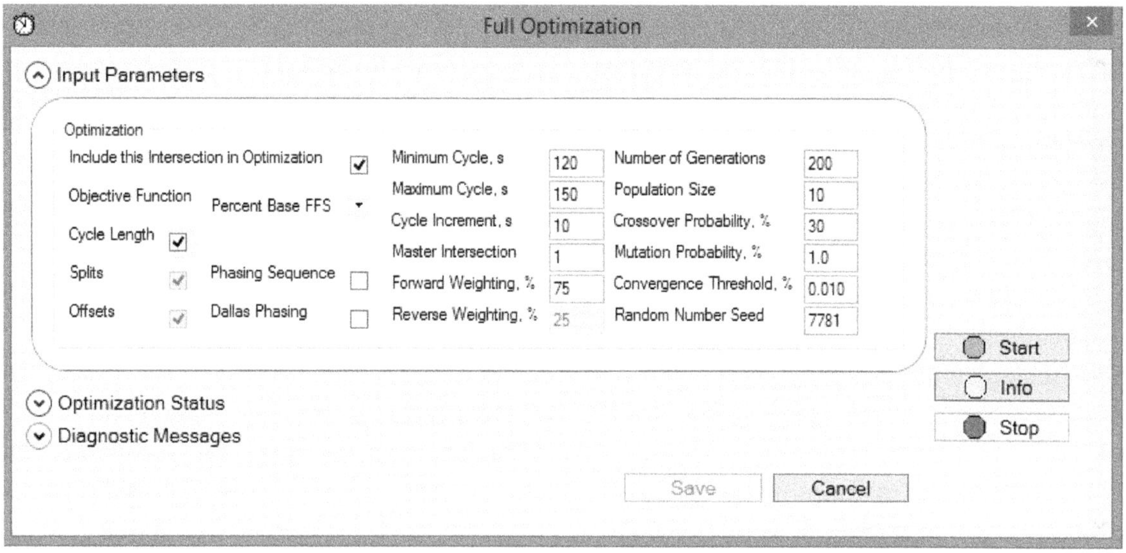

should be run. The user can request to stop and restart the process at any time, and choose to save or discard the results.

A dynamic time-space diagram is available to illustrate the progression green bands (see below). The major-street green, yellow, and

Figure 9.2 *HCS7 Optimization Status*

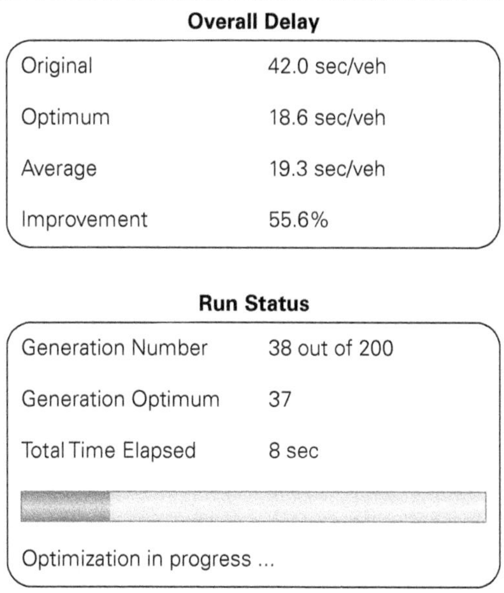

red intervals are shown vertically as the time parameter with intersections along the arterial shown by number separated by the horizontal distances. Offsets control the signal times relative to the start time to facilitate the progression bands illustrated as diagonal lines at the appropriate angle to represent the arterial speed. These can be shown in the forward, reverse, or both directions with the number of cycles also user specified to create the diagram that best suits the user's needs. The offsets can be modified by the user by cursor movement with the updated values saved to the Streets file.

A flow profile diagram is produced for visualizing the upstream departure rates or downstream arrival rates for a given segment, direction, and specified movements (see below). These rates are plotted as green when the signal is green and white when the signal is not green for the movement(s) being analyzed. The higher the proportion arriving at green, the lower the delay and travel time, the fewer the stops, and the better the level of service.

Arterials

Coordinated-actuated and pretimed operations for arterials require the optimization of several (sometimes competing) parameters, typically also by time of day, based on relative movement demands and traffic patterns. This

Figure 9.3 *HCS7 Optimization Time-Space Diagram*

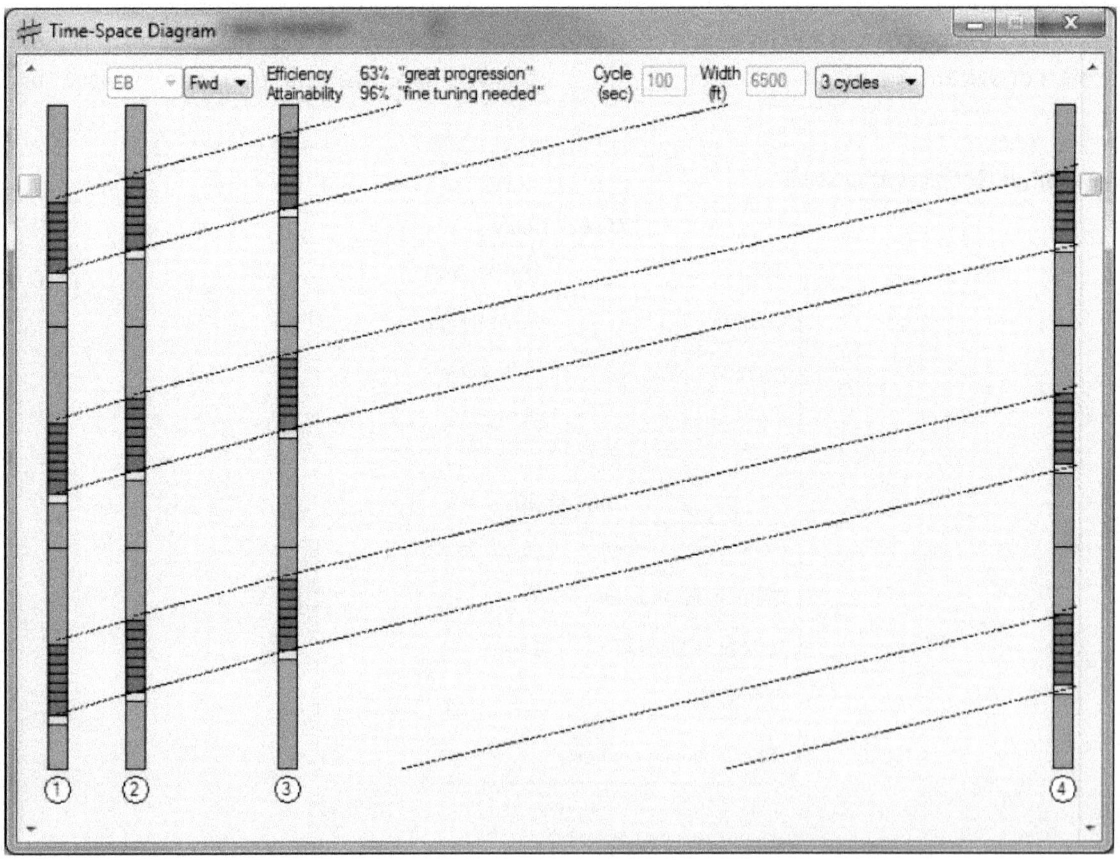

Figure 9.4 *HCS7 Flow Profile*

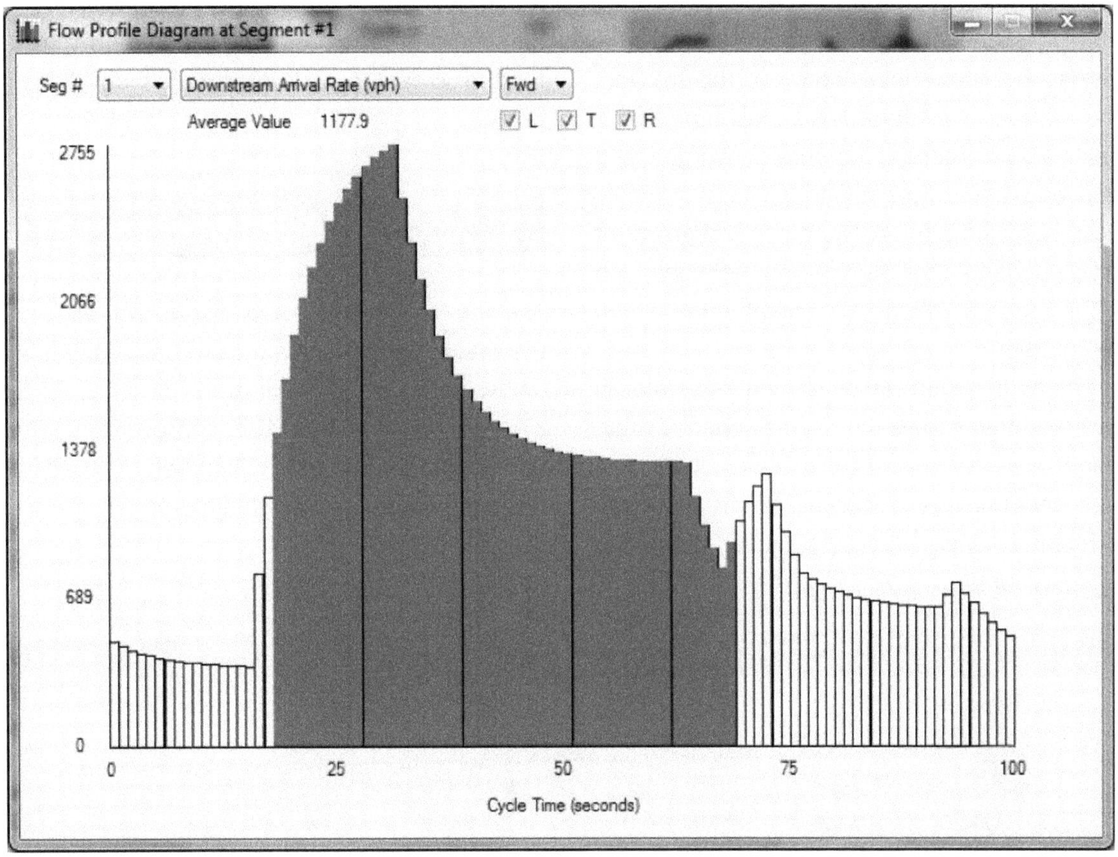

involves generating analyses for each period to be coordinated with reasonable settings, typically morning and afternoon peaks with off-peak and noon periods normally modeled as well. For major arterials, the morning, afternoon, and evening off-peak periods may be analyzed separately. Special events, work zones, or inclement weather can also dictate when special plans are needed to account for these overriding conditions.

The objective function can vary with arterial delay or overall delay most used, followed by travel time and/or arterial stops. Normally, cycle length, phase splits, and intersection offsets are included, with forward weighting used to give priority to the major direction when patterns are heavily skewed during morning or afternoon rush periods.

Think about your own driving: As you drive along an arterial, you are always aware of the upcoming signal and hoping for it to be or stay green for you to make it through without slowing or stopping. Most drivers are much more aware of how often they are stopped than they are of how many signals they clear on the green. And most drivers are more tolerant of stops during peak periods as expected than they are during off-peak periods when it seems the fault of the timing (and usually is). Most urban arterials have coordinated signals, but the efficiency often depends on the quality of the optimization and the frequency of updated timing.

After all other data are defined, the optimization can be run with Overall Delay as the Objective Function and the Splits box checked in the dashboard as shown below in Problem One. The Number of Generations should typically be set to 200 or more to be sure convergence is reached without terminating because of reaching this maximum. A Mutation Probability of 1.o is suggested for multiple intersections, and 4.0 is recommended for a single intersection optimization.

During the optimization routine, the status is displayed that shows the original, optimum, average, and improvement to the overall delay as each generation is processed as shown below in Problem One. Once the model converges on the optimum solution, the process is terminated, and the Save button will replace the maximum green times with those determined for the lowest overall intersection delay (see below).

PROBLEM ONE

The optimization setup and process screens and the results before and after screens are shown below with Overall Delay from 61 seconds to 37 seconds for a nearly 40% improvement with 200 generations in less than fifteen minutes for this nine-intersection single-time-period arterial.

Figure 9.5 *HCS7 StreetsOptInitial1*

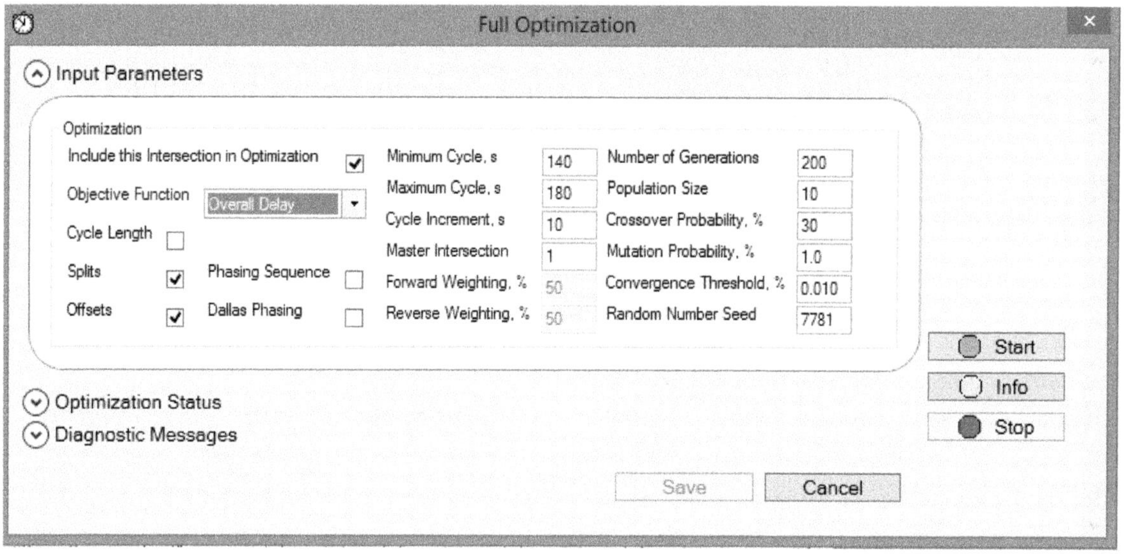

Figure 9.6 *HCS7 StreetsOptFinal1*

Overall Delay

Original	60.6 sec/veh
Optimum	36.6 sec/veh
Average	38.5 sec/veh
Improvement	39.6%

Run Status

Generation Number	200 out of 200
Generation Optimum	195
Total Time Elapsed	846 sec

Termination via max number of generations

Figure 9.7 *HCS7 StreetsOptBefore1*

Facility Output Data	Westbound	Eastbound
Facility Travel Time, s	384.95	339.61
Facility Travel Speed, mph	22.76	25.80
Facility Base Free Flow Speed, mph	41.72	41.72
Facility Percent of Base FFS	54.55	61.83
Facility Level of Service	C	C
Facility Auto Traveler Perception Score	2.47	2.42
Multimodal Results (Facility)		

Figure 9.8 *HCS7 StreetsOptAfter1*

Facility Output Data	Westbound	Eastbound
Facility Travel Time, s	412.19	369.99
Facility Travel Speed, mph	21.26	23.68
Facility Base Free Flow Speed, mph	41.72	41.72
Facility Percent of Base FFS	50.95	56.76
Facility Level of Service	C	C
Facility Auto Traveler Perception Score	2.47	2.44

PROBLEM TWO

The optimization setup and process screens and the results before and after screens are shown below with Percent BFFS from 58% (LOS C) to 81% (LOS A) for about a 28% improvement with 200 generations in less than fourteen minutes for this same nine-intersection single-time-period arterial.

Figure 9.9 *HCS7 StreetsOptInitial2*

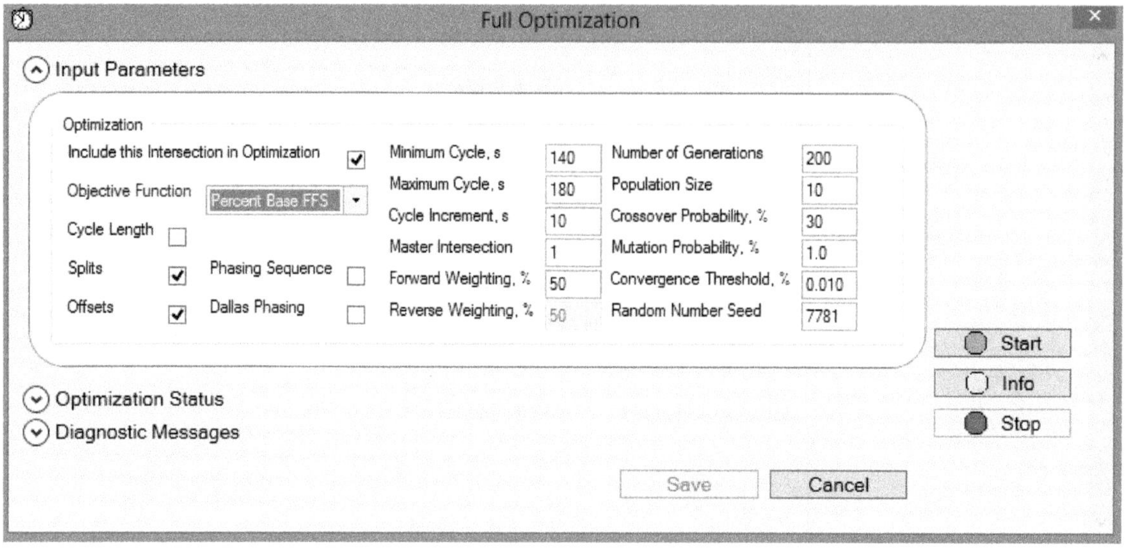

Figure 9.10 *HCS7 StreetsOptFinal2*

Percent Base FFS

Original	58.2%
Optimum	80.8%
Average	80.6%
Improvement	28.0%

Run Status

Generation Number	200 out of 200
Generation Optimum	190
Total Time Elapsed	803 sec

Termination via max number of generations

Figure 9.11 *HCS7 StreetsOptBefore2*

Facility Output Data	Westbound	Eastbound
Facility Travel Time, s	384.95	339.61
Facility Travel Speed, mph	22.76	25.80
Facility Base Free Flow Speed, mph	41.72	41.72
Facility Percent of Base FFS	54.55	61.83
Facility Level of Service	C	C
Facility Auto Traveler Perception Score	2.47	2.42

Figure 9.12 *HCS7 StreetsOptAfter2*

Facility Output Data	Westbound	Eastbound
Facility Travel Time, s	259.32	262.04
Facility Travel Speed, mph	33.79	33.43
Facility Base Free Flow Speed, mph	41.72	41.72
Facility Percent of Base FFS	80.98	80.14
Facility Level of Service	B	B
Facility Auto Traveler Perception Score	2.33	2.34

Intersections

For isolated intersections normally operating in fully actuated or semi-actuated mode, the process is simplified with the goal essentially to optimize maximum green times by time of day based on varying demands. A signalized intersection analysis is run for each time period to be optimized with reasonable maximum green times, setting the objective function as overall delay to generate the maximum green times that yield the lowest overall intersection delay. For fully actuated signal operations, the signalized intersection analysis is established as uncoordinated to set all phases to be actuated; then, the major-street through phases are set to maximum recall for semi-actuated operations.

Think about your own driving:
Waiting for a green at an isolated intersection can be frustrating if there is no demand for the phase that is green while yours is red. This can be caused by a fixed cycle length for semi-actuated operations (to guarantee a minimum time the major-street through phases regardless of demand) or a broken detector loop (which normally serves that phase much too long for demand that doesn't exist). Fully actuated signals with relatively short minimum green times are usually most efficient for these signalized intersections.

During the optimization routine, the status is displayed, showing the original, optimum, average, and improvement to the overall delay as each generation is processed as shown below in Problem One. The generations go very quickly for one intersection with one time period with 60 in less than 10 seconds. Once the model converges on the optimum solution, the process is terminated, and the Save button will replace the maximum green times with those determined for the lowest overall intersection delay (see below).

Figure 9.13 *HCS7 SignalsOptInitial1*

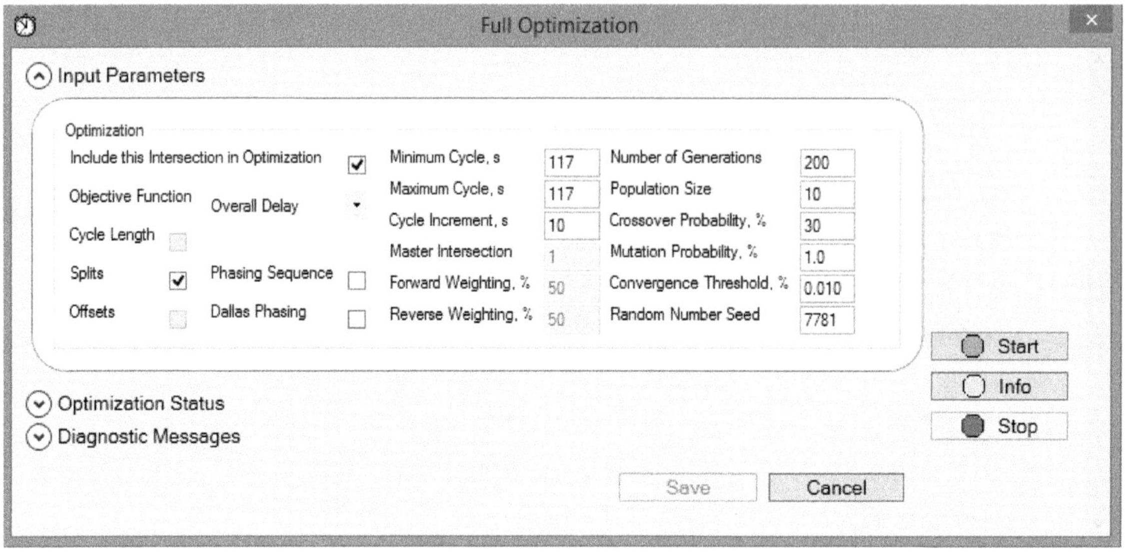

PROBLEM ONE

The before-and-after results are shown below for HCM Example Problem One that reduced this delay from 46 seconds to 28 seconds. This represents a 40% improvement in a 30-second run.

Figure 9.14 *HCS7 SignalsOptFinal1*

Overall Delay

Original	46.7 sec/veh
Optimum	28.2 sec/veh
Average	28.2 sec/veh
Improvement	39.5%

Run Status

Generation Number	61 out of 200
Generation Optimum	59
Total Time Elapsed	8 sec

Termination via convergence threshold

Figure 9.15 *HCS7 SignalsOptBefore1*

HCS7 Signalized Intersection Results Summary

General Information				Intersection Information	
Agency				Duration, h	0.25
Analyst		Analysis Date	Mar 2, 2017	Area Type	CBD
Jurisdiction		Time Period		PHF	1.00
Urban Street		Analysis Year	2017	Analysis Period	1> 7:00
Intersection		File Name	TextSignalsOptBefore1.xus		
Project Description	Chapter 31: Example Problem 1				

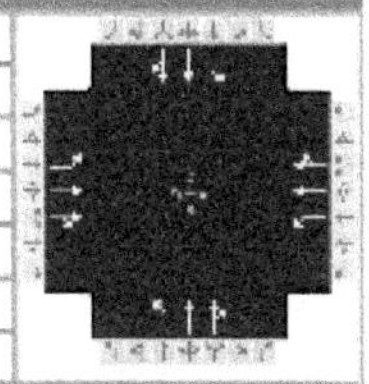

Demand Information	EB			WB			NB			SB		
Approach Movement	L	T	R	L	T	R	L	T	R	L	T	R
Demand (*v*), veh/h	71	318	106	118	600	24	133	1644	111	194	933	111

Signal Information			
Cycle, s	101.8	Reference Phase	2
Offset, s	0	Reference Point	Begin
Uncoordinated	Yes	Simult. Gap E/W	On
Force Mode	Fixed	Simult. Gap N/S	On

Green	30.0	6.2	3.6	50.0	0.0	0.0
Yellow	4.0	4.0	0.0	4.0	0.0	0.0
Red	0.0	0.0	0.0	0.0	0.0	0.0

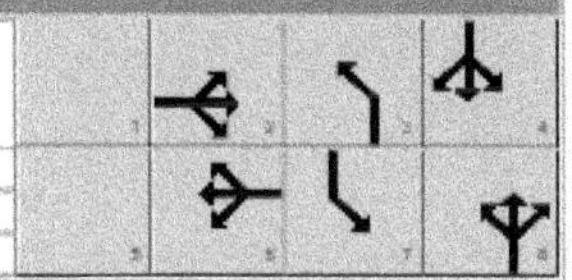

Timer Results	EBL	EBT	WBL	WBT	NBL	NBT	SBL	SBT
Assigned Phase		2		6	3	8	7	4
Case Number		6.0		6.0	1.1	4.0	1.1	4.0
Phase Duration, s		34.0		34.0	10.2	54.0	13.8	57.6
Change Period, ($Y+R_c$), s		4.0		4.0	4.0	4.0	4.0	4.0
Max Allow Headway (*MAH*), s		3.4		3.4	3.1	3.1	3.1	3.1
Queue Clearance Time (g_s), s		30.8		29.2	6.1	52.0	9.6	23.2
Green Extension Time (g_e), s		0.0		0.4	0.2	0.0	0.3	7.8
Phase Call Probability		1.00		1.00	0.98	1.00	1.00	1.00
Max Out Probability		1.00		1.00	0.00	1.00	0.00	0.18

Movement Group Results	EB			WB			NB			SB		
Approach Movement	L	T	R	L	T	R	L	T	R	L	T	R
Assigned Movement	5	2	12	1	6	16	3	8	18	7	4	14
Adjusted Flow Rate (*v*), veh/h	71	239	185	118	337	287	133	870	863	194	514	497
Adjusted Saturation Flow Rate (*s*), veh/h/ln	702	1643	1201	825	1643	1398	1603	1683	1648	1603	1683	1630
Queue Service Time (g_s), s	10.2	12.2	13.0	14.2	18.5	18.6	4.1	50.0	50.0	7.6	21.2	21.2
Cycle Queue Clearance Time (g_c), s	28.8	12.2	13.0	27.2	18.5	18.6	4.1	50.0	50.0	7.6	21.2	21.2
Green Ratio (*g/C*)	0.29	0.29	0.29	0.29	0.29	0.29	0.55	0.49	0.49	0.60	0.53	0.53
Capacity (*c*), veh/h	149	484	354	208	484	412	318	827	809	225	887	859
Volume-to-Capacity Ratio (*X*)	0.475	0.494	0.522	0.567	0.695	0.698	0.418	1.053	1.066	0.861	0.579	0.579
Back of Queue (*Q*), ft/ln (50 th percentile)	46.1	123.8	97.3	76.6	198.2	171.5	35.8	734.6	746.4	124.8	196.7	190.6
Back of Queue (*Q*), veh/ln (50 th percentile)	1.8	4.8	3.7	2.9	7.6	6.6	1.4	28.9	29.4	4.9	7.7	7.5
Queue Storage Ratio (*RQ*) (50 th percentile)	0.23	0.12	0.10	0.38	0.20	0.17	0.18	0.74	0.75	0.62	0.20	0.19
Uniform Delay (d_1), s/veh	44.6	29.6	29.9	41.3	31.8	31.9	13.7	25.9	25.9	28.9	16.4	16.4
Incremental Delay (d_2), s/veh	0.9	0.3	0.7	2.3	3.6	4.3	0.3	46.0	50.8	3.8	0.6	0.6
Initial Queue Delay (d_3), s/veh	0.0	0.0	0.0	0.0	0.0	0.0	0.0	0.0	0.0	0.0	0.0	0.0
Control Delay (*d*), s/veh	45.5	29.9	30.6	43.5	35.5	36.2	14.0	71.9	76.7	32.6	17.0	17.1
Level of Service (LOS)	D	C	C	D	D	D	B	F	F	C	B	B
Approach Delay, s/veh / LOS	32.4		C	37.0		D	70.0		E	19.6		B
Intersection Delay, s/veh / LOS	45.9						D					

Figure 9.16 *HCS7 SignalsOptAfter1*

HCS7 Signalized Intersection Results Summary

General Information				Intersection Information	
Agency				Duration, h	0.25
Analyst		Analysis Date	Mar 2, 2017	Area Type	CBD
Jurisdiction		Time Period		PHF	1.00
Urban Street		Analysis Year	2017	Analysis Period	1> 7:00
Intersection		File Name	TextSignalsOptAfter1.xus		
Project Description	Chapter 31: Example Problem 1				

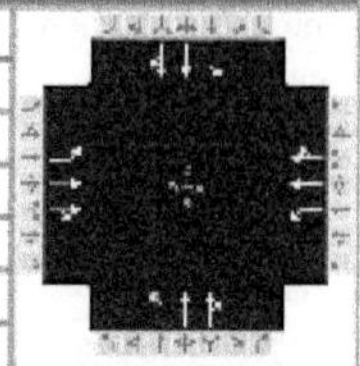

Demand Information	EB			WB			NB			SB		
Approach Movement	L	T	R	L	T	R	L	T	R	L	T	R
Demand (*v*), veh/h	71	318	106	118	600	24	133	1644	111	194	933	111

Signal Information			
Cycle, s	89.1	Reference Phase	2
Offset, s	0	Reference Point	Begin
Uncoordinated	Yes	Simult. Gap E/W	On
Force Mode	Fixed	Simult. Gap N/S	On

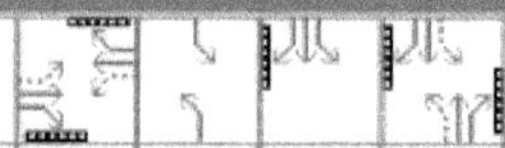

Green	20.0	4.9	2.0	50.2	0.0	0.0
Yellow	4.0	4.0	0.0	4.0	0.0	0.0
Red	0.0	0.0	0.0	0.0	0.0	0.0

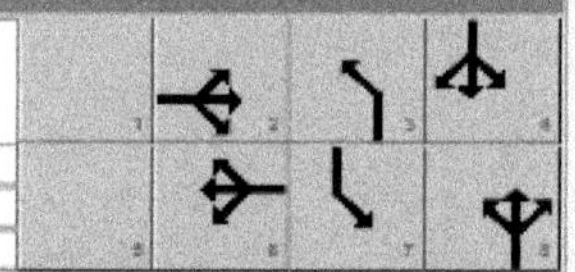

Timer Results	EBL	EBT	WBL	WBT	NBL	NBT	SBL	SBT
Assigned Phase		2		6	3	8	7	4
Case Number		6.0		6.0	1.1	4.0	1.1	4.0
Phase Duration, s		24.0		24.0	8.9	54.2	10.9	56.2
Change Period, ($Y+R_c$), s		4.0		4.0	4.0	4.0	4.0	4.0
Max Allow Headway (*MAH*), s		3.4		3.4	3.1	3.1	3.1	3.1
Queue Clearance Time (g_s), s		22.0		22.0	5.1	44.7	6.8	18.2
Green Extension Time (g_e), s		0.0		0.0	0.0	5.5	0.2	0.0
Phase Call Probability		1.00		1.00	0.96	1.00	0.99	1.00
Max Out Probability		1.00		1.00	1.00	0.00	0.05	1.00

Movement Group Results	EB			WB			NB			SB		
Approach Movement	L	T	R	L	T	R	L	T	R	L	T	R
Assigned Movement	5	2	12	1	6	16	3	8	18	7	4	14
Adjusted Flow Rate (*v*), veh/h	71	242	182	118	337	287	133	870	863	194	513	498
Adjusted Saturation Flow Rate (*s*), veh/h/ln	690	1643	1164	801	1643	1391	1603	1683	1648	1603	1683	1631
Queue Service Time (g_s), s	2.1	11.9	12.8	7.2	17.8	17.9	3.1	41.6	42.7	4.8	16.2	16.2
Cycle Queue Clearance Time (g_c), s	20.0	11.9	12.8	20.0	17.8	17.9	3.1	41.6	42.7	4.8	16.2	16.2
Green Ratio (*g/C*)	0.22	0.22	0.22	0.22	0.22	0.22	0.62	0.56	0.56	0.64	0.59	0.59
Capacity (*c*), veh/h	97	369	261	146	369	312	359	948	928	226	986	955
Volume-to-Capacity Ratio (*X*)	0.731	0.655	0.697	0.810	0.914	0.918	0.370	0.918	0.930	0.857	0.521	0.521
Back of Queue (*Q*), ft/ln (50 th percentile)	55.1	127.2	102.9	94	246.8	218.8	23.8	362.6	366.4	68.1	135.3	131.2
Back of Queue (*Q*), veh/ln (50 th percentile)	2.1	4.9	4.0	3.6	9.5	8.4	0.9	14.3	14.4	2.7	5.3	5.2
Queue Storage Ratio (*RQ*) (50 th percentile)	0.28	0.13	0.10	0.47	0.25	0.22	0.12	0.36	0.37	0.34	0.14	0.13
Uniform Delay (d_1), s/veh	44.3	31.4	31.7	42.5	33.7	33.7	9.1	17.6	17.8	21.6	11.0	11.0
Incremental Delay (d_2), s/veh	21.6	3.3	6.7	26.3	26.0	30.1	0.2	1.6	1.9	9.2	0.2	0.2
Initial Queue Delay (d_3), s/veh	0.0	0.0	0.0	0.0	0.0	0.0	0.0	0.0	0.0	0.0	0.0	0.0
Control Delay (*d*), s/veh	65.9	34.7	38.4	68.9	59.7	63.8	9.4	19.2	19.8	30.8	11.2	11.2
Level of Service (LOS)	E	C	D	E	E	E	A	B	B	C	B	B
Approach Delay, s/veh / LOS	40.5	D		62.7	E		18.8	B		14.4	B	
Intersection Delay, s/veh / LOS	27.6						C					

Figure 9.17 *HCS7 SignalsOptInitial2*

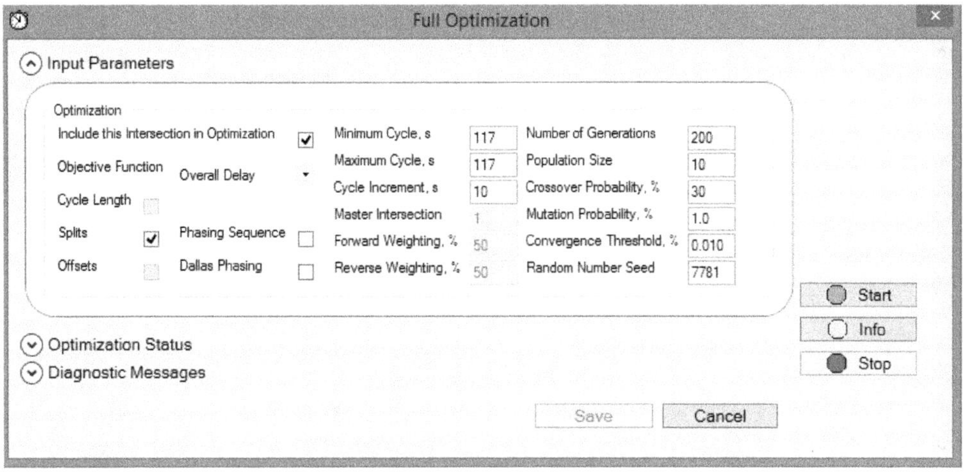

PROBLEM TWO

Semi-actuated operation would be modeled in the same way, but with Phases 2 and 6 set to maximum recall. The before-and-after results are shown below for HCM Example Problem One modified to be more appropriate for semi-actuated mode, which reduced this delay from 48 seconds to 31 seconds. This represents a very similar 36% improvement in a three-second run with semi-actuation making little difference in this case.

Figure 9.18 *HCS7 SignalsOptFinal2*

Overall Delay

Original	48.1 sec/veh
Optimum	30.8 sec/veh
Average	30.8 sec/veh
Improvement	35.9%

Run Status

Generation Number	17 out of 200
Generation Optimum	14
Total Time Elapsed	3 sec

Termination via convergence threshold

Figure 9.19 *HCS7 SignalsOptBefore2*

HCS7 Signalized Intersection Results Summary

General Information				Intersection Information	
Agency				Duration, h	0.25
Analyst		Analysis Date	Mar 2, 2017	Area Type	CBD
Jurisdiction		Time Period		PHF	1.00
Urban Street		Analysis Year	2017	Analysis Period	1> 7:00
Intersection		File Name	TextSignalsOptBefore2.xus		
Project Description	Chapter 31: Example Problem 1				

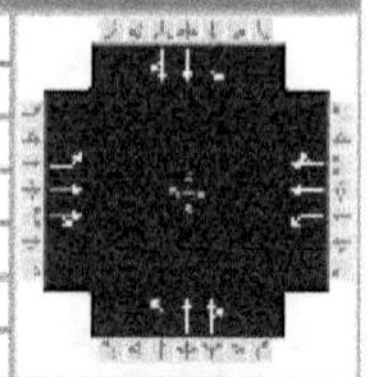

Demand Information	EB			WB			NB			SB		
Approach Movement	L	T	R	L	T	R	L	T	R	L	T	R
Demand (v), veh/h	71	318	106	118	600	24	133	1644	111	194	933	111

Signal Information										
Cycle, s	101.8	Reference Phase	2							
Offset, s	0	Reference Point	Begin	Green	30.0	6.2	3.6	50.0	0.0	0.0
Uncoordinated	Yes	Simult. Gap E/W	On	Yellow	4.0	4.0	0.0	4.0	0.0	0.0
Force Mode	Fixed	Simult. Gap N/S	On	Red	0.0	0.0	0.0	0.0	0.0	0.0

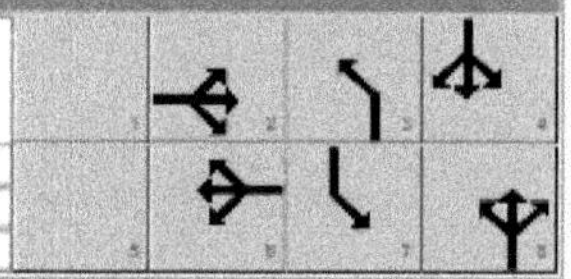

Timer Results	EBL	EBT	WBL	WBT	NBL	NBT	SBL	SBT
Assigned Phase		2		6	3	8	7	4
Case Number		6.0		6.0	1.1	4.0	1.1	4.0
Phase Duration, s		34.0		34.0	10.2	54.0	13.8	57.6
Change Period, ($Y+R_c$), s		4.0		4.0	4.0	4.0	4.0	4.0
Max Allow Headway (MAH), s		3.4		3.4	3.1	3.1	3.1	3.1
Queue Clearance Time (g_s), s		30.8		29.2	6.1	52.0	9.6	23.2
Green Extension Time (g_e), s		0.0		0.4	0.2	0.0	0.3	7.8
Phase Call Probability		1.00		1.00	0.98	1.00	1.00	1.00
Max Out Probability		1.00		1.00	0.00	1.00	0.00	0.18

Movement Group Results	EB			WB			NB			SB		
Approach Movement	L	T	R	L	T	R	L	T	R	L	T	R
Assigned Movement	5	2	12	1	6	16	3	8	18	7	4	14
Adjusted Flow Rate (v), veh/h	71	239	185	118	337	287	133	870	863	194	514	497
Adjusted Saturation Flow Rate (s), veh/h/ln	702	1643	1201	825	1643	1398	1603	1683	1648	1603	1683	1630
Queue Service Time (g_s), s	10.2	12.2	13.0	14.2	18.5	18.6	4.1	50.0	50.0	7.6	21.2	21.2
Cycle Queue Clearance Time (g_c), s	28.8	12.2	13.0	27.2	18.5	18.6	4.1	50.0	50.0	7.6	21.2	21.2
Green Ratio (g/C)	0.29	0.29	0.29	0.29	0.29	0.29	0.55	0.49	0.49	0.60	0.53	0.53
Capacity (c), veh/h	149	484	354	208	484	412	318	827	809	225	887	859
Volume-to-Capacity Ratio (X)	0.475	0.494	0.522	0.567	0.695	0.698	0.418	1.053	1.066	0.861	0.579	0.579
Back of Queue (Q), ft/ln (50 th percentile)	56.4	135.3	109.4	89.3	213.5	186.6	35.8	734.6	746.4	124.8	196.7	190.6
Back of Queue (Q), veh/ln (50 th percentile)	2.2	5.2	4.2	3.4	8.2	7.2	1.4	28.9	29.4	4.9	7.7	7.5
Queue Storage Ratio (RQ) (50 th percentile)	0.28	0.14	0.11	0.45	0.21	0.19	0.18	0.74	0.75	0.62	0.20	0.19
Uniform Delay (d_1), s/veh	44.6	29.6	29.9	41.3	31.8	31.9	13.7	25.9	25.9	28.9	16.4	16.4
Incremental Delay (d_2), s/veh	10.4	3.6	5.4	10.7	8.0	9.4	0.3	46.0	50.8	3.8	0.6	0.6
Initial Queue Delay (d_3), s/veh	0.0	0.0	0.0	0.0	0.0	0.0	0.0	0.0	0.0	0.0	0.0	0.0
Control Delay (d), s/veh	55.1	33.2	35.3	52.0	39.9	41.3	14.0	71.9	76.7	32.6	17.0	17.1
Level of Service (LOS)	E	C	D	D	D	D	B	F	F	C	B	B
Approach Delay, s/veh / LOS	37.1	D		42.4	D		70.0	E		19.6	B	
Intersection Delay, s/veh / LOS	47.4						D					

Figure 9.20 *HCS7 SignalsOptAfter2*

HCS7 Signalized Intersection Results Summary

General Information				Intersection Information	
Agency				Duration, h	0.25
Analyst		Analysis Date	Mar 2, 2017	Area Type	CBD
Jurisdiction		Time Period		PHF	1.00
Urban Street		Analysis Year	2017	Analysis Period	1> 7:00
Intersection		File Name	TextSignalsOptAfter2.xus		
Project Description	Chapter 31: Example Problem 1				

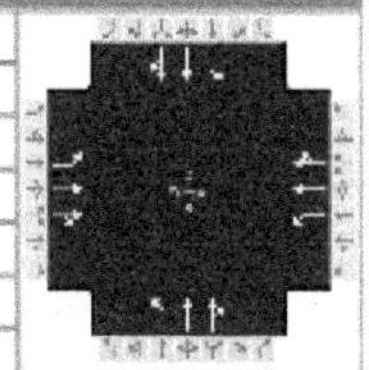

Demand Information	EB			WB			NB			SB		
Approach Movement	L	T	R	L	T	R	L	T	R	L	T	R
Demand (v), veh/h	71	318	106	118	600	24	133	1644	111	194	933	111

Signal Information			
Cycle, s	80.6	Reference Phase	2
Offset, s	0	Reference Point	Begin
Uncoordinated	Yes	Simult. Gap E/W	On
Force Mode	Fixed	Simult. Gap N/S	On

Green	18.0	4.7	0.3	45.6	0.0	0.0
Yellow	4.0	4.0	0.0	4.0	0.0	0.0
Red	0.0	0.0	0.0	0.0	0.0	0.0

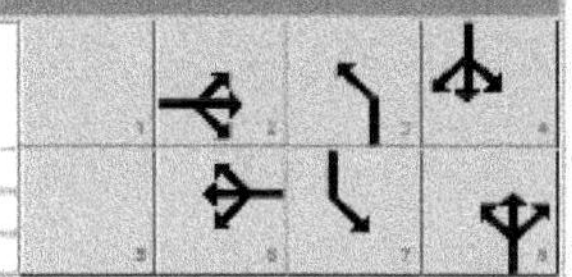

Timer Results	EBL	EBT	WBL	WBT	NBL	NBT	SBL	SBT
Assigned Phase		2		6	3	8	7	4
Case Number		6.0		6.0	1.1	4.0	1.1	4.0
Phase Duration, s		22.0		22.0	8.7	49.6	9.0	49.9
Change Period, ($Y+R_c$), s		4.0		4.0	4.0	4.0	4.0	4.0
Max Allow Headway (MAH), s		3.5		3.5	3.1	3.1	3.1	3.1
Queue Clearance Time (g_s), s		20.0		20.0	4.7	40.5	6.1	17.3
Green Extension Time (g_e), s		0.0		0.0	0.0	5.2	0.0	0.0
Phase Call Probability		1.00		1.00	0.95	1.00	0.99	1.00
Max Out Probability		1.00		1.00	1.00	0.15	1.00	1.00

Movement Group Results	EB			WB			NB			SB		
Approach Movement	L	T	R	L	T	R	L	T	R	L	T	R
Assigned Movement	5	2	12	1	6	16	3	8	18	7	4	14
Adjusted Flow Rate (v), veh/h	71	242	182	118	337	287	133	870	863	194	513	498
Adjusted Saturation Flow Rate (s), veh/h/ln	677	1643	1163	795	1643	1391	1603	1683	1648	1603	1683	1631
Queue Service Time (g_s), s	1.7	10.8	11.6	6.4	16.2	16.3	2.7	37.4	38.5	4.1	15.3	15.3
Cycle Queue Clearance Time (g_c), s	18.0	10.8	11.6	18.0	16.2	16.3	2.7	37.4	38.5	4.1	15.3	15.3
Green Ratio (g/C)	0.22	0.22	0.22	0.22	0.22	0.22	0.62	0.57	0.57	0.63	0.57	0.57
Capacity (c), veh/h	104	367	259	152	367	310	362	953	933	211	958	928
Volume-to-Capacity Ratio (X)	0.684	0.659	0.702	0.777	0.920	0.924	0.368	0.913	0.925	0.919	0.536	0.536
Back of Queue (Q), ft/ln (50 th percentile)	58.7	129.1	107.9	93.6	238.6	212.2	20	334.4	342	106	125	121.1
Back of Queue (Q), veh/ln (50 th percentile)	2.3	5.0	4.1	3.6	9.2	8.2	0.8	13.2	13.5	4.2	4.9	4.8
Queue Storage Ratio (RQ) (50 th percentile)	0.29	0.13	0.11	0.47	0.24	0.21	0.10	0.33	0.34	0.53	0.13	0.12
Uniform Delay (d_1), s/veh	40.1	28.5	28.9	38.6	30.6	30.7	8.6	15.7	15.9	18.8	10.8	10.8
Incremental Delay (d_2), s/veh	30.9	9.0	14.7	31.4	30.5	34.8	0.2	4.8	5.8	39.5	0.3	0.3
Initial Queue Delay (d_3), s/veh	0.0	0.0	0.0	0.0	0.0	0.0	0.0	0.0	0.0	0.0	0.0	0.0
Control Delay (d), s/veh	71.0	37.5	43.6	70.0	61.1	65.5	8.9	20.5	21.7	58.3	11.1	11.1
Level of Service (LOS)	E	D	D	E	E	E	A	C	C	E	B	B
Approach Delay, s/veh / LOS	44.6	D		64.2	E		20.3	C		18.7	B	
Intersection Delay, s/veh / LOS	30.2						C					

CHAPTER 10 Traffic Impact Analysis

This procedure includes analyzing existing, future, and mitigated analyses that are used to establish improvements necessary for a development to avoid negatively impacting the adjacent traffic system. The process includes estimating additional trips generated by the development, assigning those trips with growth to the nearby streets and intersections, identifying the deficiencies created, and defining the mitigating improvements required to offset the additional traffic.

Existing Conditions

The definition of the analysis of existing conditions begins with identifying the appropriate affected area to be analyzed. The size of the proposed development dictates the area that will be affected, so care must be taken to prescribe all intersections and roadways that should be included in the impact and study area. All intersections and roadways are analyzed using these procedures to determine the existing delays, queues, and levels of service.

Scope

The definition of the impacted area for a given development is normally one of the initial tasks and can be somewhat subjective with so many variables in play. The combination of the size and type of the development with the surrounding intersections and roadways becomes a bit of an art to understand what will be affected and at what level.

This can be a delicate decision to both not underestimate the limits (for fear of neglecting what should have been included) and not overestimate the reach (and increase the analysis expense unnecessarily). Typically, this is proposed by the development team and reviewed by the government entity toward discussing (and sometimes negotiating) the appropriate boundaries.

Think about your own driving:
You can probably think of a new development that generated enough traffic to affect your commute, even if it was unrelated and not even nearby your home or work. Larger developments can have far-reaching impacts that should be included in the study to ensure all affected are accommodated by the mitigating improvements.

Data

Once the scope is defined, data must be collected for the existing conditions analysis that will normally include defining all parameters for included intersections and roadways. This is usually limited to signalized and unsignalized intersections with urban street segments (and sometimes ramp terminals) included. Uninterrupted flow does not typically come into the analysis as most are more urban and at the street level.

The primary data needs include turning movement counts (which must include arrival rate for congested signalized intersections) in fifteen-minute increments (for multiple-period analyses). Lane configurations, signal control parameters, and segment lengths are also of primary importance for these analyses to be performed. Critical parameters like base saturation flow rates and headway times are frequently location specific with calibration needed, or at least agreement among all parties on the values to be used.

Depending on the type of development, the time periods to be analyzed may vary; commercial and retail would generally have different critical times from office and residential. This decision can also become part of the collaborative process between the developer and the government to ensure all agree on which time periods will be best for the overall goals of the study.

Analysis

Once the data have been collected and the parameters have been approved, the analysis follows the procedures defined in this text for the appropriate intersections and segments involved. Signalized intersections are usually the critical points that generate the most delay and queue issues when demand values are increased to account for growth and development. This makes the existing analysis important to establish the baseline for the underlying parameters as duplication field observations for maximum credibility.

All existing signalized intersections (including urban street segments, if applicable) and unsignalized intersections are analyzed to determine delay, level of service, and queues for each lane group (and segment), which are then consolidated into approaches and intersections for delay and level of service. These results serve to define the operational status of the potentially impacted area in order to quantify the increases created by the additional traffic when compared with the future analysis.

Future Conditions

Since development typically takes time to plan, design, and construct, the growth rate is applied to the existing demands to account for the time elapsed. The size of all aspects of the development must be known in order to quantify the trips that will be generated using the *ITE*

Trip Generation Manual, the accepted standard for this process. The trips are then distributed appropriately to the various roadway segments with turning movement estimates at all affected intersections as well. The capacity analysis is then repeated with the future demand values to determine the delays, queues, and LOS values with additional development demand in place.

Generation

The type and size of the development are used to estimate the generated trips using the *ITE Trip Generation Manual*. This is normally a function of square footage for retail and office buildings or number of units for residences or gas stations, but can vary widely with many land-use types covered in detail. The estimates can be by time of day and include entering and exiting percentage to facilitate quantifying demand for different scenarios.

These values might be adjusted for the effects of internal capture (for mixed-use developments) and pass-by trips (for existing adjacent traffic). Internal capture is a percentage established to represent trips that begin and end within the development and therefore are not included in the additional generated trips. Pass-by trips are trips coming from traffic already on the adjacent roadways that enter the development, but are also not additional trips generated by the development.

The growth of existing traffic independent of the development must also be considered as adding to the demand increase beyond the newly generated traffic. The area growth rate can be used to estimate this percentage applied to existing demand levels.

Distribution

The additional growth and generated trips to be added to the system are distributed to intersections and streets, usually with the help of the area planning model. The travel patterns that are predicted by this model will generally include the development being studied and can identify the origins and destinations of the new trips. These proportions are applied to the generated traffic when assigning those demands to streets and intersection turning movements.

Since it is important to run multiple-period analyses for congested conditions, the growth and generated traffic must be assigned by time period. Many times this can be done simply as proportions by period, but for developments like an office complex, the dramatic peaking of arrivals and departures as a function of start and end work hours must be considered.

Think about your own driving:
The traffic around land uses dominated by the workforce is obviously quite different approaching the times the offices open and close. You may work in a big complex and suffer through the ramifications of everyone leaving at the same time. A very minor shift in your exit time by only a few minutes can make a significant difference in your travel time home.

Analysis

All existing signalized intersection (including urban street segments, if applicable) and unsignalized intersection analyses are updated to include the increased demand from the trip generation and distribution processes. This future analysis is performed without changing any lane configurations, intersection control types, or signal timing and/or phasing for an accurate comparison as affected by the additional traffic alone.

The resulting delay, level of service, and queues for each lane group (and segment) will change to typically increase delay and queue with potentially worsened level of service. These results quantify these increases created by the additional traffic when compared with the existing analysis toward identifying strategies for mitigating the additional delay and queue to make the development travel time and expense neutral.

Mitigation

Improvements to mitigate any increased delays and queues, as well as any deterioration in LOS, are identified with cost estimates included. The local city, county, or state reviews and negotiates the proposed improvements to finalize a mitigation strategy that must be accepted by the developer for the project to move forward. This process can involve building and/or improving existing intersections and/or streets, which can require defining control type if unsignalized or signalization if warrants are met. Existing or new signals normally will be retimed (at least) or have phasing modified (sometimes) in response to the new demands.

Control

There can be smaller or internal intersections that either don't exist in the present (so control type needs to be defined) or may change purpose in the future (so control type may need to change). The process typically includes analyzing these intersections as two-way stop controlled (TWSC); all-way stop controlled (AWSC) or roundabouts if unsignalized; or as signalized intersections with a variety of phasing and/or timing options if signal warrants are met.

Many times as part of this process, an unsignalized intersection will meet the criteria for signalization as determined by the *Manual on Traffic Control Devices (MUTCD)* signal warrant procedure, so this can become part of the overall process as well. There are nine warrants that define thresholds for potential signalization with engineering judgment applied if only one or a few are met, with overall safety and efficiency as the goals.

If no warrants are met, a comparison of results among the three unsignalized intersection control options will generally show which will be best operationally as the one with the least delay and shortest queues. This can be done following the procedures defined in Chapters Four and Five.

Strategies

Comparisons between existing and future analysis results are used to identify the lanes, intersections, and/or streets where delays and queues have increased to possibly worsen level of service. An overall view of the area is needed in this process because the problems may potentially involve adjacent intersections or roadways. Once these have been identified, the strategies can be formulated to address each with the appropriate proposed improvements.

For signalized intersections, selecting the best operation mode (usually between fully actuated and coordinated-actuated) and considering interconnecting signals (if not in place) then optimizing phasing and timing, is usually the first tactic employed as the least expensive. This is necessary to adjust for a new signalized intersection or new demand conditions for existing signals. This can be accomplished using the techniques discussed in Chapter Nine for all appropriate phasing combinations to be considered—again, to find the least delay and shortest queues.

Second, increased traffic can sometimes cause turn lanes to overflow with longer queues, even after signal timing has been updated. This requires lengthening of these storage bays, since the spillover can have dramatic effects on the adjacent through-lane capacity. Sometimes, this can be accomplished with dual turn lanes (instead of longer turn bays) if the right-of-way and phasing can be accommodated.

Next—usually as a last resort because of construction and right-of-way costs—segment lanes may need to be added between intersections to meet capacity requirements. This can escalate quickly when additional lanes require appropriate transitioning to and from the smaller cross-section before and after the development impact area.

There are certainly other strategies that can come into play for specific developments and/or areas: access management techniques like closing or combining median openings to minimize conflicts on the adjacent streets. Multimodal considerations could include sidewalk buffers, bicycle lanes, or bus turnouts.

Solution

Strategies are identified and tested toward providing the most efficient improvements to mitigate the problems created by the new development. This can sometimes become somewhat contentious with the developer's desire to minimize costs and the government's need to ensure safe and efficient operations. Generally, the developer presents the proposed improvements and their analysis results to the government toward confirming the required delay, queue, and level of service goals are met. The government reviews the submission for the accuracy of the data and analysis, as well as the results that meet their requirements. It is always best for the communication between the developer and the government to be continuous throughout the process to strive toward consistency in the final proposed solution.

Project

The traffic impact analysis project is patterned after an actual study with freedom taken to simplify in some areas and make points in others. A 216,734 square-foot discount supercenter that includes a gas station with twelve fueling stations is proposed on 12th Avenue adjacent to Waldo Road in Gainesville, Florida. The scope of the impact area includes three intersections on Waldo Road and one intersection on 12th Avenue.

All existing intersection and street data are available for the four intersections, including lane configurations; signal timing and phasing parameters; and turning movement counts. Additional information is defined by the government to be used, including peak hour factor; heavy vehicle percentage; base saturation flow rate; and extension of effective green values for existing and future analyses. The trip generation results and the planning model distribution percentages are also provided, along with growth and pass-by trip percentages.

Figure 10.1 *Project Development Map*

Existing

Data are coded into *HCS7* for two signalized intersections on an urban street and one TWSC intersection. The reports for these four analyses that show input data and output results are shown below. You can see the two signals (8th and 16th) as coordinated along the urban street (Waldo) with all LOS values at D or better (except the WBL on 16th at LOS E), and all queue-storage ratios are below 1.0 (except EBR and WBL on 16th); the urban street segment analysis shows LOS B (NB) and LOS C (SB). The TWSC intersection at 12th has one lane (WBL) with LOS F.

Future

The trip generation and distribution process includes defining additional trips created by the supercenter with a gas station using the *ITE Trip Generation Manual* and assigning those trips to intersections and turning movements using the local planning model as the guide. A spreadsheet is constructed to compute entering and exiting trips for the development after accounting for pass-by trips at a typical 25% rate. From the planning model, growth is estimated at 1.31% per year to add to the generated trips before assigning to movements proportionally from the model. The calculations are shown in the spreadsheet below.

An *MUTCD* warrants analysis with future traffic satisfies Warrant 8 for projected weekday vehicular volume meeting Warrants 1 and 2 thresholds on 12th at Waldo. Data are coded into *HCS7* for three signalized intersections on an urban street and one TWSC intersection. The reports for these five analyses that show input data and output results are shown below. You can see the three signals (8th, 12th, and 16th) as coordinated along the urban street (Waldo), with all LOS values at D or better (except WBR on 12th at LOS E and EBR on 16th at LOS F) and all queue-storage ratios below 1.0 (except WBL and WBR on 12th and EBR, WBL,

Figure 10.2 *HCS7 Signal1 Existing*

General Information				Intersection Information	
Agency				Duration, h	0.25
Analyst		Analysis Date	Mar 29, 2011	Area Type	Other
Jurisdiction		Time Period		PHF	0.90
Urban Street	Waldo	Analysis Year	2011	Analysis Period	1> 7:00
Intersection	8th	File Name	WaldoExistingPM.xus		
Project Description	Existing				

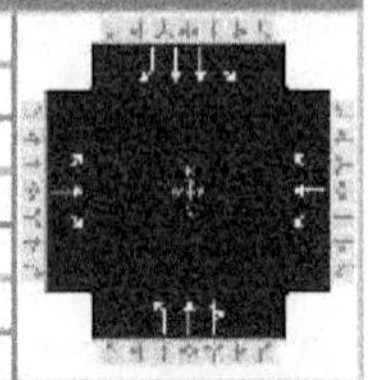

Demand Information	EB			WB			NB			SB		
Approach Movement	L	T	R	L	T	R	L	T	R	L	T	R
Demand (v), veh/h	106	196	82	41	123	173	94	831	53	211	856	70

Signal Information			
Cycle, s	120.0	Reference Phase	2
Offset, s	0	Reference Point	Begin
Uncoordinated	No	Simult. Gap E/W	On
Force Mode	Fixed	Simult. Gap N/S	On

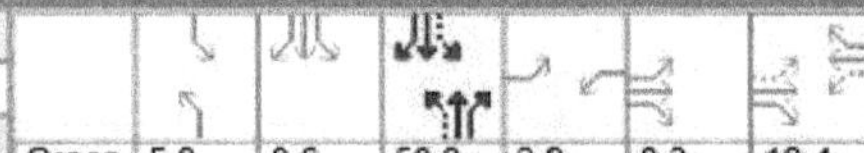

Green	5.9	0.6	59.9	3.9	0.3	19.4
Yellow	4.0	4.0	4.0	4.0	4.0	4.0
Red	1.0	1.0	1.0	1.0	1.0	1.0

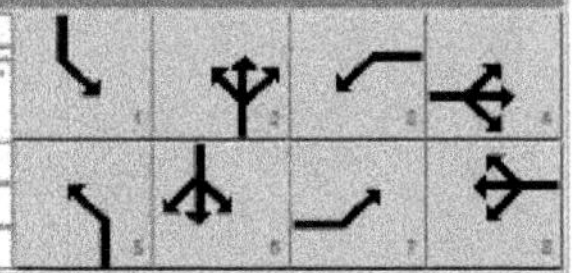

Timer Results	EBL	EBT	WBL	WBT	NBL	NBT	SBL	SBT
Assigned Phase	7	4	3	8	5	2	1	6
Case Number	1.1	3.0	1.1	3.0	1.1	4.0	1.1	3.0
Phase Duration, s	14.2	29.7	8.9	24.4	10.9	64.9	16.5	70.5
Change Period, (Y+R c), s	5.0	5.0	5.0	5.0	5.0	5.0	5.0	5.0
Max Allow Headway (MAH), s	3.1	3.2	3.1	3.2	3.1	0.0	3.1	0.0
Queue Clearance Time (g s), s	9.3	16.6	4.9	18.2	5.9		11.0	
Green Extension Time (g e), s	0.1	1.2	0.0	1.2	0.2	0.0	0.4	0.0
Phase Call Probability	0.98	1.00	0.78	1.00	0.97		1.00	
Max Out Probability	0.02	0.00	0.00	0.00	0.00		0.00	

Movement Group Results	EB			WB			NB			SB		
Approach Movement	L	T	R	L	T	R	L	T	R	L	T	R
Assigned Movement	7	4	14	3	8	18	5	2	12	1	6	16
Adjusted Flow Rate (v), veh/h	118	218	91	46	137	192	104	496	486	254	1031	84
Adjusted Saturation Flow Rate (s), veh/h/ln	1558	1636	1386	1558	1636	1386	1558	1636	1601	1558	1557	1386
Queue Service Time (g s), s	7.3	14.6	6.7	2.9	9.2	16.2	3.9	26.2	26.2	9.0	31.4	5.8
Cycle Queue Clearance Time (g c), s	7.3	14.6	6.7	2.9	9.2	16.2	3.9	26.2	26.2	9.0	31.4	5.8
Green Ratio (g/C)	0.25	0.21	0.21	0.19	0.16	0.16	0.55	0.50	0.50	0.61	0.55	0.55
Capacity (c), veh/h	264	336	285	179	264	224	265	817	799	362	1701	757
Volume-to-Capacity Ratio (X)	0.445	0.647	0.320	0.255	0.517	0.859	0.394	0.608	0.608	0.703	0.606	0.111
Back of Queue (Q), ft/ln (50 th percentile)	73.7	157	60.6	29.7	98.9	152.3	35.8	275.6	270.2	81	324.6	82.4
Back of Queue (Q), veh/ln (50 th percentile)	2.8	5.9	2.3	1.1	3.7	5.8	1.4	10.4	10.2	3.1	12.3	3.1
Queue Storage Ratio (RQ) (50 th percentile)	0.37	0.03	0.61	0.15	0.49	0.76	0.18	0.09	0.09	0.41	0.13	0.82
Uniform Delay (d 1), s/veh	36.6	43.7	40.5	40.7	46.0	49.0	17.2	21.6	21.6	17.2	25.6	23.3
Incremental Delay (d 2), s/veh	0.4	0.8	0.2	0.3	0.6	3.7	0.4	3.4	3.4	0.7	1.2	0.2
Initial Queue Delay (d 3), s/veh	0.0	0.0	0.0	0.0	0.0	0.0	0.0	0.0	0.0	0.0	0.0	0.0
Control Delay (d), s/veh	37.1	44.5	40.8	41.0	46.6	52.7	17.6	24.9	25.0	17.9	26.8	23.6
Level of Service (LOS)	D	D	D	D	D	D	B	C	C	B	C	C
Approach Delay, s/veh / LOS	41.6	D		49.0	D		24.3	C		25.0	C	
Intersection Delay, s/veh / LOS	29.7						C					

Figure 10.3 *HCS7 Signal2 Existing*

General Information				Intersection Information	
Agency				Duration, h	0.25
Analyst		Analysis Date	Mar 29, 2011	Area Type	Other
Jurisdiction		Time Period		PHF	0.90
Urban Street	Waldo	Analysis Year	2011	Analysis Period	1> 7:00
Intersection	16th	File Name	WaldoExistingPM.xus		
Project Description	Existing				

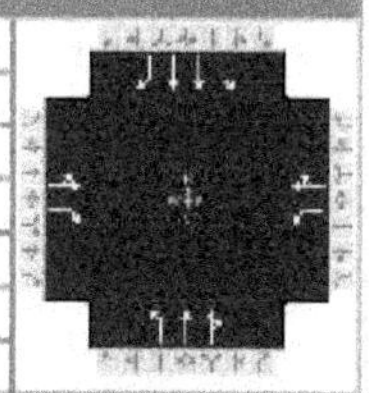

Demand Information	EB			WB			NB			SB		
Approach Movement	L	T	R	L	T	R	L	T	R	L	T	R
Demand (v), veh/h	71	10	221	100	76	52	249	1079	11	9	912	56

Signal Information			
Cycle, s	120.0	Reference Phase	2
Offset, s	0	Reference Point	Begin
Uncoordinated	No	Simult. Gap E/W	On
Force Mode	Fixed	Simult. Gap N/S	On

Green	1.4	3.2	64.4	31.0	0.0	0.0
Yellow	4.0	4.0	4.0	4.0	0.0	0.0
Red	1.0	1.0	1.0	1.0	0.0	0.0

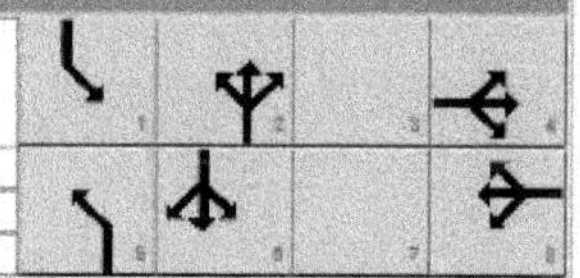

Timer Results	EBL	EBT	WBL	WBT	NBL	NBT	SBL	SBT
Assigned Phase		4		8	5	2	1	6
Case Number		7.0		6.0	1.1	4.0	1.1	3.0
Phase Duration, s		36.0		36.0	14.6	77.6	6.4	69.4
Change Period, ($Y+R_c$), s		5.0		5.0	5.0	5.0	5.0	5.0
Max Allow Headway (*MAH*), s		3.4		3.4	3.1	0.0	3.1	0.0
Queue Clearance Time (g_s), s		21.2		31.1	9.3		2.4	
Green Extension Time (g_e), s		1.1		0.0	0.3	0.0	0.0	0.0
Phase Call Probability		1.00		1.00	1.00		0.28	
Max Out Probability		0.05		1.00	0.00		0.00	

Movement Group Results	EB			WB			NB			SB		
Approach Movement	L	T	R	L	T	R	L	T	R	L	T	R
Assigned Movement	7	4	14	3	8	18	5	2	12	1	6	16
Adjusted Flow Rate (v), veh/h		90	246	111	142		229	503	501	10	1013	62
Adjusted Saturation Flow Rate (s), veh/h/ln		876	1386	1066	1524		1558	1636	1630	1558	1557	1386
Queue Service Time (g_s), s		7.4	19.2	12.3	9.2		7.3	17.5	17.5	0.4	26.8	2.6
Cycle Queue Clearance Time (g_c), s		16.6	19.2	29.1	9.2		7.3	17.5	17.5	0.4	26.8	2.6
Green Ratio (g/C)		0.26	0.26	0.26	0.26		0.63	0.60	0.60	0.55	0.54	0.54
Capacity (c), veh/h		283	358	186	394		341	989	986	314	1670	743
Volume-to-Capacity Ratio (X)		0.318	0.686	0.596	0.361		0.672	0.508	0.508	0.032	0.607	0.084
Back of Queue (Q), ft/ln (50 th percentile)		62	182	90.9	90.2		68.7	147.1	145.7	3.2	254.9	22.4
Back of Queue (Q), veh/ln (50 th percentile)		2.3	6.9	3.4	3.4		2.6	5.6	5.5	0.1	9.7	0.8
Queue Storage Ratio (RQ) (50 th percentile)		0.01	1.82	1.21	0.60		0.27	0.06	0.06	0.02	0.10	0.15
Uniform Delay (d_1), s/veh		42.7	40.1	52.0	36.4		16.6	10.0	9.9	12.9	19.1	13.5
Incremental Delay (d_2), s/veh		0.2	4.5	3.6	0.2		0.6	1.4	1.4	0.0	1.6	0.2
Initial Queue Delay (d_3), s/veh		0.0	0.0	0.0	0.0		0.0	0.0	0.0	0.0	0.0	0.0
Control Delay (d), s/veh		42.9	44.6	55.6	36.6		17.2	11.4	11.3	12.9	20.8	13.7
Level of Service (LOS)		D	D	E	D		B	B	B	B	C	B
Approach Delay, s/veh / LOS	44.2		D	44.9		D	12.4		B	20.3		C
Intersection Delay, s/veh / LOS	21.9						C					

Figure 10.4 *HCS7 Streets Existing*

General Information				Streets Information	
Agency				Number of Intersections	2
Analyst		Analysis Date	Mar 29, 2011	Number of Segments	1
Jurisdiction		Time Period		Number of Iterations	15
File Name	WaldoExistingPM.xus	Analysis Year	2011	System Cycle Length, s	120
Intersections	8th	16th		Analysis Period	1> 7:00
Project Description	Existing				

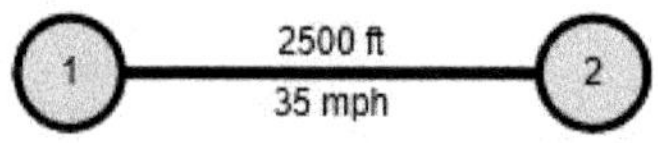

Basic Segment Information														
Segment	Speed Limit		Through Lanes		Segment Length		Intersection Wid		Length of RM		Percent Curb		Other Delay	
	NB	SB	NB	SB	NB	SB	NB	SB	NB	SB	NB	SB	NB	SB
1	35	35	2	2	2500	2500	50	50	0	0	70	70	0.0	0.0

		Northbound			Southbound		
Segment Output Data		NBL	NBT	NBR	SBL	SBT	SBR
Segment	Movement	5	2	12	1	6	16
1	Bay/Lane Spillback Time, h		never			never	
1	Shared Lane Spillback Time, h	never			never		never
1	Base Free-Flow Speed, mph		41.72			41.72	
1	Running Time, s		43.88			44.08	
1	Running Speed, mph		38.84			38.67	
1	Through Delay, s/veh		11.33			26.82	
1	Travel Time, s		55.21			70.90	
1	Travel Speed, mph		30.87			24.04	
1	Stop Rate, stops/veh		0.33			0.72	
1	Spatial Stop Rate, stops/mi		0.70			1.51	
1	Through vol/cap Ratio		0.51			0.61	
1	Percent of Base FFS		74.00			57.63	
1	Level of Service		B			C	
1	Auto Traveler Perception Score		2.24			2.37	

Figure 10.5 *HCS7 TWSC Existing*

General Information

General Information		Site Information	
Analyst		Intersection	Waldo/12th
Agency/Co.		Jurisdiction	
Date Performed	6/7/2017	East/West Street	
Analysis Year	2017	North/South Street	
Time Analyzed		Peak Hour Factor	0.90
Intersection Orientation	North-South	Analysis Time Period (hrs)	0.25
Project Description	Existing		

Lanes

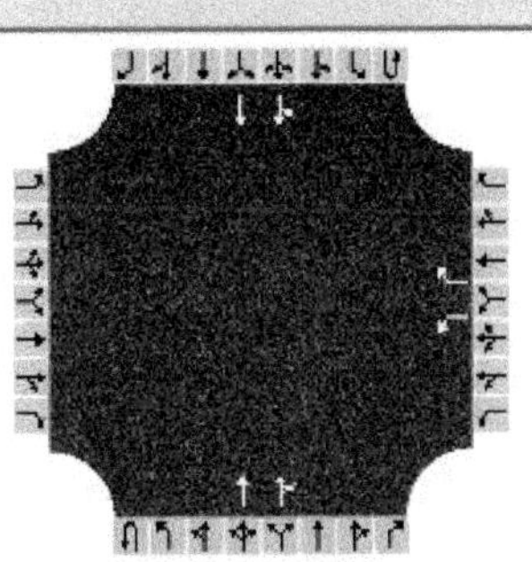

Vehicle Volumes and Adjustments

Approach	Eastbound				Westbound				Northbound				Southbound			
Movement	U	L	T	R	U	L	T	R	U	L	T	R	U	L	T	R
Priority		10	11	12		7	8	9	1U	1	2	3	4U	4	5	6
Number of Lanes		0	0	0		1	0	1	0	0	2	0	0	0	2	0
Configuration						L		R			T	TR		LT	T	
Volume, V (veh/h)						14		49			1297	26		67	1261	
Percent Heavy Vehicles (%)						7		7						7		
Proportion Time Blocked																
Percent Grade (%)					0											
Right Turn Channelized	No				No				No				No			
Median Type/Storage	Undivided															

Critical and Follow-up Headways

	U	L	T	R	U	L	T	R	U	L	T	R	U	L	T	R
Base Critical Headway (sec)																
Critical Headway (sec)																
Base Follow-Up Headway (sec)																
Follow-Up Headway (sec)																

Delay, Queue Length, and Level of Service

	U	L	T	R	U	L	T	R	U	L	T	R	U	L	T	R
Flow Rate, v (veh/h)						16		54						74		
Capacity, c (veh/h)						25		351						431		
v/c Ratio						0.64		0.15						0.17		
95% Queue Length, Q_{95} (veh)						1.9		0.5						0.6		
Control Delay (s/veh)						286.6		17.1						15.1		
Level of Service, LOS						F		C						C		
Approach Delay (s/veh)					78.7								5.8			

Figure 10.6 *FutureTrips*

Supercenter	216.734	3.87	839	Total	Passby	Net						
Gas	12	13.86	166	1005	251	754	**Supercenter**	1005 (No Passby Trips)				
Growth	Growth		Enter	Exit	Project		EBL	EBT	WBT	WBR	SBL	SBR
1.0131	0.039817		369	384	754		**451**	97	66	**44**	**46**	**464**
			49%	51%								
								0.43				
Waldo/16			0.23	0			0.23	0.2	0		0.2	
			Enter	Enter			Exit	Exit	Exit		Enter	
	EBL	EBT	**EBR**	**WBL**	WBT	WBR	**NBL**	**NBT**	**NBR**	SBL	**SBT**	SBR
Exist	71	10	221	100	76	52	246	1079	11	9	912	56
Growth	3	0	9	4	3	2	10	43	0	0	36	2
Project	0	0	85	0	0	0	88	77	0	0	74	0
Total	74	10	315	104	79	54	344	1199	11	9	1022	58
					0.88							
Waldo/12		0		0.45	0	0.43			0.45	0.43		
		Enter		Exit	Exit	Exit			Enter	Enter		
	EBL	EBT	EBR	**WBL**	WBT	**WBR**	NBL	NBT	**NBR**	**SBL**	SBT	SBR
Exist	0	0	0	14	0	49	0	1297	26	67	1261	0
Growth	0	0	0	1	0	2	0	52	1	3	50	0
Project	0	0	0	173	0	165	0	0	166	159	0	0
Passby				63		63			63	63		
Total	0	0	0	250	0	279	0	1349	256	291	1311	0
											0.45	
Waldo/8	0.1					0		0.35		0	0.35	0.1
	Enter					Enter		Enter		Exit	Exit	Exit
	EBL	EBT	EBR	WBL	WBT	**WBR**	NBL	**NBT**	NBR	**SBL**	**SBT**	**SBR**
Exist	106	196	82	41	123	173	94	831	53	211	856	70
Growth	4	8	3	2	5	7	4	33	2	8	34	3
Project	37	0	0	0	0	0	0	129	0	0	135	38
Total	147	204	85	43	128	180	98	993	55	219	1025	111

and NBL on 16th). The urban street segment analysis shows LOS D (NB) and LOS E (SB). The TWSC intersection on 12th has one lane with LOS F (SB at Supercenter).

Mitigated

The local government requires that no lane group level of service can be worse than LOS D (with three at LOS E/F the future analysis) with no lane queue-storage ratios exceeding 1.0 to prevent spillover (with five over 1.0 in the future analysis).

The first step is normally signal timing adjustments as easiest and least expensive, so the timing is optimized for cycle, splits, and offsets along the urban street. This results in lowering the cycle length from 120 seconds to 107 seconds (shorter cycles can yield lower delays if capacity is satisfied) and adjusted offsets for the updated cycle and the third signal at 50 seconds (providing for higher proportions arriving on green for lower delays). With one spillover deficiency remaining after retiming, the only solution appears to be lengthening the storage bay from 100 feet to 265 feet.

Figure 10.7 *HCS7 Signal1 Future*

General Information				Intersection Information	
Agency				Duration, h	0.25
Analyst		Analysis Date	Mar 29, 2011	Area Type	Other
Jurisdiction		Time Period		PHF	0.95
Urban Street	Waldo	Analysis Year	2011	Analysis Period	1> 7:00
Intersection	8th	File Name	WaldoFuturePM.xus		
Project Description	Future				

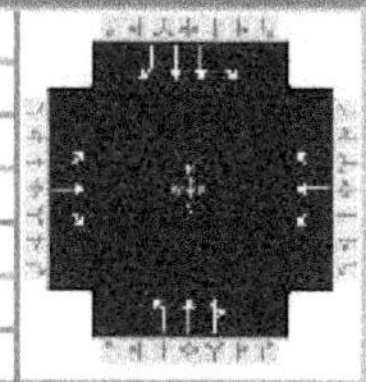

Demand Information	EB			WB			NB			SB		
Approach Movement	L	T	R	L	T	R	L	T	R	L	T	R
Demand (v), veh/h	147	204	85	43	128	180	98	993	55	219	1025	111

Signal Information										
Cycle, s	120.0	Reference Phase	2							
Offset, s	0	Reference Point	Begin	Green	7.2	2.3	54.3	5.3	1.3	19.6
Uncoordinated	No	Simult. Gap E/W	On	Yellow	4.0	4.0	4.0	4.0	4.0	4.0
Force Mode	Fixed	Simult. Gap N/S	On	Red	1.0	1.0	1.0	1.0	1.0	1.0

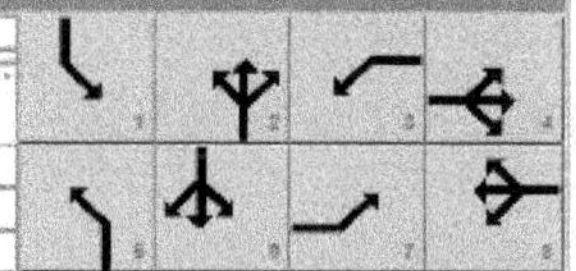

Timer Results	EBL	EBT	WBL	WBT	NBL	NBT	SBL	SBT
Assigned Phase	7	4	3	8	5	2	1	6
Case Number	1.1	3.0	1.1	3.0	1.1	4.0	1.1	3.0
Phase Duration, s	16.6	30.8	10.3	24.6	12.2	59.3	19.6	66.6
Change Period, (Y+R_c), s	5.0	5.0	5.0	5.0	5.0	5.0	5.0	5.0
Max Allow Headway (MAH), s	3.1	3.2	3.1	3.2	3.1	0.0	3.1	0.0
Queue Clearance Time (g_s), s	11.2	16.1	4.7	17.6	6.1		13.3	
Green Extension Time (g_e), s	0.1	1.2	0.0	1.1	0.1	0.0	0.4	0.0
Phase Call Probability	0.99	1.00	0.78	1.00	0.97		1.00	
Max Out Probability	0.31	0.00	0.00	0.01	0.00		0.00	

Movement Group Results	EB			WB			NB			SB		
Approach Movement	L	T	R	L	T	R	L	T	R	L	T	R
Assigned Movement	7	4	14	3	8	18	5	2	12	1	6	16
Adjusted Flow Rate (v), veh/h	155	215	89	45	135	189	103	557	546	257	1201	130
Adjusted Saturation Flow Rate (s), veh/h/ln	1558	1636	1386	1558	1636	1386	1558	1636	1605	1558	1557	1386
Queue Service Time (g_s), s	9.2	14.1	6.4	2.7	8.8	15.6	4.1	34.5	34.5	11.3	35.3	5.0
Cycle Queue Clearance Time (g_c), s	9.2	14.1	6.4	2.7	8.8	15.6	4.1	34.5	34.5	11.3	35.3	5.0
Green Ratio (g/C)	0.30	0.22	0.22	0.23	0.17	0.17	0.51	0.46	0.46	0.59	0.52	0.52
Capacity (c), veh/h	349	363	307	247	292	248	220	725	712	320	1562	695
Volume-to-Capacity Ratio (X)	0.443	0.592	0.291	0.183	0.461	0.765	0.469	0.768	0.768	0.802	0.769	0.187
Back of Queue (Q), ft/ln (50 th percentile)	91.8	150.5	0.4	27.4	95	146.3	38.2	384.4	378.5	137.3	323.1	41.9
Back of Queue (Q), veh/ln (50 th percentile)	3.5	5.7	0.0	1.0	3.6	5.5	1.4	14.6	14.3	5.2	12.2	1.6
Queue Storage Ratio (RQ) (50 th percentile)	0.46	0.03	0.00	0.14	0.47	0.73	0.19	0.13	0.13	0.69	0.13	0.42
Uniform Delay (d_1), s/veh	32.7	42.3	38.8	35.9	44.5	46.9	24.4	28.6	28.2	31.9	20.5	12.9
Incremental Delay (d_2), s/veh	0.3	0.6	0.2	0.1	0.4	3.1	0.6	7.6	7.8	6.2	3.7	0.6
Initial Queue Delay (d_3), s/veh	0.0	0.0	0.0	0.0	0.0	0.0	0.0	0.0	0.0	0.0	0.0	0.0
Control Delay (d), s/veh	33.1	42.8	39.0	36.1	45.0	50.0	24.9	36.2	36.0	38.2	24.2	13.4
Level of Service (LOS)	C	D	D	D	D	D	C	D	D	D	C	B
Approach Delay, s/veh / LOS	38.8	D		46.5	D		35.1	D		25.6	C	
Intersection Delay, s/veh / LOS	32.6						C					

Figure 10.8 *HCS7 Signal2 Future*

General Information				Intersection Information	
Agency				Duration, h	0.25
Analyst		Analysis Date	Mar 29, 2011	Area Type	Other
Jurisdiction		Time Period		PHF	0.90
Urban Street	Waldo	Analysis Year	2011	Analysis Period	1> 7:00
Intersection	12th	File Name	WaldoFuturePM.xus		
Project Description	Future				

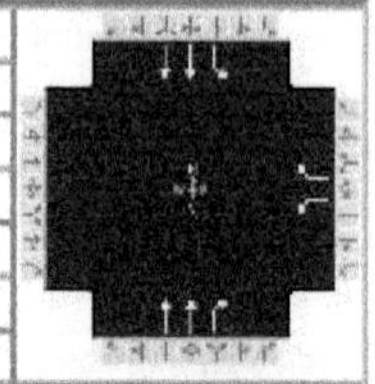

Demand Information	EB			WB			NB			SB		
Approach Movement	L	T	R	L	T	R	L	T	R	L	T	R
Demand (*v*), veh/h				250		279		1349	256	291	1311	

Signal Information										
Cycle, s	120.0	Reference Phase	2							
Offset, s	0	Reference Point	Begin	Green	11.3	64.8	28.9	0.0	0.0	0.0
Uncoordinated	No	Simult. Gap E/W	On	Yellow	4.0	4.0	4.0	0.0	0.0	0.0
Force Mode	Fixed	Simult. Gap N/S	On	Red	1.0	1.0	1.0	0.0	0.0	0.0

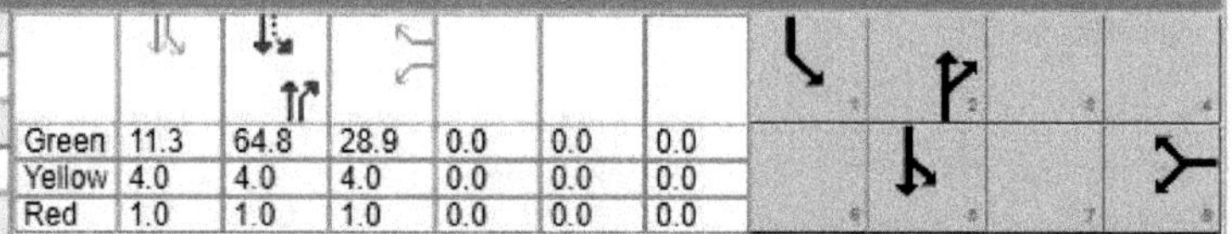

Timer Results	EBL	EBT	WBL	WBT	NBL	NBT	SBL	SBT
Assigned Phase				8		2	1	6
Case Number				9.0		7.3	1.0	4.0
Phase Duration, s				33.9		69.8	16.3	86.1
Change Period, ($Y+R_c$), s				5.0		5.0	5.0	5.0
Max Allow Headway (*MAH*), s				3.3		0.0	3.1	0.0
Queue Clearance Time (g_s), s				28.2			11.3	
Green Extension Time (g_e), s				0.5		0.0	0.1	0.0
Phase Call Probability				1.00			1.00	
Max Out Probability				1.00			1.00	

Movement Group Results	EB			WB			NB			SB		
Approach Movement	L	T	R	L	T	R	L	T	R	L	T	R
Assigned Movement				3		18		2	12	1	6	
Adjusted Flow Rate (*v*), veh/h				278		310		1168	222	291	1310	
Adjusted Saturation Flow Rate (*s*), veh/h/ln				1558		1386		1557	1386	1558	1557	
Queue Service Time (g_s), s				19.7		26.2		25.0	2.7	9.3	23.3	
Cycle Queue Clearance Time (g_c), s				19.7		26.2		25.0	2.7	9.3	23.3	
Green Ratio (*g/C*)				0.24		0.24		0.54	0.54	0.65	0.68	
Capacity (*c*), veh/h				377		336		1677	746	314	2100	
Volume-to-Capacity Ratio (*X*)				0.736		0.923		0.696	0.297	0.925	0.624	
Back of Queue (*Q*), ft/ln (50 th percentile)				213		300.3		169.4	22.3	192.7	163	
Back of Queue (*Q*), veh/ln (50 th percentile)				8.1		11.4		6.4	0.8	7.3	6.2	
Queue Storage Ratio (*RQ*) (50 th percentile)				2.84		2.00		0.07	0.01	0.96	0.07	
Uniform Delay (d_1), s/veh				41.9		44.4		11.0	3.0	22.0	7.6	
Incremental Delay (d_2), s/veh				5.5		27.2		2.4	1.0	30.3	1.4	
Initial Queue Delay (d_3), s/veh				0.0		0.0		0.0	0.0	0.0	0.0	
Control Delay (*d*), s/veh				47.4		71.6		13.5	4.0	52.4	9.0	
Level of Service (LOS)				D		E		B	A	D	A	
Approach Delay, s/veh / LOS	0.0			60.2	E		12.0	B		16.9	B	
Intersection Delay, s/veh / LOS	22.1						C					

Figure 10.9 *HCS7 Signal3 Future*

General Information				Intersection Information	
Agency				Duration, h	0.25
Analyst		Analysis Date	Mar 29, 2011	Area Type	Other
Jurisdiction		Time Period		PHF	0.90
Urban Street	Waldo	Analysis Year	2011	Analysis Period	1> 7:00
Intersection	16th	File Name	WaldoFuturePM.xus		
Project Description	Future				

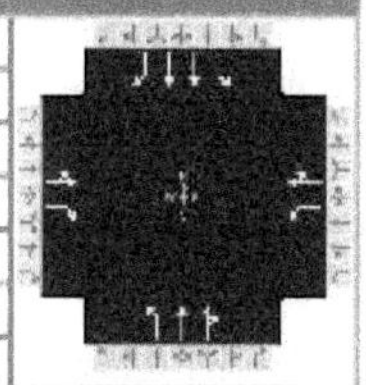

Demand Information	EB			WB			NB			SB		
Approach Movement	L	T	R	L	T	R	L	T	R	L	T	R
Demand (*v*), veh/h	74	10	315	104	79	54	344	1199	11	9	1022	58

Signal Information										
Cycle, s	120.0	Reference Phase	2							
Offset, s	0	Reference Point	Begin	Green	2.6	9.8	56.6	31.0	0.0	0.0
Uncoordinated	No	Simult. Gap E/W	On	Yellow	4.0	4.0	4.0	4.0	0.0	0.0
Force Mode	Fixed	Simult. Gap N/S	On	Red	1.0	1.0	1.0	1.0	0.0	0.0

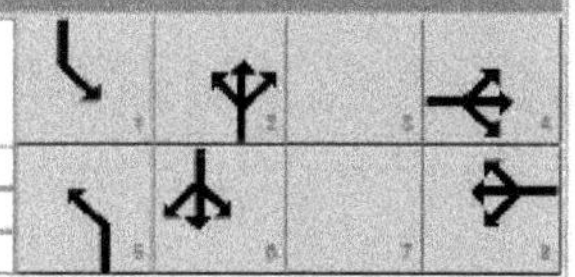

Timer Results	EBL	EBT	WBL	WBT	NBL	NBT	SBL	SBT
Assigned Phase		4		8	5	2	1	6
Case Number		7.0		6.0	1.1	4.0	1.1	3.0
Phase Duration, s		36.0		36.0	22.4	76.4	7.6	61.6
Change Period, (*Y+R* $_c$), s		5.0		5.0	5.0	5.0	5.0	5.0
Max Allow Headway (*MAH*), s		3.5		3.5	3.1	0.0	3.1	0.0
Queue Clearance Time (g_s), s		32.1		30.0	16.8		2.4	
Green Extension Time (g_e), s		0.0		0.3	0.2	0.0	0.0	0.0
Phase Call Probability		1.00		1.00	1.00		0.28	
Max Out Probability		1.00		1.00	0.84		0.00	

Movement Group Results	EB			WB			NB			SB		
Approach Movement	L	T	R	L	T	R	L	T	R	L	T	R
Assigned Movement	7	4	14	3	8	18	5	2	12	1	6	16
Adjusted Flow Rate (*v*), veh/h		93	350	116	148		327	576	574	10	1136	64
Adjusted Saturation Flow Rate (*s*), veh/h/ln		869	1386	969	1524		1558	1636	1630	1558	1557	1386
Queue Service Time (g_s), s		7.8	30.1	14.0	9.6		14.8	30.3	30.2	0.4	36.6	3.1
Cycle Queue Clearance Time (g_c), s		17.0	30.1	28.0	9.6		14.8	30.3	30.2	0.4	36.6	3.1
Green Ratio (*g/C*)		0.26	0.26	0.26	0.26		0.63	0.60	0.60	0.49	0.47	0.47
Capacity (*c*), veh/h		281	358	197	394		361	958	955	234	1461	650
Volume-to-Capacity Ratio (*X*)		0.332	0.977	0.587	0.375		0.906	0.602	0.602	0.043	0.777	0.099
Back of Queue (*Q*), ft/ln (50 th percentile)		64.5	375.3	92.8	93.9		330.7	325.1	322.7	3.6	366.5	27.5
Back of Queue (*Q*), veh/ln (50 th percentile)		2.4	14.2	3.5	3.6		12.5	12.3	12.2	0.1	13.9	1.0
Queue Storage Ratio (*RQ*) (50 th percentile)		0.01	3.75	1.24	0.63		1.32	0.13	0.13	0.02	0.15	0.18
Uniform Delay (d_1), s/veh		42.9	44.2	50.2	36.5		32.3	20.0	19.9	17.5	26.6	17.7
Incremental Delay (d_2), s/veh		0.3	41.3	3.1	0.2		21.8	2.8	2.8	0.0	4.1	0.3
Initial Queue Delay (d_3), s/veh		0.0	0.0	0.0	0.0		0.0	0.0	0.0	0.0	0.0	0.0
Control Delay (*d*), s/veh		43.1	85.4	53.3	36.8		54.1	22.8	22.7	17.5	30.8	18.0
Level of Service (LOS)		D	F	D	D		D	C	C	B	C	B
Approach Delay, s/veh / LOS	76.5		E	44.0		D	29.7		C	30.0		C
Intersection Delay, s/veh / LOS			37.0						D			

Figure 10.10 *HCS7 Streets1 Future*

General Information				Streets Information	
Agency				Number of Intersections	3
Analyst		Analysis Date	Mar 29, 2011	Number of Segments	2
Jurisdiction		Time Period		Number of Iterations	3
File Name	WaldoFuturePM.xus	Analysis Year	2011	System Cycle Length, s	120
Intersections	8th	12th		Analysis Period	1> 7:00
Project Description	Future				

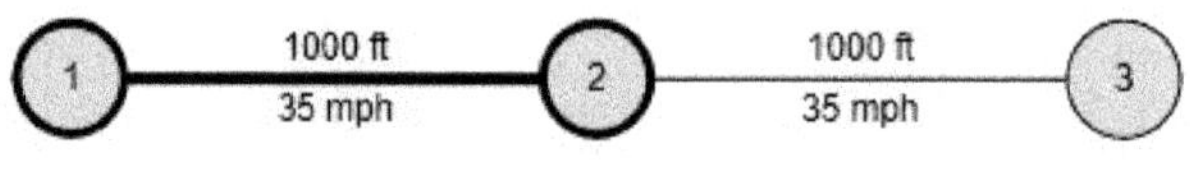

Basic Segment Information

Segment	Speed Limit		Through Lanes		Segment Length		Intersection Wid		Length of RM		Percent Curb		Other Delay	
	NB	SB	NB	SB	NB	SB	NB	SB	NB	SB	NB	SB	NB	SB
1	35	35	2	2	1000	1000	50	50	0	0	70	70	0.0	0.0

		Northbound			Southbound		
Segment Output Data		NBL	NBT	NBR	SBL	SBT	SBR
Segment	Movement	5	2	12	1	6	16
1	Bay/Lane Spillback Time, h		never		never	never	
1	Shared Lane Spillback Time, h	never			never		never
1	Base Free-Flow Speed, mph		41.72			41.72	
1	Running Time, s		20.24			20.39	
1	Running Speed, mph		33.69			33.44	
1	Through Delay, s/veh		13.46			24.24	
1	Travel Time, s		33.70			44.63	
1	Travel Speed, mph		20.23			15.28	
1	Stop Rate, stops/veh		0.33			0.61	
1	Spatial Stop Rate, stops/mi		1.74			3.23	
1	Through vol/cap Ratio		0.70			0.77	
1	Percent of Base FFS		48.50			36.62	
1	Level of Service		D			E	
1	Auto Traveler Perception Score		2.63			2.65	

Figure 10.11 *HCS7 Streets2 Future*

General Information				Streets Information	
Agency				Number of Intersections	3
Analyst		Analysis Date	Mar 29, 2011	Number of Segments	2
Jurisdiction		Time Period		Number of Iterations	3
File Name	StreetsFuture.xus	Analysis Year	2011	System Cycle Length, s	120
Intersections	12th	16th		Analysis Period	1> 7:00
Project Description	Future				

Basic Segment Information															
Segment	Speed Limit		Through Lanes		Segment Length		Intersection Wid		Length of RM		Percent Curb		Other Delay		
	NB	SB	NB	SB	NB	SB	NB	SB	NB	SB	NB	SB	NB	SB	
2	35	35	2	2	1000	1000	50	50	0	0	70	70	0.0	0.0	

		Northbound			Southbound		
Segment Output Data		NBL	NBT	NBR	SBL	SBT	SBR
Segment	Movement		2	12	1	6	
2	Bay/Lane Spillback Time, h		never		3.24	never	
2	Shared Lane Spillback Time, h	never			5.82		never
2	Base Free-Flow Speed, mph		41.72			41.72	
2	Running Time, s		20.30			20.40	
2	Running Speed, mph		33.58			33.43	
2	Through Delay, s/veh		22.75			9.02	
2	Travel Time, s		43.05			29.42	
2	Travel Speed, mph		15.84			23.17	
2	Stop Rate, stops/veh		0.64			0.28	
2	Spatial Stop Rate, stops/mi		3.38			1.49	
2	Through vol/cap Ratio		0.60			0.62	
2	Percent of Base FFS		37.96			55.55	
2	Level of Service		E			C	
2	Auto Traveler Perception Score		2.68			2.37	

Figure 10.12 *HCS7 TWSC Future*

General Information		Site Information	
Analyst		Intersection	12th/Supercenter
Agency/Co.		Jurisdiction	
Date Performed	4/10/2011	East/West Street	
Analysis Year	2017	North/South Street	
Time Analyzed		Peak Hour Factor	0.95
Intersection Orientation	East-West	Analysis Time Period (hrs)	0.25
Project Description	Furture		

Lanes

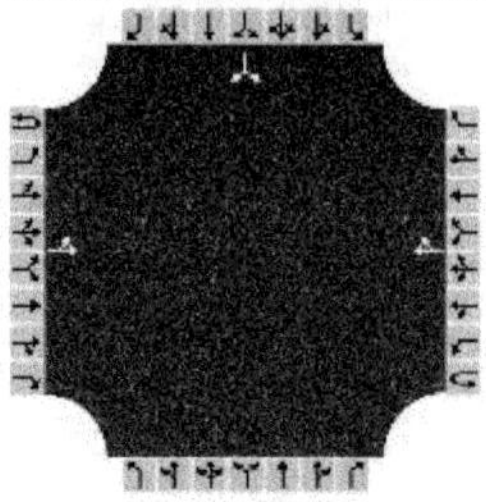

Vehicle Volumes and Adjustments																
Approach	Eastbound				Westbound				Northbound				Southbound			
Movement	U	L	T	R	U	L	T	R	U	L	T	R	U	L	T	R
Priority	1U	1	2	3	4U	4	5	6		7	8	9		10	11	12
Number of Lanes	0	0	1	0	0	0	1	0		0	0	0		0	1	0
Configuration		LT						TR							LR	
Volume, V (veh/h)		451	97				66	44						46		464
Percent Heavy Vehicles (%)		7												7		7
Proportion Time Blocked																
Percent Grade (%)													0			
Right Turn Channelized	No				No				No				No			
Median Type/Storage	Undivided															
Critical and Follow-up Headways																
Base Critical Headway (sec)																
Critical Headway (sec)																
Base Follow-Up Headway (sec)																
Follow-Up Headway (sec)																
Delay, Queue Length, and Level of Service																
Flow Rate, v (veh/h)		475													536	
Capacity, c (veh/h)		1420													634	
v/c Ratio		0.33													0.85	
95% Queue Length, Q_{95} (veh)		1.5													9.3	
Control Delay (s/veh)		8.8													34.0	
Level of Service, LOS		A													D	
Approach Delay (s/veh)	7.8												34.0			

The internal intersection (12th at Supercenter) is modeled as AWSC and a roundabout to compare results with the TWSC analysis. While AWSC generated a lower overall delay, the roundabout solution was even lower—comparing both average delay in seconds per vehicle and vehicle-hours of delay accounting for demands. However, with TWSC at LOS D to satisfy the requirement, building a roundabout would not be cost effective, leaving all criteria satisfied with construction limited to lengthening one turn lane.

Figure 10.13 *HCS7 TWSC Mitigated*

General Information		Site Information	
Analyst		Intersection	12th/Supercenter
Agency/Co.		Jurisdiction	
Date Performed		East/West Street	
Analysis Year	2017	North/South Street	
Time Analyzed		Peak Hour Factor	0.95
Intersection Orientation	East-West	Analysis Time Period (hrs)	0.25
Project Description	Mitigated		

Lanes

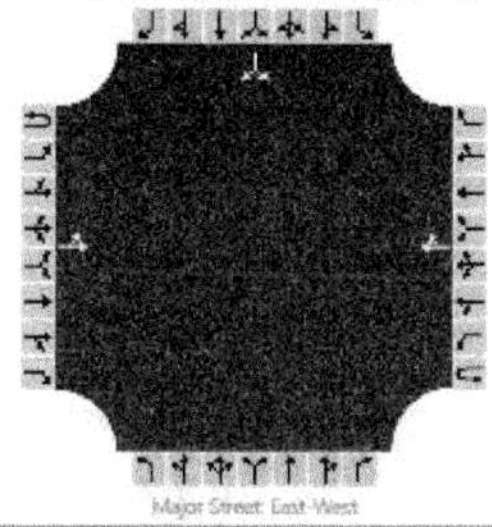

Vehicle Volumes and Adjustments																
Approach	Eastbound				Westbound				Northbound				Southbound			
Movement	U	L	T	R	U	L	T	R	U	L	T	R	U	L	T	R
Priority	1U	1	2	3	4U	4	5	6		7	8	9		10	11	12
Number of Lanes	0	0	1	0	0	0	1	0		0	0	0		0	1	0
Configuration		LT						TR							LR	
Volume, V (veh/h)		451	97				66	44						46		464
Percent Heavy Vehicles (%)		7												7		7
Proportion Time Blocked																
Percent Grade (%)													0			
Right Turn Channelized	No				No				No				No			
Median Type/Storage	Undivided															
Critical and Follow-up Headways																
Base Critical Headway (sec)																
Critical Headway (sec)																
Base Follow-Up Headway (sec)																
Follow-Up Headway (sec)																
Delay, Queue Length, and Level of Service																
Flow Rate, v (veh/h)		475													536	
Capacity, c (veh/h)		1420													634	
v/c Ratio		0.33													0.85	
95% Queue Length, Q_{95} (veh)		1.5													9.3	
Control Delay (s/veh)		8.8													34.0	
Level of Service, LOS		A													D	
Approach Delay (s/veh)	7.8												34.0			
Approach LOS													D			

Figure 10.14 *HCS7 AWSC Mitigated*

General Information		Site Information	
Analyst		Intersection	
Agency/Co.		Jurisdiction	
Date Performed		East/West Street	12th
Analysis Year		North/South Street	Supercenter
Analysis Time Period (hrs)	0.25	Peak Hour Factor	0.95
Time Analyzed			
Project Description	Mitigated		

Lanes

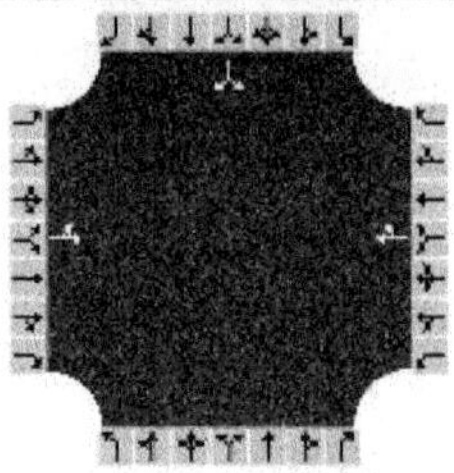

Vehicle Volume and Adjustments

Approach	Eastbound			Westbound			Northbound			Southbound		
Movement	L	T	R	L	T	R	L	T	R	L	T	R
Volume	325	204			209	44				46		0
% Thrus in Shared Lane												
Lane	L1	L2	L3	L1	L2	L3	L1	L2	L3	L1	L2	L3
Configuration	LT			TR						LR		
Flow Rate, v (veh/h)	557			266						48		
Percent Heavy Vehicles	7			7						7		
Departure Headway and Service Time												
Initial Departure Headway, hd (s)	3.20			3.20						3.20		
Initial Degree of Utilization, x	0.495			0.237						0.043		
Final Departure Headway, hd (s)	4.60			4.68						6.09		
Final Degree of Utilization, x	0.711			0.346						0.082		
Move-Up Time, m (s)	2.0			2.0						2.0		
Service Time, ts (s)	2.60			2.68						4.09		
Capacity, Delay and Level of Service												
Flow Rate, v (veh/h)	557			266						48		
Capacity	783			769						591		
95% Queue Length, Q_{95} (veh)	6.1			1.6						0.3		
Control Delay (s/veh)	18.1			10.1						9.6		
Level of Service, LOS	C			B						A		
Approach Delay (s/veh)	18.1			10.1						9.6		
Approach LOS	C			B						A		
Intersection Delay, s/veh | LOS	15.2						C					

Figure 10.15 *HCS7 Roundabout Mitigated*

General Information		Site Information	
Analyst		Intersection	Internal
Agency or Co.		E/W Street Name	12th
Date Performed	8/28/2012	N/S Street Name	Supercenter
Analysis Year	2017	Analysis Time Period (hrs)	0.25
Time Analyzed		Peak Hour Factor	0.95
Project Description	Mitigated	Jurisdiction	

Volume Adjustments and Site Characteristics

Approach	EB				WB				NB				SB			
Movement	U	L	T	R	U	L	T	R	U	L	T	R	U	L	T	R
Number of Lanes (N)	0	0	1	0	0	0	1	0	0	0	0	0	0	0	1	0
Lane Assignment			LT				TR								LR	
Volume (V), veh/h	0	325	204		0		209	44					0	46		338
Percent Heavy Vehicles, %	7	7	7		7		7	7					7	7		7
Flow Rate ($v_{i,PCE}$), pc/h	0	366	230		0		235	50					0	52		381
Right-Turn Bypass	None				None				None				None			
Conflicting Lanes	1				1								1			
Pedestrians Crossing, p/h	0				0								0			

Critical and Follow-Up Headway Adjustment

Approach	EB			WB			NB			SB		
Lane	Left	Right	Bypass	Left	Right	Bypass	Left	Right	Bypass	Left	Right	Bypass
Critical Headway (s)		4.9763			4.9763						4.9763	
Follow-Up Headway (s)		2.6087			2.6087						2.6087	

Flow Computations, Capacity and v/c Ratios

Approach	EB			WB			NB			SB		
Lane	Left	Right	Bypass	Left	Right	Bypass	Left	Right	Bypass	Left	Right	Bypass
Entry Flow (v_e), pc/h		596			285						433	
Entry Volume veh/h		557			266						405	
Circulating Flow (v_c), pc/h		52			366			648			235	
Exiting Flow (v_{ex}), pc/h		282			616			416			0	
Capacity (c_{pce}), pc/h		1309			950						1086	
Capacity (c), veh/h		1223			888						1015	
v/c Ratio (x)		0.46			0.30						0.40	

Delay and Level of Service

Approach	EB			WB			NB			SB		
Lane	Left	Right	Bypass	Left	Right	Bypass	Left	Right	Bypass	Left	Right	Bypass
Lane Control Delay (d), s/veh		7.7			7.3						7.9	
Lane LOS		A			A						A	
95% Queue, veh		2.4			1.3						1.9	
Approach Delay, s/veh		7.7			7.3						7.9	
Approach LOS		A			A						A	
Intersection Delay, s/veh | LOS		7.6						A				

Figure 10.16 *HCS7 Signal1 Mitigated*

General Information				Intersection Information	
Agency				Duration, h	0.25
Analyst		Analysis Date	Mar 29, 2011	Area Type	Other
Jurisdiction		Time Period		PHF	0.95
Urban Street	Waldo	Analysis Year	2011	Analysis Period	1> 7:00
Intersection	8th	File Name	StreetsMitigated.xus		
Project Description	Future				

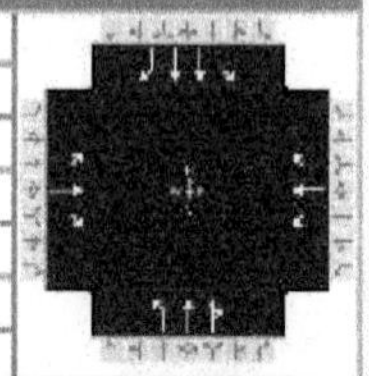

Demand Information	EB			WB			NB			SB		
Approach Movement	L	T	R	L	T	R	L	T	R	L	T	R
Demand (v), veh/h	147	204	85	43	128	180	98	993	55	219	1025	111

Signal Information										
Cycle, s	107.0	Reference Phase	2							
Offset, s	0	Reference Point	Begin							
Uncoordinated	No	Simult. Gap E/W	On	Green	7.0	0.3	46.2	5.2	0.4	18.0
Force Mode	Fixed	Simult. Gap N/S	On	Yellow	4.0	4.0	4.0	4.0	4.0	4.0
				Red	1.0	1.0	1.0	1.0	1.0	1.0

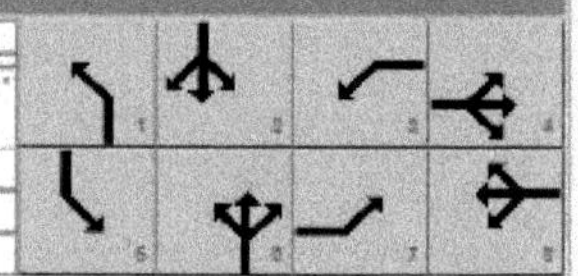

Timer Results	EBL	EBT	WBL	WBT	NBL	NBT	SBL	SBT
Assigned Phase	7	4	3	8	1	6	5	2
Case Number	1.1	3.0	1.1	3.0	1.1	4.0	1.1	3.0
Phase Duration, s	15.5	28.3	10.2	23.0	12.0	51.2	17.3	56.5
Change Period, ($Y+R_c$), s	5.0	5.0	5.0	5.0	5.0	5.0	5.0	5.0
Max Allow Headway (MAH), s	3.1	3.2	3.1	3.2	3.1	0.0	3.1	0.0
Queue Clearance Time (g_s), s	10.1	14.5	4.4	15.8	5.8		11.0	
Green Extension Time (g_e), s	0.1	1.2	0.0	1.2	0.2	0.0	0.4	0.0
Phase Call Probability	0.99	1.00	0.74	1.00	0.95		1.00	
Max Out Probability	0.08	0.00	0.00	0.00	0.00		0.00	

Movement Group Results	EB			WB			NB			SB		
Approach Movement	L	T	R	L	T	R	L	T	R	L	T	R
Assigned Movement	7	4	14	3	8	18	1	6	16	5	2	12
Adjusted Flow Rate (v), veh/h	155	215	89	45	135	189	103	557	546	246	1149	124
Adjusted Saturation Flow Rate (s), veh/h/ln	1558	1636	1386	1558	1636	1386	1558	1636	1605	1558	1557	1386
Queue Service Time (g_s), s	8.1	12.5	5.7	2.4	7.8	13.8	3.8	32.1	32.1	9.0	34.9	7.0
Cycle Queue Clearance Time (g_c), s	8.1	12.5	5.7	2.4	7.8	13.8	3.8	32.1	32.1	9.0	34.9	7.0
Green Ratio (g/C)	0.31	0.23	0.23	0.24	0.18	0.18	0.49	0.44	0.44	0.56	0.49	0.49
Capacity (c), veh/h	379	373	316	278	306	259	230	685	672	308	1447	644
Volume-to-Capacity Ratio (X)	0.408	0.576	0.283	0.163	0.441	0.731	0.449	0.813	0.813	0.796	0.794	0.193
Back of Queue (Q), ft/ln (50 th percentile)	78.5	131	50.5	23.4	82.4	124.8	34.5	366.7	360.8	80.7	358.4	82.6
Back of Queue (Q), veh/ln (50 th percentile)	3.0	5.0	1.9	0.9	3.1	4.7	1.3	13.9	13.7	3.1	13.6	3.1
Queue Storage Ratio (RQ) (50 th percentile)	0.39	0.03	0.50	0.12	0.41	0.62	0.17	0.12	0.12	0.40	0.24	0.83
Uniform Delay (d_1), s/veh	27.9	37.1	34.1	30.8	39.0	41.0	22.4	27.8	27.4	21.9	28.4	21.8
Incremental Delay (d_2), s/veh	0.3	0.5	0.2	0.1	0.4	1.5	0.5	10.2	10.4	1.9	3.4	0.5
Initial Queue Delay (d_3), s/veh	0.0	0.0	0.0	0.0	0.0	0.0	0.0	0.0	0.0	0.0	0.0	0.0
Control Delay (d), s/veh	28.2	37.7	34.3	30.9	39.3	42.5	23.0	38.0	37.8	23.8	31.8	22.3
Level of Service (LOS)	C	D	C	C	D	D	C	D	D	C	C	C
Approach Delay, s/veh / LOS	33.8	C		39.9	D		36.6	D		29.7	C	
Intersection Delay, s/veh / LOS	33.6						C					

Figure 10.17 *HCS7 Signal2 Mitigated*

General Information				Intersection Information	
Agency				Duration, h	0.25
Analyst		Analysis Date	Mar 29, 2011	Area Type	Other
Jurisdiction		Time Period		PHF	0.90
Urban Street	Waldo	Analysis Year	2011	Analysis Period	1> 7:00
Intersection	12th	File Name	StreetsMitigated.xus		
Project Description	Future				

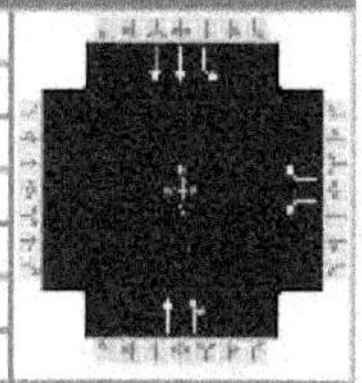

Demand Information	EB			WB			NB			SB		
Approach Movement	L	T	R	L	T	R	L	T	R	L	T	R
Demand (*v*), veh/h				250		279		1349	256	291	1311	

Signal Information										
Cycle, s	107.0	Reference Phase	2							
Offset, s	50	Reference Point	Begin	Green	15.9	48.8	27.3	0.0	0.0	0.0
Uncoordinated	No	Simult. Gap E/W	On	Yellow	4.0	4.0	4.0	0.0	0.0	0.0
Force Mode	Fixed	Simult. Gap N/S	On	Red	1.0	1.0	1.0	0.0	0.0	0.0

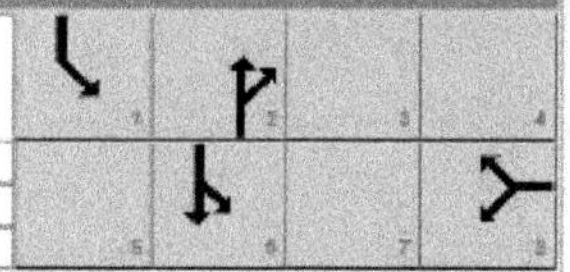

Timer Results	EBL	EBT	WBL	WBT	NBL	NBT	SBL	SBT
Assigned Phase				8		2	1	6
Case Number				9.0		8.3	1.0	4.0
Phase Duration, s				32.3		53.8	20.9	74.7
Change Period, (*Y+R* $_c$), s				5.0		5.0	5.0	5.0
Max Allow Headway (*MAH*), s				3.3		0.0	3.1	0.0
Queue Clearance Time (g_s), s				24.5			15.7	
Green Extension Time (g_e), s				1.2		0.0	0.5	0.0
Phase Call Probability				1.00			1.00	
Max Out Probability				0.00			0.00	

Movement Group Results	EB			WB			NB			SB		
Approach Movement	L	T	R	L	T	R	L	T	R	L	T	R
Assigned Movement				3		18		2	12	1	6	
Adjusted Flow Rate (*v*), veh/h				278		310		711	678	276	1241	
Adjusted Saturation Flow Rate (*s*), veh/h/ln				1558		1386		1636	1545	1558	1557	
Queue Service Time (g_s), s				17.0		22.5		94.5	46.6	13.7	20.1	
Cycle Queue Clearance Time (g_c), s				17.0		22.5		94.5	46.6	13.7	20.1	
Green Ratio (*g/C*)				0.25		0.25		0.46	0.46	0.61	0.65	
Capacity (*c*), veh/h				419		373		726	686	296	1985	
Volume-to-Capacity Ratio (*X*)				0.662		0.831		0.979	0.989	0.931	0.625	
Back of Queue (*Q*), ft/ln (50 th percentile)				168.1		204.8		550.6	542.6	181.4	127.1	
Back of Queue (*Q*), veh/ln (50 th percentile)				6.4		7.8		20.9	20.6	6.9	4.8	
Queue Storage Ratio (*RQ*) (50 th percentile)				0.67		0.82		0.37	0.36	0.91	0.13	
Uniform Delay (d_1), s/veh				34.8		36.8		28.6	28.9	27.5	7.4	
Incremental Delay (d_2), s/veh				0.7		3.2		20.7	23.5	10.4	1.0	
Initial Queue Delay (d_3), s/veh				0.0		0.0		0.0	0.0	0.0	0.0	
Control Delay (*d*), s/veh				35.4		40.0		49.3	52.5	38.0	8.3	
Level of Service (LOS)				D		D		D	D	D	A	
Approach Delay, s/veh / LOS	0.0			37.8		D	50.8		D	13.7		B
Intersection Delay, s/veh / LOS			32.5						C			

Figure 10.18 *HCS7 Signal3 Mitigated*

General Information				Intersection Information	
Agency				Duration, h	0.25
Analyst		Analysis Date	Mar 29, 2011	Area Type	Other
Jurisdiction		Time Period		PHF	0.95
Urban Street	Waldo	Analysis Year	2011	Analysis Period	1> 7:00
Intersection	16th	File Name	StreetsMitigated.xus		
Project Description	Future				

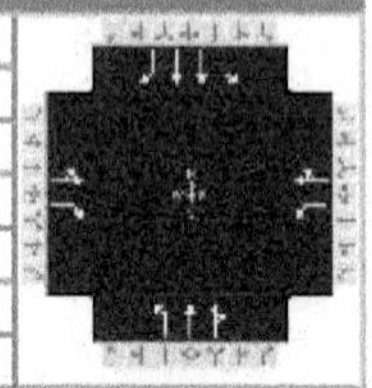

Demand Information	EB			WB			NB			SB		
Approach Movement	L	T	R	L	T	R	L	T	R	L	T	R
Demand (*v*), veh/h	74	10	315	104	79	54	344	1199	11	9	1022	58

Signal Information			
Cycle, s	107.0	Reference Phase	2
Offset, s	0	Reference Point	Begin
Uncoordinated	No	Simult. Gap E/W	On
Force Mode	Fixed	Simult. Gap N/S	On

Green	2.4	5.7	51.3	27.7	0.0	0.0
Yellow	4.0	4.0	4.0	4.0	0.0	0.0
Red	1.0	1.0	1.0	1.0	0.0	0.0

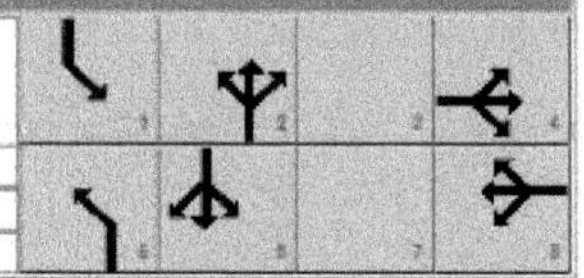

Timer Results	EBL	EBT	WBL	WBT	NBL	NBT	SBL	SBT
Assigned Phase		4		8	5	2	1	6
Case Number		7.0		6.0	1.1	4.0	1.1	3.0
Phase Duration, s		32.7		32.7	18.1	66.9	7.4	56.3
Change Period, (*Y+R* c), s		5.0		5.0	5.0	5.0	5.0	5.0
Max Allow Headway (*MAH*), s		3.5		3.5	3.1	0.0	3.1	0.0
Queue Clearance Time (*g* s), s		26.5		22.7	13.2		2.3	
Green Extension Time (*g* e), s		0.8		1.2	0.5	0.0	0.0	0.0
Phase Call Probability		1.00		1.00	1.00		0.25	
Max Out Probability		0.66		0.14	0.00		0.00	

Movement Group Results	EB			WB			NB			SB		
Approach Movement	L	T	R	L	T	R	L	T	R	L	T	R
Assigned Movement	7	4	14	3	8	18	5	2	12	1	6	16
Adjusted Flow Rate (*v*), veh/h		88	332	109	140		312	549	547	9	1076	61
Adjusted Saturation Flow Rate (*s*), veh/h/ln		964	1386	986	1524		1558	1636	1630	1558	1557	1386
Queue Service Time (*g* s), s		6.2	24.5	11.0	7.9		11.2	25.4	25.5	0.3	28.8	2.5
Cycle Queue Clearance Time (*g* c), s		12.7	24.5	20.7	7.9		11.2	25.4	25.5	0.3	28.8	2.5
Green Ratio (*g/C*)		0.27	0.27	0.27	0.27		0.64	0.59	0.59	0.52	0.49	0.49
Capacity (*c*), veh/h		324	375	245	413		342	939	936	266	1528	680
Volume-to-Capacity Ratio (*X*)		0.273	0.884	0.448	0.339		0.912	0.584	0.584	0.036	0.704	0.090
Back of Queue (*Q*), ft/ln (50 th percentile)		51.2	261.2	70	76.1		98	256.1	257.6	2.7	276.9	21.6
Back of Queue (*Q*), veh/ln (50 th percentile)		1.9	9.9	2.7	2.9		3.7	9.7	9.8	0.1	10.5	0.8
Queue Storage Ratio (*RQ*) (50 th percentile)		0.01	0.99	0.93	0.51		0.39	0.26	0.26	0.01	0.11	0.14
Uniform Delay (*d* 1), s/veh		35.4	37.4	40.5	31.3		17.4	18.1	18.0	14.3	21.5	14.5
Incremental Delay (*d* 2), s/veh		0.2	17.3	0.5	0.2		5.5	0.9	0.9	0.0	2.7	0.3
Initial Queue Delay (*d* 3), s/veh		0.0	0.0	0.0	0.0		0.0	0.0	0.0	0.0	0.0	0.0
Control Delay (*d*), s/veh		35.6	54.7	41.0	31.5		22.9	19.0	18.9	14.4	24.3	14.8
Level of Service (LOS)		D	D	D	C		C	B	B	B	C	B
Approach Delay, s/veh / LOS	50.7		D	35.7		D	19.8		B	23.7		C
Intersection Delay, s/veh / LOS	26.4						C					

Figure 10.19 *HCS7 Streets1 Mitigated*

General Information				Streets Information	
Agency				Number of Intersections	3
Analyst		Analysis Date	Mar 29, 2011	Number of Segments	2
Jurisdiction		Time Period		Number of Iterations	3
File Name	StreetsMitigated.xus	Analysis Year	2011	System Cycle Length, s	107
Intersections	8th	12th		Analysis Period	1> 7:00
Project Description	Future				

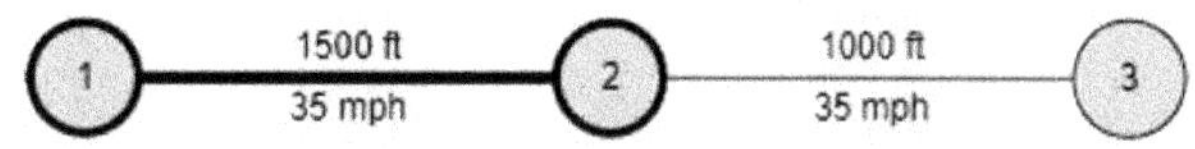

Basic Segment Information															
Segment	Speed Limit		Through Lanes		Segment Length		Intersection Wid		Length of RM		Percent Curb		Other Delay		
	NB	SB	NB	SB	NB	SB	NB	SB	NB	SB	NB	SB	NB	SB	
1	35	35	2	2	1500	1500	50	50	0	0	70	70	0.0	0.0	

		Northbound			Southbound		
Segment Output Data		NBL	NBT	NBR	SBL	SBT	SBR
Segment	Movement		2	12	1	6	
1	Bay/Lane Spillback Time, h	never	never	never			
1	Shared Lane Spillback Time, h	never		never			
1	Base Free-Flow Speed, mph	41.72			41.72		
1	Running Time, s	27.95			28.08		
1	Running Speed, mph	36.59			36.42		
1	Through Delay, s/veh	50.53			31.81		
1	Travel Time, s	78.48			59.89		
1	Travel Speed, mph	13.03			17.08		
1	Stop Rate, stops/veh	1.00			0.80		
1	Spatial Stop Rate, stops/mi	3.52			2.80		
1	Through vol/cap Ratio	0.98			0.79		
1	Percent of Base FFS	31.24			40.93		
1	Level of Service	D			D		
1	Auto Traveler Perception Score	2.94			2.58		

Figure 10.20 *HCS7 Streets2 Mitigated*

General Information				Streets Information	
Agency				Number of Intersections	3
Analyst		Analysis Date	Mar 29, 2011	Number of Segments	2
Jurisdiction		Time Period		Number of Iterations	3
File Name	StreetsMitigated.xus	Analysis Year	2011	System Cycle Length, s	107
Intersections	12th	16th		Analysis Period	1> 7:00
Project Description	Future				

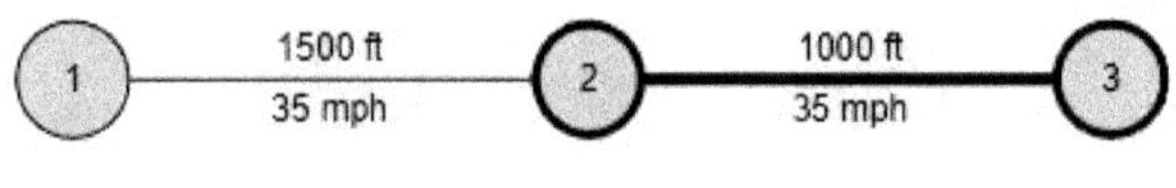

Basic Segment Information																
Segment	Speed Limit		Through Lanes		Segment Length		Intersection Wid		Length of RM		Percent Curb		Other Delay	
	NB	SB	NB	SB	NB	SB	NB	SB	NB	SB	NB	SB	NB	SB
2	35	35	2	2	1000	1000	50	50	0	0	70	70	0.0	0.0

		Northbound			Southbound		
Segment Output Data		NBL	NBT	NBR	SBL	SBT	SBR
Segment	Movement		2	12	1	6	
2	Bay/Lane Spillback Time, h		never		never	never	
2	Shared Lane Spillback Time, h	never			never		never
2	Base Free-Flow Speed, mph		41.72			41.72	
2	Running Time, s		20.25			20.33	
2	Running Speed, mph		33.66			33.53	
2	Through Delay, s/veh		18.94			8.34	
2	Travel Time, s		39.20			28.67	
2	Travel Speed, mph		17.39			23.78	
2	Stop Rate, stops/veh		0.60			0.26	
2	Spatial Stop Rate, stops/mi		3.15			1.38	
2	Through vol/cap Ratio		0.58			0.63	
2	Percent of Base FFS		41.69			56.99	
2	Level of Service		D			C	
2	Auto Traveler Perception Score		2.64			2.35	

Glossary

TERMS

95th Percentile Queue—Queue lengths for all yielding movements in the analysis period to the 95th percentile rate

Acceleration Lane Length—Distance from the gore to the end of the ramp taper connecting to the freeway

Access Point Density—Number of right-side access points per mile used to adjust free-flow speed

Access Points—Unsignalized side streets and driveways that can enter and exit the segment

Actuated—Signal phases with variable times that serve a minimum if vehicles are present and end when demand ends

Additional Lost Time—Green time with vehicles not moving due to downstream queues or demand starvation

Adjacent Ramp—Upstream or downstream ramp that affect the lanes 1 and 2 computation if within the equilibrium length

All-Way Stop Control—Intersections with stop control on all approaches

Alternative Intersections—Innovative designs that provide options for improving overall intersection results

Alternative Tools—Guidance for using simulation and other tools to generate HCM-compliant delay or density results

Analysis Period—The (typically) fifteen-minute period being analyzed expressed as 0.25 hours

Area Type—Reduction in saturation flow rate due to CBD or similar conditions that generate distraction and caution

Arterial Delay—Average delay weighted by demand of the major-street through lane groups for all included intersections

Auxiliary Lane—Additional freeway lane connection an on-ramp with an off-ramp to form a weaving segment

Average Travel Speed—Segment length divided by the average travel time during the analysis period

Back of Queue—Maximum length of queued vehicles during a typical cycle expressed as an average or a percentile.

Barrier—One or two managed lanes with access to and from the general purpose lanes controlled by a barrier

Base Free-Flow Speed—Starting value to compute free-flow speed with a default of 75.4 mi/h as 5 mi/h or 7 mi/h over the speed limit. Computed as a function of speed limit, access points, median type, curbs and parking.

Basic Freeway Segment—Section of freeway outside the demand influence of ramps

Bottleneck—One or more segments with flow rate exceeding capacity creating delay and queue affecting other segments

Break Point—Flow rate beyond which the free-flow speed is less than the travel speed as influenced by demand

Buffer—One or two Managed lanes with access to and from the general purpose lanes controlled by two solid lines

Capacity—Maximum number of vehicles passing a defined point in a specified time period that is usually fifteen minutes. Adjusted saturation flow rate times the green-to-cycle ratio computed in vehicles per hour by lane group.

Capacity Adjustment Factor—Adjusts capacity for weather events, driver familiarity, work zones and incidents

Capacity Checks—Segment flow rate compared with capacity as function of demand and density including ramps

Channelized Right—Exclusive right-turn lane with a triangular island and a stop or yield sign at the minor street

Circulating Flow—Traffic circulating within the roundabout becoming conflicting flow to entering traffic

Circulating Speed—Speed computed from the average radius of the circulating path through the roundabout

Climbing Lane—Passing lane on a specific grade that is designated for climbing for adjusting flow rates

Conflicting Flow—Sum of the flow rates of all movements for a given entry movement that must yield

Continuous Access—One managed lane with access to and from the general purpose lanes allowed continuously

Control Delay—Delay to yielding movements that includes time in queue as well as deceleration and acceleration delay. Sum of delay terms that includes slowed and stopped delay by lane group, approach and intersection.

Coordination—Signal timing that maximizes the proportion arriving on green to minimize downstream delay

Crawl Speed—Speeds for trucks on steep downhill grades forces to gear down to maintain control at lower speeds

Critical Headway—Minimum headway (front bumper to front bumper) time for a given movement to proceed

Cross Weave—On-ramp and/or off-ramp vehicles crossing freeway traffic to enter and/or exit the managed lanes

Crossover Intersections—Supplemental intersections that facilitate crossover or U-turn movements

d/c Ratio—Demand divided by capacity for the analysis conditions that define oversaturated conditions and LOS F

Dallas Phasing—Lead-lag with protected-permitted phasing with the leading left-turn permitted portion also lagging

Deceleration Lane Length—Distance from the end of the ramp taper connecting to the freeway to the gore

De-Facto Lanes—Share lanes that function as exclusive turn lanes due to unequal flow rates

Degree-of Conflict—Defined as a function of the possible scenarios for a given geometric design

Degree of Utilization—Lane flow rate times the departure headway divided by 3600 for use in the delay computation

Delay—Computed as an average in seconds per vehicle usually by lane or lane group for determining level of service

Demand—Number of vehicles per hour by movement not restricted by capacity as collected in the field

Demand Starvation—Signal timing preventing upstream flows from proceeding to the downstream green signal

Density—Computed from the flow rate and speed in passenger cars per mile per lane for determining level of service

Departure Headway—Average time between the departure of successive vehicles on the subject approach

Distributed Intersections—Expanding a conventional intersection to include supplemental adjacent intersections

Distribution—Process to assign generated trips to streets and intersections normally by using the area planning model

Downstream Queue—Queuing between the terminal intersection that generate additional lost time upstream

Driver Familiarity—Adjusts free-flow speed and capacity for the relative familiarity of the demand being analyzed

Effective Green Time—Green time less start-up and clearance lost times to represent the time vehicles are moving

Entering Flow—Traffic entering the roundabout from one- or two-lane approaches

Existing—Analysis of current traffic, lane, control and signal conditions within development impact scope area

Exiting Flow—Demand exiting the one- or two-lane roundabout as free movements

Facilities—Series of segments analyzed together for composite results that can be affected by system coordination

Flared Right—Geometry within which a right-turning vehicle can turn past a queue without a right-turn lane

Flow Balancing—Process to ensure flow arriving at the downstream intersection match those leaving upstream

Flow Profile—Model for the platoon from the upstream signal including platoon dispersion and access point effects

Flow Rate—Demand adjusting for peaking characteristics and the presence of heavy vehicles in passenger cars per hour

Follow-Up Headway—Time for second vehicle in queue to accept the same headway as the preceding vehicle

Free-Flow Speed—Average speed under low volume conditions usually controlled by geometry. Computed for both subsegments as a function of influence area and geometric conditions.

Freeway Facilities—Analysis of multiple freeway segments for the facility results and/or to model oversaturated conditions

Future—Analysis of current lane, control and signal conditions including growth and development generated traffic

Gap Acceptance—Vehicles yielding to other traffic must select appropriate gaps in those conflicting flows

Generation—Conversion of development parameters into estimated trips generated using the ITE Trip Generation Manual

Genetic Algorithm—Optimization process used in HCS to converge on the solution based on mutation and selection

Geometric Delay—Delay additional to control delay to account for slowing down and navigating the roundabout

Geometry Group—Definition of approach as a function of number of approaches and lanes available at each

Growth—Increase in background traffic to account for the growth rate that will add trips prior to the development opening

Heavy Vehicles—Proportion of heavy vehicles in the traffic stream divided between tractor trailers and single-unit trucks

Highway Capacity Manual—Transportation Research Board publication defining highway capacity analysis methods

Highway Capacity Software—University of Florida McTrans Center software application implementing HCM procedures

Highway Class—Classes I, II and III are defined as the function of the highway that dictates the service measure for each

Impedance—Conflicting flow that develops a queue generates this further reduction in capacity

Incident—Adjusts the capacity of open lanes for the effects of closing the shoulder, lane or lanes

Incremental Delay—Second term of the delay equation to estimate random arrival fluctuation and oversaturated periods

Influence Area—Distance of 1500 feet from the gore including the outer two freeway lanes for computing density

Initial Queue—Unmet demand from the previous time period by movement that generates initial queue delay

Initial Queue Delay—Third term of the delay equation estimating delay caused by unmet demand from the previous period

Inscribed Diameter—Diameter of the largest circle that can be inscribed across the roundabout

Interchange Ramp Terminals—Procedure to analyze the two signalized intersections controlling freeway ramps

Internal Capture—Percentage of trips that remain internal with both origin and destination within mixed-use development

Interrupted Flow—Procedures for Stop Control, Roundabouts, Signals, Streets, Interchanges and Alternative Intersections

Intersection Offsets—Time between signals referenced to a given phase to control timing among successive intersections

Jam Density—Maximum density in queuing condition dependent on vehicle spacing in essentially stop-and-go traffic

Lane Balancing—Check ensuring the outer freeway lanes don't exceed 2700/hour/lane after lanes 1 and 2 computed

Lane-Changing Rates—Flow rate multiplied by the lane changes for weaving and non-weaving vehicles

Lane Utilization—Distribution of traffic in multiple lanes serving the same movement that can be quite unequal

Lane Width—Average lane width across all lanes in the analysis direction if less than 12 feet

Lanes 1 and 2—Outer two freeway lanes closest to the on-ramp or off-ramp included in the influence area

Lead-Lag Phasing—Design with left-turn phases before through phases in one direction and after in the other direction

Level of Service—Determined from the service measure as defined by each methodology on a scale for A through F

Major Diverge—Freeway segment that diverges to two freeway segments each with multiple lanes

Major Merge—Two freeways each with multiple lanes merge to form one freeway segment

Managed Lanes—Including high-occupancy vehicle, high-occupancy toll or express toll lanes for five design types

Maximum Weaving Length—Distance for analysis conditions beyond which weaving is optional without friction

Median Type—Divided with a median or two-way left-turn lane used to adjust free-flow speed

Merge or Diverge Segment—Single-lane or two-lane on-ramp or off-ramp on the right or left side of the freeway

Mitigated—Strategy analysis to determine changes and improvement to lanes, control and signals to offset added traffic

Mixed-Flow Model—Required for segments with steep grades and/or high truck proportions redefining speed-flow curves

Multilane Highway—Four- or six-lane highway with signalized intersections more than two miles apart

Multiple-Period Analysis—Sequential analysis to facilitate unmet demand to build and dissipate over time

NEMA—Standard to which signal phases must adhere to be consistent in control for safety and consistency

Non-Yielding Bypass Lanes—Normally right-turn lanes that don't enter the roundabout and merge downstream

Number of Weaving Lanes—Total lanes accomplishing an entering or exiting movement with zero or one lane change

Objective Function—Focus of the optimization process defined typically as overall delay, arterial delay or level of service

Optimization—Iterative process to generate the optimum cycle, splits and offsets for a particular objective function. Process to converge on the optimum value of a specified objective function for a given set of conditions.

Overall Delay—Average delay weighted by demand of all lane groups, approaches and intersections in the analysis

Pass-By Trips—Trips that were already on the adjacent roadway that accessed the development with no generated traffic

Passenger-Car Equivalents—Conversion of heavy vehicles as a function of general terrain or specific grade parameters

Passing Lane—Region with a temporary additional lane to facilitate passing without conflicting traffic

Passing Zones—Portion of the segment where vehicles are permitted to pass slower vehicles in front of them

Peak Hour Factor—Adjusts the hourly volume to a peak 15-minute flow rate expressed in passenger cars per hour

Percent Time-Spent-Following—Average percent of time vehicles are in platoon following slower vehicles

Phase Duration—Effective green time plus start-up and clearance lost times to represent the phase portion of the cycle

Platoon Decay—Part of the flow profile model that accounts for platoon segregating due to access point activity

Potential Capacity—Capacity computed before upstream signal platooning and/or impedance is considered

Pretimed—Signal phases with preset times that do not vary as functions of demand by movement or approach

Proportion Arriving on Green—Proportion of the vehicles arriving downstream during the green

Proportion Time Blocked—Parameter from the urban streets analysis used to model upstream signal platooning effects

Protected-Permitted—Left-turn phases that split time between protected and permitted modes either leading or lagging

Regions—Four parts of the analysis segment before, with and after a passing lane including effective downstream length

Restrictive Median—Percent of the segment with median access controlled usually by a raised curb

Right-Side Lateral Clearance—Distance to the nearest obstruction impacting drivers if less than 6 feet

Roundabout—Intersections with yield-controlled entries into a circular pattern around a central island

Roundabout Segments—Segments between roundabouts including the through movement downstream

Running Time—Time along the segment between intersections considering lost time, proximity and access points

Saturation Flow Rate—Number of vehicles per hour per lane that can cross the stop bar on a continuous green

Saturation Headway—Average time between the departure of successive vehicles for a given case

Segment—Segments between signalized intersections including the through movement downstream

Service Measure—Performance measure used to determine level of service for each method, usually density or delay

Service Time—Average time spent to depart after becoming the first vehicle in queue

Short Length—Distance between the on-ramp and off-ramp within delineation if it exists or the gore if not

Signal Operation—Fully-actuated, semi-actuated, coordinated-actuated or pretimed modes of signal operations

Single-Unit Trucks—Heavy vehicles including buses, recreational vehicles and two- or three-axle trucks

Speed—Mean speed computed from the adjusted free-flow speed and flow rate according to the speed-flow curve

Speed Adjustment Factor—Adjusts free-flow speed for weather events, driver familiarity and work zones

Split Phasing—Typically side streets serving all movements on each approach independently one after the other

Subsegment—Two subsegments defined to compute free-flow speed and geometric delay for the segment

Supplemental Intersections—Adjacent intersections to manage certain movements separate from the main intersection

Sustained Spillback—Queue from the downstream signal that extends to the upstream signal not cyclical in nature

Through Delay—Control delay plus geometric delay for both subsegments for the downstream through movement

Time-Space Diagram—Graphic progression representation of signal timing and intersection distances for the travel speed

Timing Plan—Set of controller parameters typically for a given time of day to accommodate those particular conditions

Total Lateral Clearance—Sum of distances on both sides to the nearest obstruction impacting drivers if less than 6 feet

Total Ramp Density—Average number of ramps three miles upstream and downstream of analysis segment midpoint

Traffic Pressure—Saturation flow adjustment to account for headway variance due to driver aggression

Travel Speed—The sum of running time and through delay with the segment length converted to speed.

Truncated—Situation with the analysis segment ending before the end of the effective downstream length

Turn Radius—Saturation flow adjustment to model the effects of turning as a function of the turn radius

Two-Lane Highway—Two-lane highway operating as uninterrupted flow without traffic control

Two-Stage Gap Acceptance—Major-street refuge design for left-turns and through movements to move in two stages

Two-Way Stop Control—Intersections with minor-street stop controlled and major-street movements uncontrolled

Uniform Delay—First term of the delay equation using arrival and departure rates to estimate delay associated with cycle

Uninterrupted Flow—Procedures for Freeway Segments, Freeway Facilities, Multilane Highways and Two-Lane Highways

Unmet Demand—Demand not served in a particular time period by movement that spills to the subsequent period

Upstream Signal—Signalized intersection that creates platoons to increase headways for minor-street gap acceptance

Urban Street—Segments and facilities connecting signalized intersections with or without access points v/c Ratio—Volume-to-capacity ratio used as the primary parameter in computing control delay and 95th percentile queue

Volume—Demand in vehicles per hour under prevailing conditions normally from field data collection

Volume Ratio—Entering plus exiting weaving flow rates divided by the total segment flow rate

Warrants—Following the MUTCD to test for intersections meeting the threshold for signalizing unsignalized intersections

Weather Event—Adjusts free-flow speed and capacity for the effects of rain, snow, extreme cold or low visibility

Weaving Intensity Factor—Function of total lane changing rates and the short length used to compute the weaving speed

Weaving Segment—On-ramp followed by an off-ramp with a continuous auxiliary lane between the ramps

Work Zones—Adjusts speed and capacity as a function of barrier type and placement, speed limits, area type and lighting

Yielding Bypass Lanes—Normally right-turn lanes that don't enter the roundabout but yield to exiting flow

ACRONYMS

ATDM—Active Transportation Demand Management

ATS—Average Travel Speed

AWSC—All-Way Stop Control

CAF—Capacity Adjustment Factor

CBB—Central Business District

CRF—Capacity Reduction Factor

DDI—Diverging Diamond Interchange

DLT—Displaced Left Turn

EBL—Eastbound Left

EBR—Eastbound Right

EBT—Eastbound Through

EDTT—Extra Distance Travel Time

ETT—Experience Travel Time

FFS—Free-Flow Speed

GA—Genetic Algorithm

HCM—Highway Capacity Manual

HCQS—Highway Capacity and Quality of Service

HCS—Highway Capacity Software

ITE—Institute of Transportation Engineers

LCSI—Lane Closure Severity Index

LOS—Level of Service

ML—Managed Lanes

MUT—Median U-Turn

MUTCD—Manual on Uniform Traffic Control Device

NBL—Northbound Left

NBR—Northbound Right

NBT—Northbound Through

NEMA—National Electrical Manufacturer's Association

PCE—Passenger-Car Equivalents

PHF—Peak Hour Factor

PTB—Proportion Time Blocked

PTSF—Percent Time-Spent-Following

RCUT—Restricted Crossing U-Turn

SAF—Speed Adjustment Factor

SBL—Southbound Left

SBR—Southbound Right

SBT—Southbound Through

SUT—Single-Unit Trucks

TRB—Transportation Research Board

TRD—Total Ramp Density

TT—Tractor Trailers

TTI—Travel Time Index

TTR—Travel Time Reliability

TWLTL—Two-Way Left-Turn Lane

TWSC—Two-Way Stop Control

WBL—Westbound Left

WBR—Westbound Right

WBT—Westbound Through

Figure Credits

Design image: Copyright © Depositphotos/EpicStockMedia.
Fig. 0.1: Copyright © 2016 by the National Academy of Sciences. Reprinted with permission by the National Academies Press.
Fig. 0.2: Copyright © 2016 by University of Florida Board of Trustees. Reprinted with permission.
Fig. 1.1: Copyright © 2016 by the National Academy of Sciences. Reprinted with permission by the National Academies Press.
Fig. 1.1: Copyright © 2016 by the National Academy of Sciences. Reprinted with permission by the National Academies Press.
Fig. 1.2: Copyright © 2016 by the National Academy of Sciences. Reprinted with permission by the National Academies Press.
Fig. 1.3: Copyright © 2016 by the National Academy of Sciences. Reprinted with permission by the National Academies Press.
Fig. 1.4: Copyright © 2016 by the National Academy of Sciences. Reprinted with permission by the National Academies Press.
Fig. 1.5: Copyright © 2016 by the National Academy of Sciences. Reprinted with permission by the National Academies Press.
Fig. 1.6: Copyright © 2016 by the National Academy of Sciences. Reprinted with permission by the National Academies Press.
Fig. 1.7: Copyright © 2016 by the National Academy of Sciences. Reprinted with permission by the National Academies Press.
Fig. 1.8: Copyright © 2016 by the National Academy of Sciences. Reprinted with permission by the National Academies Press.
Fig. 1.9: Copyright © 2016 by the National Academy of Sciences. Reprinted with permission by the National Academies Press.
Fig. 1.10: Copyright © 2016 by the National Academy of Sciences. Reprinted with permission by the National Academies Press.
Fig. 1.11: Copyright © 2016 by the National Academy of Sciences. Reprinted with permission by the National Academies Press.
Fig. 1.12: Copyright © 2016 by the National Academy of Sciences. Reprinted with permission by the National Academies Press.
Fig. 1.13: Copyright © 2016 by the National Academy of Sciences. Reprinted with permission by the National Academies Press.
Fig. 1.14: Copyright © 2016 by the National Academy of Sciences. Reprinted with permission by the National Academies Press.
Fig. 1.15: Copyright © 2016 by the National Academy of Sciences. Reprinted with permission by the National Academies Press.
Fig. 1.16: Copyright © 2016 by the National Academy of Sciences. Reprinted with permission by the National Academies Press.
Fig. 1.17: Copyright © 2016 by the National Academy of Sciences. Reprinted with permission by the National Academies Press.
Fig. 1.18: Copyright © 2016 by the National Academy of Sciences. Reprinted with permission by the National Academies Press.
Fig. 1.19: Copyright © 2016 by the National Academy of Sciences. Reprinted with permission by the National Academies Press.
Fig. 1.20: Copyright © 2016 by the National Academy of Sciences. Reprinted with permission by the National Academies Press.
Fig. 1.21: Copyright © 2016 by the National Academy of Sciences. Reprinted with permission by the National Academies Press.
Fig. 1.22: Copyright © 2016 by the National Academy of Sciences. Reprinted with permission by the National Academies Press.
Fig. 1.23: Copyright © 2016 by the National Academy of Sciences. Reprinted with permission by the National Academies Press.
Fig. 1.24: Copyright © 2016 by the National Academy of Sciences. Reprinted with permission by the National Academies Press.
Fig. 1.25: Copyright © 2016 by the National Academy of Sciences. Reprinted with permission by the National Academies Press.
Fig. 1.26: Copyright © 2016 by the National Academy of Sciences. Reprinted with permission by the National Academies Press.
Fig. 1.27: Copyright © 2016 by the National Academy of Sciences. Reprinted with permission by the National Academies Press.
Fig. 1.28: Copyright © 2016 by the National Academy of Sciences. Reprinted with permission by the National Academies Press.
Fig. 1.29: Copyright © 2016 by the National Academy of Sciences. Reprinted with permission by the National Academies Press.
Fig. 1.30: Copyright © 2016 by the National Academy of Sciences. Reprinted with permission by the National Academies Press.
Fig. 1.31: Copyright © 2016 by the National Academy of Sciences. Reprinted with permission by the National Academies Press.
Fig. 1.32: Copyright © 2016 by the National Academy of Sciences. Reprinted with permission by the National Academies Press.
Fig. 1.33: Copyright © 2016 by the National Academy of Sciences. Reprinted with permission by the National Academies Press.
Fig. 1.34: Copyright © 2016 by the National Academy of Sciences. Reprinted with permission by the National Academies Press.
Fig. 1.35: Copyright © 2016 by the National Academy of Sciences. Reprinted with permission by the National Academies Press.
Fig. 1.36: Copyright © 2016 by the National Academy of Sciences. Reprinted with permission by the National Academies Press.
Fig. 1.37: Copyright © 2016 by University of Florida Board of Trustees. Reprinted with permission.

Fig. 1.38: Copyright © 2016 by University of Florida Board of Trustees. Reprinted with permission.
Fig. 1.39: Copyright © 2016 by University of Florida Board of Trustees. Reprinted with permission.
Fig. 1.40: Copyright © 2016 by University of Florida Board of Trustees. Reprinted with permission.
Fig. 1.41: Copyright © 2016 by the National Academy of Sciences. Reprinted with permission by the National Academies Press.
Fig. 1.42: Copyright © 2016 by the National Academy of Sciences. Reprinted with permission by the National Academies Press.
Fig. 1.43: Copyright © 2016 by the National Academy of Sciences. Reprinted with permission by the National Academies Press.
Fig. 1.44: Copyright © 2016 by the National Academy of Sciences. Reprinted with permission by the National Academies Press.
Fig. 1.45: Copyright © 2016 by the National Academy of Sciences. Reprinted with permission by the National Academies Press.
Fig. 1.46: Copyright © 2016 by the National Academy of Sciences. Reprinted with permission by the National Academies Press.
Fig. 1.47: Copyright © 2016 by the National Academy of Sciences. Reprinted with permission by the National Academies Press.
Fig. 1.48: Copyright © 2016 by the National Academy of Sciences. Reprinted with permission by the National Academies Press.
Fig. 1.49: Copyright © 2016 by the National Academy of Sciences. Reprinted with permission by the National Academies Press.
Fig. 1.50: Copyright © 2016 by the National Academy of Sciences. Reprinted with permission by the National Academies Press.
Fig. 1.51: Copyright © 2016 by the National Academy of Sciences. Reprinted with permission by the National Academies Press.
Fig. 1.52: Copyright © 2016 by the National Academy of Sciences. Reprinted with permission by the National Academies Press.
Fig. 1.53: Copyright © 2016 by the National Academy of Sciences. Reprinted with permission by the National Academies Press.
Fig. 1.54: Copyright © 2016 by the National Academy of Sciences. Reprinted with permission by the National Academies Press.
Fig. 1.55: Copyright © 2016 by the National Academy of Sciences. Reprinted with permission by the National Academies Press.
Fig. 1.56: Copyright © 2016 by the National Academy of Sciences. Reprinted with permission by the National Academies Press.
Fig. 1.57: Copyright © 2016 by the National Academy of Sciences. Reprinted with permission by the National Academies Press.
Fig. 1.58: Copyright © 2016 by the National Academy of Sciences. Reprinted with permission by the National Academies Press.
Fig. 1.59: Copyright © 2016 by the National Academy of Sciences. Reprinted with permission by the National Academies Press.
Fig. 1.60: Copyright © 2016 by the National Academy of Sciences. Reprinted with permission by the National Academies Press.
Fig. 1.61: Copyright © 2016 by the National Academy of Sciences. Reprinted with permission by the National Academies Press.
Fig. 1.62: Copyright © 2016 by the National Academy of Sciences. Reprinted with permission by the National Academies Press.
Fig. 1.63: Copyright © 2016 by the National Academy of Sciences. Reprinted with permission by the National Academies Press.
Fig. 1.64: Copyright © 2016 by the National Academy of Sciences. Reprinted with permission by the National Academies Press.
Fig. 1.65: Copyright © 2016 by the National Academy of Sciences. Reprinted with permission by the National Academies Press.
Fig. 1.66: Copyright © 2016 by the National Academy of Sciences. Reprinted with permission by the National Academies Press.
Fig. 1.67: Copyright © 2016 by the National Academy of Sciences. Reprinted with permission by the National Academies Press.
Fig. 1.68: Copyright © 2016 by the National Academy of Sciences. Reprinted with permission by the National Academies Press.
Fig. 1.69: Copyright © 2016 by University of Florida Board of Trustees. Reprinted with permission.
Fig. 1.70: Copyright © 2016 by University of Florida Board of Trustees. Reprinted with permission.
Fig. 1.71: Copyright © 2016 by University of Florida Board of Trustees. Reprinted with permission.
Fig. 1.72: Copyright © 2016 by University of Florida Board of Trustees. Reprinted with permission.
Fig. 1.73: Copyright © 2016 by University of Florida Board of Trustees. Reprinted with permission.
Fig. 1.74: Copyright © 2016 by University of Florida Board of Trustees. Reprinted with permission.
Fig. 2.1: Copyright © 2016 by the National Academy of Sciences. Reprinted with permission by the National Academies Press.
Fig. 2.2: Copyright © 2016 by the National Academy of Sciences. Reprinted with permission by the National Academies Press.
Fig. 2.3: Copyright © 2016 by the National Academy of Sciences. Reprinted with permission by the National Academies Press.
Fig. 2.4: Copyright © 2016 by the National Academy of Sciences. Reprinted with permission by the National Academies Press.
Fig. 2.5: Copyright © 2016 by the National Academy of Sciences. Reprinted with permission by the National Academies Press.
Fig. 2.6: Copyright © 2016 by the National Academy of Sciences. Reprinted with permission by the National Academies Press.
Fig. 2.7: Copyright © 2016 by the National Academy of Sciences. Reprinted with permission by the National Academies Press.
Fig. 2.8: Copyright © 2016 by the National Academy of Sciences. Reprinted with permission by the National Academies Press.

Fig. 2.9: Copyright © 2016 by the National Academy of Sciences. Reprinted with permission by the National Academies Press.
Fig. 2.10: Copyright © 2016 by the National Academy of Sciences. Reprinted with permission by the National Academies Press.
Fig. 2.11: Copyright © 2016 by the National Academy of Sciences. Reprinted with permission by the National Academies Press.
Fig. 2.12: Copyright © 2016 by the National Academy of Sciences. Reprinted with permission by the National Academies Press.
Fig. 2.13: Copyright © 2016 by the National Academy of Sciences. Reprinted with permission by the National Academies Press.
Fig. 2.14: Copyright © 2016 by the National Academy of Sciences. Reprinted with permission by the National Academies Press.
Fig. 2.15: Copyright © 2016 by the National Academy of Sciences. Reprinted with permission by the National Academies Press.
Fig. 2.16: Copyright © 2016 by the National Academy of Sciences. Reprinted with permission by the National Academies Press.
Fig. 2.17: Copyright © 2016 by the National Academy of Sciences. Reprinted with permission by the National Academies Press.
Fig. 2.18: Copyright © 2016 by the National Academy of Sciences. Reprinted with permission by the National Academies Press.
Fig. 2.19: Copyright © 2016 by the National Academy of Sciences. Reprinted with permission by the National Academies Press.
Fig. 2.20: Copyright © 2016 by the National Academy of Sciences. Reprinted with permission by the National Academies Press.
Fig. 2.21: Copyright © 2016 by the National Academy of Sciences. Reprinted with permission by the National Academies Press.
Fig. 2.22: Copyright © 2016 by the National Academy of Sciences. Reprinted with permission by the National Academies Press.
Fig. 2.23: Copyright © 2016 by the National Academy of Sciences. Reprinted with permission by the National Academies Press.
Fig. 2.24: Copyright © 2016 by the National Academy of Sciences. Reprinted with permission by the National Academies Press.
Fig. 2.25: Copyright © 2016 by the National Academy of Sciences. Reprinted with permission by the National Academies Press.
Fig. 2.26: Copyright © 2016 by the National Academy of Sciences. Reprinted with permission by the National Academies Press.
Fig. 2.27: Copyright © 2016 by the National Academy of Sciences. Reprinted with permission by the National Academies Press.
Fig. 2.28: Copyright © 2016 by the National Academy of Sciences. Reprinted with permission by the National Academies Press.
Fig. 2.29: Copyright © 2016 by the National Academy of Sciences. Reprinted with permission by the National Academies Press.
Fig. 2.30: Copyright © 2016 by the National Academy of Sciences. Reprinted with permission by the National Academies Press.
Fig. 2.31: Copyright © 2016 by the National Academy of Sciences. Reprinted with permission by the National Academies Press.
Fig. 2.32: Copyright © 2016 by the National Academy of Sciences. Reprinted with permission by the National Academies Press.
Fig. 2.33: Copyright © 2016 by the National Academy of Sciences. Reprinted with permission by the National Academies Press.
Fig. 2.34: Copyright © 2016 by the National Academy of Sciences. Reprinted with permission by the National Academies Press.
Fig. 2.35: Copyright © 2016 by the National Academy of Sciences. Reprinted with permission by the National Academies Press.
Fig. 2.36: Copyright © 2016 by the National Academy of Sciences. Reprinted with permission by the National Academies Press.
Fig. 2.37: Copyright © 2016 by the National Academy of Sciences. Reprinted with permission by the National Academies Press.
Fig. 2.38: Copyright © 2016 by University of Florida Board of Trustees. Reprinted with permission.
Fig. 2.39: Copyright © 2016 by University of Florida Board of Trustees. Reprinted with permission.
Fig. 2.40: Copyright © 2016 by University of Florida Board of Trustees. Reprinted with permission.
Fig. 2.41: Copyright © 2016 by University of Florida Board of Trustees. Reprinted with permission.
Fig. 2.42: Copyright © 2016 by the National Academy of Sciences. Reprinted with permission by the National Academies Press.
Fig. 2.43: Copyright © 2016 by the National Academy of Sciences. Reprinted with permission by the National Academies Press.
Fig. 2.44: Copyright © 2016 by the National Academy of Sciences. Reprinted with permission by the National Academies Press.
Fig. 2.45: Copyright © 2016 by the National Academy of Sciences. Reprinted with permission by the National Academies Press.
Fig. 2.46: Copyright © 2016 by the National Academy of Sciences. Reprinted with permission by the National Academies Press.
Fig. 2.47: Copyright © 2016 by the National Academy of Sciences. Reprinted with permission by the National Academies Press.
Fig. 2.48: Copyright © 2016 by the National Academy of Sciences. Reprinted with permission by the National Academies Press.
Fig. 2.49: Copyright © 2016 by the National Academy of Sciences. Reprinted with permission by the National Academies Press.
Fig. 2.50: Copyright © 2016 by the National Academy of Sciences. Reprinted with permission by the National Academies Press.
Fig. 2.51: Copyright © 2016 by the National Academy of Sciences. Reprinted with permission by the National Academies Press.

Fig. 2.52: Copyright © 2016 by the National Academy of Sciences. Reprinted with permission by the National Academies Press.
Fig. 2.53: Copyright © 2016 by the National Academy of Sciences. Reprinted with permission by the National Academies Press.
Fig. 2.54: Copyright © 2016 by the National Academy of Sciences. Reprinted with permission by the National Academies Press.
Fig. 2.55: Copyright © 2016 by the National Academy of Sciences. Reprinted with permission by the National Academies Press.
Fig. 2.56: Copyright © 2016 by the National Academy of Sciences. Reprinted with permission by the National Academies Press.
Fig. 2.57: Copyright © 2016 by the National Academy of Sciences. Reprinted with permission by the National Academies Press.
Fig. 2.58: Copyright © 2016 by the National Academy of Sciences. Reprinted with permission by the National Academies Press.
Fig. 2.59: Copyright © 2016 by the National Academy of Sciences. Reprinted with permission by the National Academies Press.
Fig. 2.60: Copyright © 2016 by the National Academy of Sciences. Reprinted with permission by the National Academies Press.
Fig. 2.61: Copyright © 2016 by the National Academy of Sciences. Reprinted with permission by the National Academies Press.
Fig. 2.62: Copyright © 2016 by the National Academy of Sciences. Reprinted with permission by the National Academies Press.
Fig. 2.63: Copyright © 2016 by the National Academy of Sciences. Reprinted with permission by the National Academies Press.
Fig. 2.64: Copyright © 2016 by the National Academy of Sciences. Reprinted with permission by the National Academies Press.
Fig. 2.65: Copyright © 2016 by the National Academy of Sciences. Reprinted with permission by the National Academies Press.
Fig. 2.66: Copyright © 2016 by the National Academy of Sciences. Reprinted with permission by the National Academies Press.
Fig. 2.67: Copyright © 2016 by the National Academy of Sciences. Reprinted with permission by the National Academies Press.
Fig. 2.68: Copyright © 2016 by the National Academy of Sciences. Reprinted with permission by the National Academies Press.
Fig. 2.69: Copyright © 2016 by the National Academy of Sciences. Reprinted with permission by the National Academies Press.
Fig. 2.70: Copyright © 2016 by the National Academy of Sciences. Reprinted with permission by the National Academies Press.
Fig. 2.71: Copyright © 2016 by the National Academy of Sciences. Reprinted with permission by the National Academies Press.
Fig. 2.72: Copyright © 2016 by the National Academy of Sciences. Reprinted with permission by the National Academies Press.
Fig. 2.73: Copyright © 2016 by the National Academy of Sciences. Reprinted with permission by the National Academies Press.
Fig. 2.74: Copyright © 2016 by the National Academy of Sciences. Reprinted with permission by the National Academies Press.
Fig. 2.75: Copyright © 2016 by the National Academy of Sciences. Reprinted with permission by the National Academies Press.
Fig. 2.76: Copyright © 2016 by the National Academy of Sciences. Reprinted with permission by the National Academies Press.
Fig. 2.77: Copyright © 2016 by the National Academy of Sciences. Reprinted with permission by the National Academies Press.
Fig. 2.78: Copyright © 2016 by the National Academy of Sciences. Reprinted with permission by the National Academies Press.
Fig. 2.79: Copyright © 2016 by the National Academy of Sciences. Reprinted with permission by the National Academies Press.
Fig. 2.80: Copyright © 2016 by the National Academy of Sciences. Reprinted with permission by the National Academies Press.
Fig. 2.81: Copyright © 2016 by the National Academy of Sciences. Reprinted with permission by the National Academies Press.
Fig. 2.82: Copyright © 2016 by the National Academy of Sciences. Reprinted with permission by the National Academies Press.
Fig. 2.83: Copyright © 2016 by the National Academy of Sciences. Reprinted with permission by the National Academies Press.
Fig. 2.84: Copyright © 2016 by the National Academy of Sciences. Reprinted with permission by the National Academies Press.
Fig. 2.85: Copyright © 2016 by the National Academy of Sciences. Reprinted with permission by the National Academies Press.
Fig. 2.86: Copyright © 2016 by the National Academy of Sciences. Reprinted with permission by the National Academies Press.
Fig. 2.87: Copyright © 2016 by the National Academy of Sciences. Reprinted with permission by the National Academies Press.
Fig. 2.88: Copyright © 2016 by the National Academy of Sciences. Reprinted with permission by the National Academies Press.
Fig. 2.89: Copyright © 2016 by the National Academy of Sciences. Reprinted with permission by the National Academies Press.
Fig. 2.90: Copyright © 2016 by the National Academy of Sciences. Reprinted with permission by the National Academies Press.
Fig. 2.91: Copyright © 2016 by the National Academy of Sciences. Reprinted with permission by the National Academies Press.
Fig. 2.92: Copyright © 2016 by the National Academy of Sciences. Reprinted with permission by the National Academies Press.
Fig. 2.93: Copyright © 2016 by University of Florida Board of Trustees. Reprinted with permission.

Fig. 2.94: Copyright © 2016 by University of Florida Board of Trustees. Reprinted with permission.
Fig. 2.95: Copyright © 2016 by University of Florida Board of Trustees. Reprinted with permission.
Fig. 2.96: Copyright © 2016 by University of Florida Board of Trustees. Reprinted with permission.
Fig. 3.1: Copyright © 2016 by the National Academy of Sciences. Reprinted with permission by the National Academies Press.
Fig. 3.2: Copyright © 2016 by the National Academy of Sciences. Reprinted with permission by the National Academies Press.
Fig. 3.3: Copyright © 2016 by the National Academy of Sciences. Reprinted with permission by the National Academies Press.
Fig. 3.4: Copyright © 2016 by the National Academy of Sciences. Reprinted with permission by the National Academies Press.
Fig. 3.5: Copyright © 2016 by the National Academy of Sciences. Reprinted with permission by the National Academies Press.
Fig. 3.6: Copyright © 2016 by the National Academy of Sciences. Reprinted with permission by the National Academies Press.
Fig. 3.7: Copyright © 2016 by the National Academy of Sciences. Reprinted with permission by the National Academies Press.
Fig. 3.8: Copyright © 2016 by the National Academy of Sciences. Reprinted with permission by the National Academies Press.
Fig. 3.9: Copyright © 2016 by the National Academy of Sciences. Reprinted with permission by the National Academies Press.
Fig. 3.10: Copyright © 2016 by the National Academy of Sciences. Reprinted with permission by the National Academies Press.
Fig. 3.11: Copyright © 2016 by the National Academy of Sciences. Reprinted with permission by the National Academies Press.
Fig. 3.12: Copyright © 2016 by the National Academy of Sciences. Reprinted with permission by the National Academies Press.
Fig. 3.13: Copyright © 2016 by the National Academy of Sciences. Reprinted with permission by the National Academies Press.
Fig. 3.14: Copyright © 2016 by the National Academy of Sciences. Reprinted with permission by the National Academies Press.
Fig. 3.15: Copyright © 2016 by the National Academy of Sciences. Reprinted with permission by the National Academies Press.
Fig. 3.16: Copyright © 2016 by the National Academy of Sciences. Reprinted with permission by the National Academies Press.
Fig. 3.17: Copyright © 2016 by the National Academy of Sciences. Reprinted with permission by the National Academies Press.
Fig. 3.18: Copyright © 2016 by the National Academy of Sciences. Reprinted with permission by the National Academies Press.
Fig. 3.19: Copyright © 2016 by the National Academy of Sciences. Reprinted with permission by the National Academies Press.
Fig. 3.20: Copyright © 2016 by the National Academy of Sciences. Reprinted with permission by the National Academies Press.
Fig. 3.21: Copyright © 2016 by the National Academy of Sciences. Reprinted with permission by the National Academies Press.
Fig. 3.22: Copyright © 2016 by the National Academy of Sciences. Reprinted with permission by the National Academies Press.
Fig. 3.23: Copyright © 2016 by the National Academy of Sciences. Reprinted with permission by the National Academies Press.
Fig. 3.24: Copyright © 2016 by the National Academy of Sciences. Reprinted with permission by the National Academies Press.
Fig. 3.25: Copyright © 2016 by the National Academy of Sciences. Reprinted with permission by the National Academies Press.
Fig. 3.26: Copyright © 2016 by University of Florida Board of Trustees. Reprinted with permission.
Fig. 3.27: Copyright © 2016 by University of Florida Board of Trustees. Reprinted with permission.
Fig. 3.28: Copyright © 2016 by University of Florida Board of Trustees. Reprinted with permission.
Fig. 3.29: Copyright © 2016 by University of Florida Board of Trustees. Reprinted with permission.
Fig. 3.30: Copyright © 2016 by the National Academy of Sciences. Reprinted with permission by the National Academies Press.
Fig. 3.31: Copyright © 2016 by the National Academy of Sciences. Reprinted with permission by the National Academies Press.
Fig. 3.32: Copyright © 2016 by the National Academy of Sciences. Reprinted with permission by the National Academies Press.
Fig. 3.33: Copyright © 2016 by the National Academy of Sciences. Reprinted with permission by the National Academies Press.
Fig. 3.34: Copyright © 2016 by the National Academy of Sciences. Reprinted with permission by the National Academies Press.
Fig. 3.35: Copyright © 2016 by the National Academy of Sciences. Reprinted with permission by the National Academies Press.
Fig. 3.36: Copyright © 2016 by the National Academy of Sciences. Reprinted with permission by the National Academies Press.
Fig. 3.37: Copyright © 2016 by the National Academy of Sciences. Reprinted with permission by the National Academies Press.
Fig. 3.38: Copyright © 2016 by the National Academy of Sciences. Reprinted with permission by the National Academies Press.
Fig. 3.39: Copyright © 2016 by the National Academy of Sciences. Reprinted with permission by the National Academies Press.
Fig. 3.40: Copyright © 2016 by the National Academy of Sciences. Reprinted with permission by the National Academies Press.
Fig. 3.41: Copyright © 2016 by the National Academy of Sciences. Reprinted with permission by the National Academies Press.

Fig. 4.90: Copyright © 2016 by the National Academy of Sciences. Reprinted with permission by the National Academies Press.
Fig. 4.91: Copyright © 2016 by the National Academy of Sciences. Reprinted with permission by the National Academies Press.
Fig. 4.92: Copyright © 2016 by the National Academy of Sciences. Reprinted with permission by the National Academies Press.
Fig. 4.93: Copyright © 2016 by the National Academy of Sciences. Reprinted with permission by the National Academies Press.
Fig. 4.94: Copyright © 2016 by the National Academy of Sciences. Reprinted with permission by the National Academies Press.
Fig. 4.95: Copyright © 2016 by the National Academy of Sciences. Reprinted with permission by the National Academies Press.
Fig. 4.96: Copyright © 2016 by the National Academy of Sciences. Reprinted with permission by the National Academies Press.
Fig. 4.97: Copyright © 2016 by the National Academy of Sciences. Reprinted with permission by the National Academies Press.
Fig. 4.98: Copyright © 2016 by the National Academy of Sciences. Reprinted with permission by the National Academies Press.
Fig. 4.99: Copyright © 2016 by the National Academy of Sciences. Reprinted with permission by the National Academies Press.
Fig. 4.100: Copyright © 2016 by University of Florida Board of Trustees. Reprinted with permission.
Fig. 4.101: Copyright © 2016 by University of Florida Board of Trustees. Reprinted with permission.
Fig. 4.102: Copyright © 2016 by University of Florida Board of Trustees. Reprinted with permission.
Fig. 4.103: Copyright © 2016 by University of Florida Board of Trustees. Reprinted with permission.
Fig. 5.1: Copyright © 2016 by the National Academy of Sciences. Reprinted with permission by the National Academies Press.
Fig. 5.2: Copyright © 2016 by the National Academy of Sciences. Reprinted with permission by the National Academies Press.
Fig. 5.3: Copyright © 2016 by the National Academy of Sciences. Reprinted with permission by the National Academies Press.
Fig. 5.4: Copyright © 2016 by the National Academy of Sciences. Reprinted with permission by the National Academies Press.
Fig. 5.5: Copyright © 2016 by the National Academy of Sciences. Reprinted with permission by the National Academies Press.
Fig. 5.6: Copyright © 2016 by the National Academy of Sciences. Reprinted with permission by the National Academies Press.
Fig. 5.7: Copyright © 2016 by the National Academy of Sciences. Reprinted with permission by the National Academies Press.
Fig. 5.8: Copyright © 2016 by the National Academy of Sciences. Reprinted with permission by the National Academies Press.
Fig. 5.9: Copyright © 2016 by the National Academy of Sciences. Reprinted with permission by the National Academies Press.
Fig. 5.10: Copyright © 2016 by the National Academy of Sciences. Reprinted with permission by the National Academies Press.
Fig. 5.11: Copyright © 2016 by the National Academy of Sciences. Reprinted with permission by the National Academies Press.
Fig. 5.12: Copyright © 2016 by the National Academy of Sciences. Reprinted with permission by the National Academies Press.
Fig. 5.13: Copyright © 2016 by the National Academy of Sciences. Reprinted with permission by the National Academies Press.
Fig. 5.14: Copyright © 2016 by the National Academy of Sciences. Reprinted with permission by the National Academies Press.
Fig. 5.15: Copyright © 2016 by the National Academy of Sciences. Reprinted with permission by the National Academies Press.
Fig. 5.16: Copyright © 2016 by the National Academy of Sciences. Reprinted with permission by the National Academies Press.
Fig. 5.17: Copyright © 2016 by the National Academy of Sciences. Reprinted with permission by the National Academies Press.
Fig. 5.18: Copyright © 2016 by the National Academy of Sciences. Reprinted with permission by the National Academies Press.
Fig. 5.19: Copyright © 2016 by the National Academy of Sciences. Reprinted with permission by the National Academies Press.
Fig. 5.20: Copyright © 2016 by the National Academy of Sciences. Reprinted with permission by the National Academies Press.
Fig. 5.21: Copyright © 2016 by the National Academy of Sciences. Reprinted with permission by the National Academies Press.
Fig. 5.22: Copyright © 2016 by the National Academy of Sciences. Reprinted with permission by the National Academies Press.
Fig. 5.23: Copyright © 2016 by the National Academy of Sciences. Reprinted with permission by the National Academies Press.
Fig. 5.24: Copyright © 2016 by the National Academy of Sciences. Reprinted with permission by the National Academies Press.
Fig. 5.25: Copyright © 2016 by the National Academy of Sciences. Reprinted with permission by the National Academies Press.
Fig. 5.26: Copyright © 2016 by the National Academy of Sciences. Reprinted with permission by the National Academies Press.
Fig. 5.27: Copyright © 2016 by the National Academy of Sciences. Reprinted with permission by the National Academies Press.
Fig. 5.28: Copyright © 2016 by the National Academy of Sciences. Reprinted with permission by the National Academies Press.
Fig. 5.29: Copyright © 2016 by the National Academy of Sciences. Reprinted with permission by the National Academies Press.

Fig. 6.8: Copyright © 2016 by the National Academy of Sciences. Reprinted with permission by the National Academies Press.
Fig. 6.9: Copyright © 2016 by the National Academy of Sciences. Reprinted with permission by the National Academies Press.
Fig. 6.10: Copyright © 2016 by the National Academy of Sciences. Reprinted with permission by the National Academies Press.
Fig. 6.11: Copyright © 2016 by the National Academy of Sciences. Reprinted with permission by the National Academies Press.
Fig. 6.12: Copyright © 2016 by the National Academy of Sciences. Reprinted with permission by the National Academies Press.
Fig. 6.13: Copyright © 2016 by the National Academy of Sciences. Reprinted with permission by the National Academies Press.
Fig. 6.14: Copyright © 2016 by the National Academy of Sciences. Reprinted with permission by the National Academies Press.
Fig. 6.15: Copyright © 2016 by the National Academy of Sciences. Reprinted with permission by the National Academies Press.
Fig. 6.16: Copyright © 2016 by the National Academy of Sciences. Reprinted with permission by the National Academies Press.
Fig. 6.17: Copyright © 2016 by the National Academy of Sciences. Reprinted with permission by the National Academies Press.
Fig. 6.18: Copyright © 2016 by the National Academy of Sciences. Reprinted with permission by the National Academies Press.
Fig. 6.19: Copyright © 2016 by the National Academy of Sciences. Reprinted with permission by the National Academies Press.
Fig. 6.20: Copyright © 2016 by the National Academy of Sciences. Reprinted with permission by the National Academies Press.
Fig. 6.21: Copyright © 2016 by the National Academy of Sciences. Reprinted with permission by the National Academies Press.
Fig. 6.22: Copyright © 2016 by the National Academy of Sciences. Reprinted with permission by the National Academies Press.
Fig. 6.23: Copyright © 2016 by the National Academy of Sciences. Reprinted with permission by the National Academies Press.
Fig. 6.24: Copyright © 2016 by the National Academy of Sciences. Reprinted with permission by the National Academies Press.
Fig. 6.25: Copyright © 2016 by the National Academy of Sciences. Reprinted with permission by the National Academies Press.
Fig. 6.26: Copyright © 2016 by the National Academy of Sciences. Reprinted with permission by the National Academies Press.
Fig. 6.27: Copyright © 2016 by the National Academy of Sciences. Reprinted with permission by the National Academies Press.
Fig. 6.28: Copyright © 2016 by the National Academy of Sciences. Reprinted with permission by the National Academies Press.
Fig. 6.29: Copyright © 2016 by the National Academy of Sciences. Reprinted with permission by the National Academies Press.
Fig. 6.30: Copyright © 2016 by the National Academy of Sciences. Reprinted with permission by the National Academies Press.
Fig. 6.31: Copyright © 2016 by the National Academy of Sciences. Reprinted with permission by the National Academies Press.
Fig. 6.32: Copyright © 2016 by the National Academy of Sciences. Reprinted with permission by the National Academies Press.
Fig. 6.33: Copyright © 2016 by the National Academy of Sciences. Reprinted with permission by the National Academies Press.
Fig. 6.34: Copyright © 2016 by the National Academy of Sciences. Reprinted with permission by the National Academies Press.
Fig. 6.35: Copyright © 2016 by the National Academy of Sciences. Reprinted with permission by the National Academies Press.
Fig. 6.36: Copyright © 2016 by the National Academy of Sciences. Reprinted with permission by the National Academies Press.
Fig. 6.37: Copyright © 2016 by the National Academy of Sciences. Reprinted with permission by the National Academies Press.
Fig. 6.38: Copyright © 2016 by the National Academy of Sciences. Reprinted with permission by the National Academies Press.
Fig. 6.39: Copyright © 2016 by the National Academy of Sciences. Reprinted with permission by the National Academies Press.
Fig. 6.40: Copyright © 2016 by the National Academy of Sciences. Reprinted with permission by the National Academies Press.
Fig. 6.41: Copyright © 2016 by the National Academy of Sciences. Reprinted with permission by the National Academies Press.
Fig. 6.42: Copyright © 2016 by the National Academy of Sciences. Reprinted with permission by the National Academies Press.
Fig. 6.43: Copyright © 2016 by the National Academy of Sciences. Reprinted with permission by the National Academies Press.
Fig. 6.44: Copyright © 2016 by the National Academy of Sciences. Reprinted with permission by the National Academies Press.
Fig. 6.45: Copyright © 2016 by the National Academy of Sciences. Reprinted with permission by the National Academies Press.
Fig. 6.46: Copyright © 2016 by the National Academy of Sciences. Reprinted with permission by the National Academies Press.
Fig. 6.47: Copyright © 2016 by the National Academy of Sciences. Reprinted with permission by the National Academies Press.
Fig. 6.48: Copyright © 2016 by the National Academy of Sciences. Reprinted with permission by the National Academies Press.
Fig. 6.49: Copyright © 2016 by the National Academy of Sciences. Reprinted with permission by the National Academies Press.

Fig. 6.50: Copyright © 2016 by the National Academy of Sciences. Reprinted with permission by the National Academies Press.
Fig. 6.51: Copyright © 2016 by the National Academy of Sciences. Reprinted with permission by the National Academies Press.
Fig. 6.52: Copyright © 2016 by the National Academy of Sciences. Reprinted with permission by the National Academies Press.
Fig. 6.53: Copyright © 2016 by the National Academy of Sciences. Reprinted with permission by the National Academies Press.
Fig. 6.54: Copyright © 2016 by the National Academy of Sciences. Reprinted with permission by the National Academies Press.
Fig. 6.55: Copyright © 2016 by the National Academy of Sciences. Reprinted with permission by the National Academies Press.
Fig. 6.56: Copyright © 2016 by the National Academy of Sciences. Reprinted with permission by the National Academies Press.
Fig. 6.57: Copyright © 2016 by the National Academy of Sciences. Reprinted with permission by the National Academies Press.
Fig. 6.58: Copyright © 2016 by the National Academy of Sciences. Reprinted with permission by the National Academies Press.
Fig. 6.59: Copyright © 2016 by the National Academy of Sciences. Reprinted with permission by the National Academies Press.
Fig. 6.60: Copyright © 2016 by the National Academy of Sciences. Reprinted with permission by the National Academies Press.
Fig. 6.61: Copyright © 2016 by University of Florida Board of Trustees. Reprinted with permission.
Fig. 6.62: Copyright © 2016 by University of Florida Board of Trustees. Reprinted with permission.
Fig. 6.63: Copyright © 2016 by University of Florida Board of Trustees. Reprinted with permission.
Fig. 6.64: Copyright © 2016 by University of Florida Board of Trustees. Reprinted with permission.
Fig. 7.1: Copyright © 2016 by the National Academy of Sciences. Reprinted with permission by the National Academies Press.
Fig. 7.2: Copyright © 2016 by the National Academy of Sciences. Reprinted with permission by the National Academies Press.
Fig. 7.3: Copyright © 2016 by the National Academy of Sciences. Reprinted with permission by the National Academies Press.
Fig. 7.4: Copyright © 2016 by the National Academy of Sciences. Reprinted with permission by the National Academies Press.
Fig. 7.5: Copyright © 2016 by the National Academy of Sciences. Reprinted with permission by the National Academies Press.
Fig. 7.6: Copyright © 2016 by the National Academy of Sciences. Reprinted with permission by the National Academies Press.
Fig. 7.7: Copyright © 2016 by the National Academy of Sciences. Reprinted with permission by the National Academies Press.
Fig. 7.8: Copyright © 2016 by the National Academy of Sciences. Reprinted with permission by the National Academies Press.
Fig. 7.9: Copyright © 2016 by the National Academy of Sciences. Reprinted with permission by the National Academies Press.
Fig. 7.10: Copyright © 2016 by the National Academy of Sciences. Reprinted with permission by the National Academies Press.
Fig. 7.11: Copyright © 2016 by the National Academy of Sciences. Reprinted with permission by the National Academies Press.
Fig. 7.12: Copyright © 2016 by the National Academy of Sciences. Reprinted with permission by the National Academies Press.
Fig. 7.13: Copyright © 2016 by the National Academy of Sciences. Reprinted with permission by the National Academies Press.
Fig. 7.14: Copyright © 2016 by the National Academy of Sciences. Reprinted with permission by the National Academies Press.
Fig. 7.15: Copyright © 2016 by the National Academy of Sciences. Reprinted with permission by the National Academies Press.
Fig. 7.16: Copyright © 2016 by the National Academy of Sciences. Reprinted with permission by the National Academies Press.
Fig. 7.17: Copyright © 2016 by the National Academy of Sciences. Reprinted with permission by the National Academies Press.
Fig. 7.18: Copyright © 2016 by the National Academy of Sciences. Reprinted with permission by the National Academies Press.
Fig. 7.19: Copyright © 2016 by the National Academy of Sciences. Reprinted with permission by the National Academies Press.
Fig. 7.20: Copyright © 2016 by the National Academy of Sciences. Reprinted with permission by the National Academies Press.
Fig. 7.21: Copyright © 2016 by the National Academy of Sciences. Reprinted with permission by the National Academies Press.
Fig. 7.22: Copyright © 2016 by the National Academy of Sciences. Reprinted with permission by the National Academies Press.
Fig. 7.23: Copyright © 2016 by the National Academy of Sciences. Reprinted with permission by the National Academies Press.
Fig. 7.24: Copyright © 2016 by the National Academy of Sciences. Reprinted with permission by the National Academies Press.
Fig. 7.25: Copyright © 2016 by the National Academy of Sciences. Reprinted with permission by the National Academies Press.
Fig. 7.26: Copyright © 2016 by the National Academy of Sciences. Reprinted with permission by the National Academies Press.
Fig. 7.27: Copyright © 2016 by the National Academy of Sciences. Reprinted with permission by the National Academies Press.
Fig. 7.28: Copyright © 2016 by the National Academy of Sciences. Reprinted with permission by the National Academies Press.

Fig. 7.29: Copyright © 2016 by the National Academy of Sciences. Reprinted with permission by the National Academies Press.
Fig. 7.30: Copyright © 2016 by the National Academy of Sciences. Reprinted with permission by the National Academies Press.
Fig. 7.31: Copyright © 2016 by the National Academy of Sciences. Reprinted with permission by the National Academies Press.
Fig. 7.32: Copyright © 2016 by the National Academy of Sciences. Reprinted with permission by the National Academies Press.
Fig. 7.33: Copyright © 2016 by the National Academy of Sciences. Reprinted with permission by the National Academies Press.
Fig. 7.34: Copyright © 2016 by the National Academy of Sciences. Reprinted with permission by the National Academies Press.
Fig. 7.35: Copyright © 2016 by the National Academy of Sciences. Reprinted with permission by the National Academies Press.
Fig. 7.36: Copyright © 2016 by the National Academy of Sciences. Reprinted with permission by the National Academies Press.
Fig. 7.37: Copyright © 2016 by the National Academy of Sciences. Reprinted with permission by the National Academies Press.
Fig. 7.38: Copyright © 2016 by the National Academy of Sciences. Reprinted with permission by the National Academies Press.
Fig. 7.39: Copyright © 2016 by the National Academy of Sciences. Reprinted with permission by the National Academies Press.
Fig. 7.40: Copyright © 2016 by the National Academy of Sciences. Reprinted with permission by the National Academies Press.
Fig. 7.41: Copyright © 2016 by the National Academy of Sciences. Reprinted with permission by the National Academies Press.
Fig. 7.42: Copyright © 2016 by the National Academy of Sciences. Reprinted with permission by the National Academies Press.
Fig. 7.43: Copyright © 2016 by the National Academy of Sciences. Reprinted with permission by the National Academies Press.
Fig. 7.44: Copyright © 2016 by the National Academy of Sciences. Reprinted with permission by the National Academies Press.
Fig. 7.45: Copyright © 2016 by the National Academy of Sciences. Reprinted with permission by the National Academies Press.
Fig. 7.46: Copyright © 2016 by the National Academy of Sciences. Reprinted with permission by the National Academies Press.
Fig. 7.47: Copyright © 2016 by the National Academy of Sciences. Reprinted with permission by the National Academies Press.
Fig. 7.48: Copyright © 2016 by the National Academy of Sciences. Reprinted with permission by the National Academies Press.
Fig. 7.49: Copyright © 2016 by the National Academy of Sciences. Reprinted with permission by the National Academies Press.
Fig. 7.50: Copyright © 2016 by the National Academy of Sciences. Reprinted with permission by the National Academies Press.
Fig. 7.51: Copyright © 2016 by the National Academy of Sciences. Reprinted with permission by the National Academies Press.
Fig. 7.52: Copyright © 2016 by the National Academy of Sciences. Reprinted with permission by the National Academies Press.
Fig. 7.53: Copyright © 2016 by the National Academy of Sciences. Reprinted with permission by the National Academies Press.
Fig. 7.54: Copyright © 2016 by the National Academy of Sciences. Reprinted with permission by the National Academies Press.
Fig. 7.55: Copyright © 2016 by the National Academy of Sciences. Reprinted with permission by the National Academies Press.
Fig. 7.56: Copyright © 2016 by the National Academy of Sciences. Reprinted with permission by the National Academies Press.
Fig. 7.57: Copyright © 2016 by the National Academy of Sciences. Reprinted with permission by the National Academies Press.
Fig. 7.58: Copyright © 2016 by the National Academy of Sciences. Reprinted with permission by the National Academies Press.
Fig. 7.59: Copyright © 2016 by the National Academy of Sciences. Reprinted with permission by the National Academies Press.
Fig. 7.60: Copyright © 2016 by the National Academy of Sciences. Reprinted with permission by the National Academies Press.
Fig. 7.61: Copyright © 2016 by the National Academy of Sciences. Reprinted with permission by the National Academies Press.
Fig. 7.62: Copyright © 2016 by the National Academy of Sciences. Reprinted with permission by the National Academies Press.
Fig. 7.63: Copyright © 2016 by the National Academy of Sciences. Reprinted with permission by the National Academies Press.
Fig. 7.64: Copyright © 2016 by the National Academy of Sciences. Reprinted with permission by the National Academies Press.
Fig. 7.65: Copyright © 2016 by the National Academy of Sciences. Reprinted with permission by the National Academies Press.
Fig. 7.66: Copyright © 2016 by the National Academy of Sciences. Reprinted with permission by the National Academies Press.
Fig. 7.67: Copyright © 2016 by the National Academy of Sciences. Reprinted with permission by the National Academies Press.
Fig. 7.68: Copyright © 2016 by University of Florida Board of Trustees. Reprinted with permission.
Fig. 7.69: Copyright © 2016 by University of Florida Board of Trustees. Reprinted with permission.
Fig. 7.70: Copyright © 2016 by University of Florida Board of Trustees. Reprinted with permission.
Fig. 7.71: Copyright © 2016 by University of Florida Board of Trustees. Reprinted with permission.

Fig. 8.1: Copyright © 2016 by the National Academy of Sciences. Reprinted with permission by the National Academies Press.
Fig. 8.2: Copyright © 2016 by the National Academy of Sciences. Reprinted with permission by the National Academies Press.
Fig. 8.3: Copyright © 2016 by the National Academy of Sciences. Reprinted with permission by the National Academies Press.
Fig. 8.4: Copyright © 2016 by the National Academy of Sciences. Reprinted with permission by the National Academies Press.
Fig. 8.5: Copyright © 2016 by the National Academy of Sciences. Reprinted with permission by the National Academies Press.
Fig. 8.6: Copyright © 2016 by the National Academy of Sciences. Reprinted with permission by the National Academies Press.
Fig. 8.7: Copyright © 2016 by the National Academy of Sciences. Reprinted with permission by the National Academies Press.
Fig. 8.8: Copyright © 2016 by the National Academy of Sciences. Reprinted with permission by the National Academies Press.
Fig. 8.9: Copyright © 2016 by the National Academy of Sciences. Reprinted with permission by the National Academies Press.
Fig. 8.10: Copyright © 2016 by the National Academy of Sciences. Reprinted with permission by the National Academies Press.
Fig. 8.11: Copyright © 2016 by the National Academy of Sciences. Reprinted with permission by the National Academies Press.
Fig. 8.12: Copyright © 2016 by the National Academy of Sciences. Reprinted with permission by the National Academies Press.
Fig. 8.13: Copyright © 2016 by the National Academy of Sciences. Reprinted with permission by the National Academies Press.
Fig. 8.14: Copyright © 2016 by the National Academy of Sciences. Reprinted with permission by the National Academies Press.
Fig. 8.15: Copyright © 2016 by the National Academy of Sciences. Reprinted with permission by the National Academies Press.
Fig. 8.16: Copyright © 2016 by the National Academy of Sciences. Reprinted with permission by the National Academies Press.
Fig. 8.17: Copyright © 2016 by the National Academy of Sciences. Reprinted with permission by the National Academies Press.
Fig. 8.18: Copyright © 2016 by the National Academy of Sciences. Reprinted with permission by the National Academies Press.
Fig. 8.19: Copyright © 2016 by the National Academy of Sciences. Reprinted with permission by the National Academies Press.
Fig. 8.20: Copyright © 2016 by the National Academy of Sciences. Reprinted with permission by the National Academies Press.
Fig. 8.21: Copyright © 2016 by the National Academy of Sciences. Reprinted with permission by the National Academies Press.
Fig. 8.22: Copyright © 2016 by the National Academy of Sciences. Reprinted with permission by the National Academies Press.
Fig. 8.23: Copyright © 2016 by the National Academy of Sciences. Reprinted with permission by the National Academies Press.
Fig. 8.24: Copyright © 2016 by the National Academy of Sciences. Reprinted with permission by the National Academies Press.
Fig. 8.25: Copyright © 2016 by the National Academy of Sciences. Reprinted with permission by the National Academies Press.
Fig. 8.26: Copyright © 2016 by the National Academy of Sciences. Reprinted with permission by the National Academies Press.
Fig. 8.27: Copyright © 2016 by the National Academy of Sciences. Reprinted with permission by the National Academies Press.
Fig. 8.28: Copyright © 2016 by the National Academy of Sciences. Reprinted with permission by the National Academies Press.
Fig. 8.29: Copyright © 2016 by the National Academy of Sciences. Reprinted with permission by the National Academies Press.
Fig. 8.30: Copyright © 2016 by the National Academy of Sciences. Reprinted with permission by the National Academies Press.
Fig. 8.31: Copyright © 2016 by the National Academy of Sciences. Reprinted with permission by the National Academies Press.
Fig. 8.32: Copyright © 2016 by the National Academy of Sciences. Reprinted with permission by the National Academies Press.
Fig. 8.33: Copyright © 2016 by the National Academy of Sciences. Reprinted with permission by the National Academies Press.
Fig. 8.34: Copyright © 2016 by the National Academy of Sciences. Reprinted with permission by the National Academies Press.
Fig. 8.35: Copyright © 2016 by University of Florida Board of Trustees. Reprinted with permission.
Fig. 8.36: Copyright © 2016 by the National Academy of Sciences. Reprinted with permission by the National Academies Press.
Fig. 8.37: Copyright © 2016 by the National Academy of Sciences. Reprinted with permission by the National Academies Press.
Fig. 8.38: Copyright © 2016 by the National Academy of Sciences. Reprinted with permission by the National Academies Press.
Fig. 8.39: Copyright © 2016 by the National Academy of Sciences. Reprinted with permission by the National Academies Press.
Fig. 8.40: Copyright © 2016 by the National Academy of Sciences. Reprinted with permission by the National Academies Press.
Fig. 8.41: Copyright © 2016 by the National Academy of Sciences. Reprinted with permission by the National Academies Press.
Fig. 8.42: Copyright © 2016 by the National Academy of Sciences. Reprinted with permission by the National Academies Press.

Fig. 8.43: Copyright © 2016 by the National Academy of Sciences. Reprinted with permission by the National Academies Press.
Fig. 8.44: Copyright © 2016 by the National Academy of Sciences. Reprinted with permission by the National Academies Press.
Fig. 8.45: Copyright © 2016 by the National Academy of Sciences. Reprinted with permission by the National Academies Press.
Fig. 8.46: Copyright © 2016 by the National Academy of Sciences. Reprinted with permission by the National Academies Press.
Fig. 8.47: Copyright © 2016 by the National Academy of Sciences. Reprinted with permission by the National Academies Press.
Fig. 8.48: Copyright © 2016 by the National Academy of Sciences. Reprinted with permission by the National Academies Press.
Fig. 8.49: Copyright © 2016 by the National Academy of Sciences. Reprinted with permission by the National Academies Press.
Fig. 8.50: Copyright © 2016 by the National Academy of Sciences. Reprinted with permission by the National Academies Press.
Fig. 8.51: Copyright © 2016 by the National Academy of Sciences. Reprinted with permission by the National Academies Press.
Fig. 8.52: Copyright © 2016 by the National Academy of Sciences. Reprinted with permission by the National Academies Press.
Fig. 8.53: Copyright © 2016 by the National Academy of Sciences. Reprinted with permission by the National Academies Press.
Fig. 8.54: Copyright © 2016 by the National Academy of Sciences. Reprinted with permission by the National Academies Press.
Fig. 8.55: Copyright © 2016 by the National Academy of Sciences. Reprinted with permission by the National Academies Press.
Fig. 8.56: Copyright © 2016 by the National Academy of Sciences. Reprinted with permission by the National Academies Press.
Fig. 8.57: Copyright © 2016 by the National Academy of Sciences. Reprinted with permission by the National Academies Press.
Fig. 8.58: Copyright © 2016 by the National Academy of Sciences. Reprinted with permission by the National Academies Press.
Fig. 8.59: Copyright © 2016 by the National Academy of Sciences. Reprinted with permission by the National Academies Press.
Fig. 8.60: Copyright © 2016 by University of Florida Board of Trustees. Reprinted with permission.
Fig. 8.61: Copyright © 2016 by the National Academy of Sciences. Reprinted with permission by the National Academies Press.
Fig. 8.62: Copyright © 2016 by the National Academy of Sciences. Reprinted with permission by the National Academies Press.
Fig. 8.63: Copyright © 2016 by the National Academy of Sciences. Reprinted with permission by the National Academies Press.
Fig. 8.64: Copyright © 2016 by the National Academy of Sciences. Reprinted with permission by the National Academies Press.
Fig. 8.65: Copyright © 2016 by the National Academy of Sciences. Reprinted with permission by the National Academies Press.
Fig. 8.66: Copyright © 2016 by the National Academy of Sciences. Reprinted with permission by the National Academies Press.
Fig. 8.67: Copyright © 2016 by the National Academy of Sciences. Reprinted with permission by the National Academies Press.
Fig. 8.68: Copyright © 2016 by the National Academy of Sciences. Reprinted with permission by the National Academies Press.
Fig. 8.69: Copyright © 2016 by University of Florida Board of Trustees. Reprinted with permission.
Fig. 8.70: Copyright © 2016 by University of Florida Board of Trustees. Reprinted with permission.
Fig. 8.71: Copyright © 2016 by the National Academy of Sciences. Reprinted with permission by the National Academies Press.
Fig. 8.72: Copyright © 2016 by the National Academy of Sciences. Reprinted with permission by the National Academies Press.
Fig. 8.73: Copyright © 2016 by the National Academy of Sciences. Reprinted with permission by the National Academies Press.
Fig. 8.74: Copyright © 2016 by the National Academy of Sciences. Reprinted with permission by the National Academies Press.
Fig. 8.75: Copyright © 2016 by University of Florida Board of Trustees. Reprinted with permission.
Fig. 8.76: Copyright © 2016 by University of Florida Board of Trustees. Reprinted with permission.
Fig. 8.77: Copyright © 2016 by the National Academy of Sciences. Reprinted with permission by the National Academies Press.
Fig. 8.78: Copyright © 2016 by the National Academy of Sciences. Reprinted with permission by the National Academies Press.
Fig. 8.79: Copyright © 2016 by the National Academy of Sciences. Reprinted with permission by the National Academies Press.
Fig. 8.80: Copyright © 2016 by the National Academy of Sciences. Reprinted with permission by the National Academies Press.
Fig. 8.81: Copyright © 2016 by the National Academy of Sciences. Reprinted with permission by the National Academies Press.
Fig. 8.82: Copyright © 2016 by the National Academy of Sciences. Reprinted with permission by the National Academies Press.
Fig. 8.83: Copyright © 2016 by the National Academy of Sciences. Reprinted with permission by the National Academies Press.
Fig. 8.84: Copyright © 2016 by the National Academy of Sciences. Reprinted with permission by the National Academies Press.
Fig. 8.85: Copyright © 2016 by the National Academy of Sciences. Reprinted with permission by the National Academies Press.

Fig. 8.86: Copyright © 2016 by the National Academy of Sciences. Reprinted with permission by the National Academies Press.

Fig. 8.87: Copyright © 2016 by the National Academy of Sciences. Reprinted with permission by the National Academies Press.

Fig. 8.88: Copyright © 2016 by the National Academy of Sciences. Reprinted with permission by the National Academies Press.

Fig. 8.89: Copyright © 2016 by the National Academy of Sciences. Reprinted with permission by the National Academies Press.

Fig. 8.90: Copyright © 2016 by the National Academy of Sciences. Reprinted with permission by the National Academies Press.

Fig. 8.91: Copyright © 2016 by the National Academy of Sciences. Reprinted with permission by the National Academies Press.

Fig. 8.92: Copyright © 2016 by the National Academy of Sciences. Reprinted with permission by the National Academies Press.

Fig. 8.93: Copyright © 2016 by the National Academy of Sciences. Reprinted with permission by the National Academies Press.

Fig. 8.94: Copyright © 2016 by the National Academy of Sciences. Reprinted with permission by the National Academies Press.

Fig. 8.95: Copyright © 2016 by the National Academy of Sciences. Reprinted with permission by the National Academies Press.

Fig. 8.96: Copyright © 2016 by University of Florida Board of Trustees. Reprinted with permission.

Fig. 8.97: Copyright © 2016 by University of Florida Board of Trustees. Reprinted with permission.

Fig. 8.98: Copyright © 2016 by the National Academy of Sciences. Reprinted with permission by the National Academies Press.

Fig. 8.99: Copyright © 2016 by the National Academy of Sciences. Reprinted with permission by the National Academies Press.

Fig. 8.100: Copyright © 2016 by the National Academy of Sciences. Reprinted with permission by the National Academies Press.

Fig. 8.101: Copyright © 2016 by the National Academy of Sciences. Reprinted with permission by the National Academies Press.

Fig. 8.102: Copyright © 2016 by the National Academy of Sciences. Reprinted with permission by the National Academies Press.

Fig. 8.103: Copyright © 2016 by University of Florida Board of Trustees. Reprinted with permission.

Fig. 8.104: Copyright © 2016 by University of Florida Board of Trustees. Reprinted with permission.

Fig. 8.105: Copyright © 2016 by the National Academy of Sciences. Reprinted with permission by the National Academies Press.

Fig. 8.106: Copyright © 2016 by the National Academy of Sciences. Reprinted with permission by the National Academies Press.

Fig. 8.107: Copyright © 2016 by the National Academy of Sciences. Reprinted with permission by the National Academies Press.

Fig. 8.108: Copyright © 2016 by the National Academy of Sciences. Reprinted with permission by the National Academies Press.

Fig. 8.109: Copyright © 2016 by the National Academy of Sciences. Reprinted with permission by the National Academies Press.

Fig. 8.110: Copyright © 2016 by University of Florida Board of Trustees. Reprinted with permission.

Fig. 8.111: Copyright © 2016 by University of Florida Board of Trustees. Reprinted with permission.

Fig. 9.1: Copyright © 2016 by University of Florida Board of Trustees. Reprinted with permission.

Fig. 9.2: Copyright © 2016 by University of Florida Board of Trustees. Reprinted with permission.

Fig. 9.3: Copyright © 2016 by University of Florida Board of Trustees. Reprinted with permission.

Fig. 9.4: Copyright © 2016 by University of Florida Board of Trustees. Reprinted with permission.

Fig. 9.5: Copyright © 2016 by University of Florida Board of Trustees. Reprinted with permission.

Fig. 9.6: Copyright © 2016 by University of Florida Board of Trustees. Reprinted with permission.

Fig. 9.7: Copyright © 2016 by University of Florida Board of Trustees. Reprinted with permission.

Fig. 9.8: Copyright © 2016 by University of Florida Board of Trustees. Reprinted with permission.

Fig. 9.9: Copyright © 2016 by University of Florida Board of Trustees. Reprinted with permission.

Fig. 9.10: Copyright © 2016 by University of Florida Board of Trustees. Reprinted with permission.

Fig. 9.11: Copyright © 2016 by University of Florida Board of Trustees. Reprinted with permission.

Fig. 9.12: Copyright © 2016 by University of Florida Board of Trustees. Reprinted with permission.

Fig. 9.13: Copyright © 2016 by University of Florida Board of Trustees. Reprinted with permission.

Fig. 9.14: Copyright © 2016 by University of Florida Board of Trustees. Reprinted with permission.

Fig. 9.15: Copyright © 2016 by University of Florida Board of Trustees. Reprinted with permission.

Fig. 9.16: Copyright © 2016 by University of Florida Board of Trustees. Reprinted with permission.

Fig. 9.17: Copyright © 2016 by University of Florida Board of Trustees. Reprinted with permission.

Fig. 9.18: Copyright © 2016 by University of Florida Board of Trustees. Reprinted with permission.

Fig. 9.19: Copyright © 2016 by University of Florida Board of Trustees. Reprinted with permission.

Fig. 9.20: Copyright © 2016 by University of Florida Board of Trustees. Reprinted with permission.

Fig. 10.1: Copyright © 2016 by University of Florida Board of Trustees. Reprinted with permission.

Fig. 10.2: Copyright © 2016 by University of Florida Board of Trustees. Reprinted with permission.

Fig. 10.3: Copyright © 2016 by University of Florida Board of Trustees. Reprinted with permission.

Fig. 10.4: Copyright © 2016 by University of Florida Board of Trustees. Reprinted with permission.

Fig. 10.5: Copyright © 2016 by University of Florida Board of Trustees. Reprinted with permission.

Fig. 10.6: Copyright © 2016 by University of Florida Board of Trustees. Reprinted with permission.

CPSIA information can be obtained
at www.ICGtesting.com
Printed in the USA
LVHW050332221122
733733LV00003B/14